THE

PHILADELPHIA BOARD OF TRADE,

BELIEVING that this volume on "PHILADELPHIA AND ITS MANUFACTURES," contains matter entirely reliable, of much value to every Business Community, as demonstrating the important fact that Philadelphia is the chief seat of the Manufacturing and Industrial Resources of the Country, and therefore the best and cheapest Market for buyers, have taken the liberty of mailing it to you, with the request that you will publish the accompanying Abstract of its contents, or such extracts as you may see proper; and call to the whole subject the attention of your readers. The Board would also be obliged by your addressing a copy of your paper, with such notice, to their Rooms.

BOARD OF TRADE ROOMS,
　　　　Philadelphia, Penn'a.

From the Public Ledger.

PHILADELPHIA MANUFACTURES.

The *Ledger*, ever since its existence, has repeatedly advocated the importance of ascertaining and making known the development which manufacturing industry has attained in Philadelphia. Our mechanics were entitled, we believe, to have their important achievements placed upon permanent record; our position as a producing market, we were convinced, was sufficiently well defined to entitle us to rank among the great manufacturing cities of the world. It was evident that our neighbors in the East, by superior activity in inviting public attention to their manufactures, not only attracted the trade which belonged to them, but a portion of that which in reason and right belonged to Philadelphia. It is a notorious fact, that not only have buyers from distant sections passed through this city to purchase in other places Philadelphia made goods, but our own merchants have repeatedly gone to other cities to buy the identical goods which were made within a few squares of their own stores. More accurate information as to the extent, variety, and characteristics of our manufacturing resources and development would prevent such mistakes as these, benefit our manufacturers, save money to our merchants, attract new customers from abroad, and strengthen the confidence of those who are already our customers. The public may be said to have coincided in our view of its importance, but the execution of the task was deferred, mainly, it would seem, because no one possessing the requisite qualifications was willing to undertake the vast amount of labor necessary to its accomplishment. Committees of the Board of Trade, of the Statistical Society, and of other Associations, attempted an investigation of this sort, but with no appreciable result. About one year ago, the *Ledger* announced that a work was in course of preparation on the Manufactures of Philadelphia, by EDWIN T. FREEDLEY, who was well and favorably known as the author of two or three very successful books. Various circumstances have postponed its publication, but the work is now before us, entitled, *Philadelphia and its Manufactures : A Hand-book exhibiting the Development, Variety, and Statistics of the Manufacturing Industry of Philadelphia in 1857, together with Sketches of Remarkable Manufactories ; and a List of Articles now made in Philadelphia. By Edwin T. Freedley,* Author of "A Practical Treatise on Business,', "The Legal Adviser," etc. Philadelphia, Edward Young, 333 Walnut Street, 1858. It is a large 12mo., 490 pages, handsomely bound, good paper, beautiful typography, and illustrated with engravings of some of the most noted of our manufacturing establishments.

The contents are divided into four parts :—Part I. being a well-written Essay on the causes of eminence in Manufactures. Part II. shows that Philadelphia possesses all the essential, and some extraordinary advantages for manufacturing. Part III. illustrates the present development of manufactures in Philadelphia; and Part IV., which, though nominally an Index to contents, is a minute and valuable *List of Articles now made in Philadelphia, with the address of one or more manufacturers of each.* We shall subjoin a comprehensive summary of its statistical results; but it is proper to state, that the value and interest of the volume do not consist chiefly in the statistical details. Descriptions of the classes of goods made, and of leading establishments, together with peculiar characteristic incidents connected with the various branches, preponderate largely over the statistics. The author has aimed to make a *readable* exhibit; and in this respect, the volume before us is far superior to the dry reports of Boards of Trade, or any thing of the kind ever before attempted. The book is in fact *sui generis*—no other city either in Europe or America having as yet, by its productive industry, become the subject of a volume so well entitled to popularity, for its style and general interest, besides fulfilling the special object for which it was written.

Agricultural Implements, Fertilizers, &c.—The manufacture of Agricultural Implements is comparatively a new business in Philadelphia. There are, however, several extensive establishments, each of which turns out admirable machines, both as respects workmanship and materials. *Of Garden and Field Seeds*, Philadelphia is a principal distributing point. Mr. Landreth's seed-grounds embrace nearly 400 acres, requiring a large force of hands, and a steam-engine for threshing and cleaning seeds. The value of *Fertilizers* made in Philadelphia is stated at $503,000—the principal being *Super-Phosphate of Lime,* of which 7000 tons, or 55,000 bbls. are made annually, worth $45 per ton.

Alcohol, Burning Fluid, and Camphene.—Nine establishments in this branch produced in 1857, 395,000 gallons alcohol; 147,250 gallons pine oil; 1,112,000 gallons burning fluid—the estimated value of the whole being $1,022,149. The quality of the burning fluid made in Philadelphia is very superior.

Books, Magazines, and Newspapers.—The chapter on books, in the volume before us, is particularly deserving of attention. It relates the principal events in the history of the publishing interest from the establishing of the first printing-press in Philadelphia—being the second set up in North America—to the present time. *Nine-tenths* of the *medical books* issued in the United States are published in Philadelphia; while in law books, religious books, school books, and other standard literature, the publishers of this city take a leading rank. The capital invested in the business is stated at $2,500,000, and the value of the books published annually at $3,690,000. There are 12 newspapers published daily, over 40 weekly, and about 50 periodicals, including four literary magazines. There are 4 type foundries, having a capital invested of $500,000, and producing annually $420,000; seven stereotype foundries, employing 180 hands, and having a capital invested of $150,000. There are about 50 printing offices, employing from 3 to 100 persons each; 60 book-binderies and blank-book manufactories, producing annually a value of $1,210,000, and furnishing employment to 1700 persons, male and female; engravers of all kinds, and lithographers whose work has attracted marked attention in England; manufacturers of marble paper, maps, printing inks, stationery, &c. There are 9 paper-mills in the city, producing annually $1,250,000, and 35 houses for the sale of paper. The annual product of the book manufacture and its kindred branches is stated at $5,593,000.

Boots and Shoes.—The two principal centres of the boot and shoe manufacture in the United States, are Philadelphia, and Lynn, Mass., the latter being the centre for the common, cheap pegged work, the former of the fine, sewed work. The product of this manufacture in Philadelphia is $4,141,000, while the entire trade of the city in boots and shoes approximates $15,000,000.

Brass and Copper.—The manufactures in these metals are extremely varied, embracing almost every description of article. The product of the brass foundries is stated at $830,000 ; of copper at $400,000.

Brewing and Distilling.—The breweries of Philadelphia have long been celebrated. Of *Ale* and *Porter* there are 9 extensive breweries, having a capital invested of $1,500,000, and producing annually $1,020,000. Of *Lager Beer,* there are about thirty brewers, employing a capital of $1,200,000, and producing 180,000 barrels of Lager, worth $1,080,000. There are 5 distilleries in the city, which produced in 1857, 2,100,000 gallons of Whiskey, worth $630,000. There are 8 firms extensively engaged in *rectifying Whiskey,* having a capital invested of $1,250,000, and whose business in 1857 amounted to $2,524,500.

Bricks, Fire Bricks, Earthenware, &c.—There are fifty brick yards which produce annually about 100,000,000 of common bricks, worth $7 per thousand, and 8,000,000 pressed bricks, worth $14. Total value, $812,000. The manufacture of *Fire Bricks, Chemical Ware, Terra Cotta,* and miscellaneous manufactures in clay, furnish an aggregate product of $647,000.

Carriages and Wagons.—Philadelphia is unrivalled in the manufacture of both light carriage and government wagons. The growth of hickory, oak and ash in the vicinity of Philadelphia is especially suited for carriage-making purposes. There are thirty manufacturers of light carriages, having a capital invested of about $500,000, who turn out an average annual product of $900,000. There are also about 45 manufactories of wagons, carts, drays, &c., three of which are very extensive establishments—one covering six acres. Total product, $815,000.

Chemicals, Paints, Glue, &c.—The chemical establishments of Philadelphia are among the largest and most important in the Union. The statistical summary of the products of pharmaceutical processes is as follows: Chemicals, dye stuffs, &c., $3,335,000 ; medicines, $1,300,000 ; white lead, $960,000 ; paints, zinc, &c., $770,000 ; glue, &c., $775,000 ; varnishes, $230,000. Total, $7,370,000.

Clothing, Ready-made.—The capital invested in the manufacture of ready-made clothing, shirts, collars, bosoms, mantillas and corsets, is stated at $3,300,000, and the product at $11,157,500. There are 67 firms that each make annually over $40,000 of clothing.

Confectionery.—The value of sugar confectionery, pieces montées, &c., made annually, is $1,020,000. The chapter on this subject is particularly amusing and interesting.

Dry Goods.—To one who knows Philadelphia only as a clean, quiet and beautiful city, the statement seems incredible that it is the centre of upward of 260 cotton and woolen factories, besides hand-looms equal in force and production to 70 additional factories of ordinary size. In the volume before us, the names of over two hundred manufacturers of textile fabrics are given, all within the limits of the city; while in the adjacent counties, there are over fifty factories, some of them very large and fine establishments. *Tickings*, of unusual excellence, are made largely. Of *Apron* and *Furniture Checks*, Philadelphia may be said to have the monopoly in the manufacture. Of *Ginghams, Cottonades, Kentucky Jeans, Ducks, Osnaburgs, &c.*, the production is enormous. The manufacture of *Ingrain* and *Rag Carpetings*, employs 2060 looms, and amounts to 8,160,000 yards, worth $3,096,000. Of *Woolen Hosiery* or *Fancy Knit Work*, the goods known as Germantown goods, are almost as celebrated as Nottingham or Leicester hosiery. In the manufacture of *Trimmings* or *Narrow Textile Fabrics*, there are twenty establishments, which produce annually to the value of about $1,600,000. The Messrs. Hortsmann have a capital invested of $400,000, employing 400 hands, who receive annually $100,000 in wages, and produce an average annual product of $400,000. Of *Hand-Loom Weaving and Manufacturing*, Philadelphia is now the great seat in America. There are now 4760 looms in the city, and 7180 hand-loom weavers, who produce annually in checks, ginghams, carpetings and hosiery, a value of $5,754,150. The following is the summary of production in Dry Goods:

Woolen and cotton goods, by power,					....	$13,163,968
"	"	"	"	hand looms (exclusive of hosiery),	....	4,746,000
Hosiery and fancy woolen goods, hand power,				....................$1,008,150		
"	"	"	"	" factories,	 800,000	1,808,150
Narrow textile fabrics, sewing silks, &c.,				....		1,600,000
Total annual product in Philadelphia of Dry Goods,				....		$21,318,110

Flour, Provisions, &c.—There are 22 flour mills in Philadelphia, having 90 run of stones, which produced in 1857 over 400,000 barrels of flour, worth $3,000,000. The *Curing of Meats* and *Packing of Beef*, constitutes a very extensive business, employing a capital of over $2,000,000. The value of the provisions cured, &c., annually, exceeds $4,000,000. Among the Philadelphia brands of flour and provisions there are several that are noted in all the markets of Europe. The value of *Furniture* manufactured annually, is estimated at $2,500,000, and including the *Upholstery*, $3,000,000. There are 13 manufacturers of *Glass*, whose headquarters are in this city. Value of product, $1,600,000.

The *Hat, Cap, and Fur Manufacture*, yields a product of $1,900,000.

Iron Manufactures.—It is demonstrated that Philadelphia is the commercial centre of the district in the United States, which produces the best and cheapest iron and coal. The development which has been attained in the production of the various articles manufactured from iron, is therefore less astonishing, and we are prepared to learn that the stoves, saws, shovels and spades, forks, cutlery, bolts, nuts and washers, locomotives and steam-engines, and machinery generally, made in this city, are unsurpassed in quality and cheapness. The chapters on this subject in the volume before us, occupy over 60 pages; and the extraordinary machines that have been constructed in Philadelphia, and the machine-making establishments as described, possessing, as some of them do, tools and appliances not elsewhere in use in this country or in Europe, are deserving objects of pride to all who are interested in American manufactures. The summary of production in iron and its manufactures, states the hands at 10,410, and the product at $12,852,150, while the commercial value of finished work in the vicinity of the city is at least *three millions* more.

Jewelry, Silver Ware, &c.—All branches of the working in precious metals, and their imitations, are carried on very largely and successfully here. The Mint, since its establishment in 1793 to the close of 1856, coined $391,730,571 86; while in the manufacture of fine jewelry, pure silver-ware, surpassing coin, and gold and silver plate, watch-cases, silver-plated wares, &c., this is a leading market. Hands employed, 1700; product, viz.:

Gold jewelry, pens, spectacles, &c.,	$1,275,000
" watch-cases,	942,000
" leaf and foil,	325,000
Silver-ware,	450,000
Plated and Britannia-ware,	380,000
Total,	$3,372,000

Lamps, Chandeliers, Gas Fixtures.—The Philadelphia houses in this branch, it is well known, are unequaled by any others in this country. They employ 1250 hands, and produce $1,300,000.

Leather Manufactures.—There are 10 tanneries in the city, 35 curriers, 25 morocco manufacturers, several buckskin and kid glove makers, the annual product being $3,091,250. The morocco leather and buckskin goods made here are especially excellent.

Marble, &c.—There are a number of quarries, producing excellent building marble, in the vicinity of Philadelphia, and one soap-stone quarry. There are about 60 marble yards in the city, employing 840 hands, and executing work to the amount of $860,000 annually, or, including brown stone, granite, &c., $1,160,000. The chapter on this subject is very comprehensive and interesting.

Oils.—There are 24 establishments engaged in producing various oils, having a capital invested of $1,040,000, employing 245 hands, and yielding an annual product of $2,131,230.

Paper Hangings.—This is another manufacture in which Philadelphia has deservedly a high reputation. There are 6 principal manufactories, employing 456 hands, and producing a value of $800,000. One of the firms occupies a brick factory 396 feet long, 80 wide, and 4 stories high, believed to be the largest in the world.

Rope and Cordage are made to the amount of $810,000 annually, consuming about 1500 tons of Manilla hemp, and 1250 tons Western hemp. *Jute Rope* is also made largely.

The *Saddle and Harness* manufacture employs 960 hands, and yields a product of $1,500,000; while *Whips, Trunks and Portmanteaus* add a half million more. In all these branches the Philadelphia establishments take first rank.

Ship and Boat Building constitutes a very interesting chapter, to which we must refer.

The *Soap and Candle Factories* produce $2,057,600 annually. The *Sugar Refineries*, of which there are five, besides two extensive establishments that extract sugar from molasses, are very remarkable, and have a capacity for producing $10,000,000 annually. The value of the sugar refined in 1857 was about $6,000,000. The manufactures of *Tobacco* are extremely important—the quantity of cigars made annually being 312,000,000, worth $3,000,000. The cigar, snuff, and leaf tobacco trade employs a capital of $3,000,000, turned twice a year, which produces an annual business of $6,000,000. The *Umbrella and Parasol Factories* of Philadelphia are the largest in the United States, and the quality of their products the best. About 2500 persons are employed in this business, producing an average annual value of $1,275,000, though in 1853 it amounted to $2,000,000. For the details of these and other manufactures—artificial teeth, brushes, buttons, combs, musical instruments, oil cloths, perfumery and fancy soaps, surgical instruments, straw and millinery goods, tinware, &c.,—we must refer to the volume.

The following is the Author's summary of

Aggregate Value of Articles produced in Philadelphia, for the year ending June 30th, 1857.

Agricultural Implements, Seeds, &c., (estimated)...................................$500,000
Alcohol, Burning Fluid, and Camphene 1,022,140
Ale, Porter, and Brown Stout 1,020,000
Artificial Flowers 85,000
Awnings, Bags, &c. 91,750
Assaying and Refining Precious Metals, including actual expenses of U. S. Mint, $130,000.........................850,000
Barrels, Casks, Shooks, and Vats......715,000
Beer, Lager and Small..................1,280,000
Blacking, Ink, and Lampblack, (estimated)..500,000
Bolts, Nuts, Screws, &c.....................411,000
Book and Periodical Publishing, *exclusive of Paper, Printing, Binding, &c.*......................................818,000
Book Binding, Blank Books, and Marble Paper.............................1,230,000
Boots and Shoes............................4,141,000
Boxes, Packing, (estimated)..............500,000
Brass Articles 830,000

Bread, Bakers, (including Crackers,) Ship Bread, &c.........................$5,600,000
Bricks, Common and Pressed............812,000
Britannia and Plated Wares............380,000
Brooms, Corn and other...................104,000
Brushes...225,000
Candles, Adamantine & Oleine Oils....570,000
Caps...400,000
Cards, Playing 118,000
Carpeting, Ingrain.......................2,592,000
Carpeting, Rag................................504,000
Carriages and Coaches.....................900,000
Cars and Car Wheels.......................550,000
Chemicals, Dye-Stuffs, Chrome Colors, and Extracts............ 3,335,000
Clothing 9,640,000
Coffins, Ready-made.......................219,000
Combs...150,000
Confectionery, &c.........................1,020,000
Copper Work....................................400,000
Cordials, Bay Water, &c...................200,000
Cotton and Woolen Goods, exclusive of Hosiery, Carpetings, &c.........14,813,968

Cordage, Twines, &c.................................$810,000
Cutlery, Skates, &c.............................150,000
Daguerreotypes, Cases, and Materials, (estimated).............................600,000
Edge Tools, Hammers, &c....................127,000
Earthenware, Fire-Bricks, &c 647,000
Engines, Locomotive, Stationary and Fire...3,428,000
Engraving and Lithography..............570,000
Envelopes and Fancy Stationery.......150,000
Flooring and Planed Lumber...........370,000
Flour..3,200,000
Fertilizers............................ 503,000
Fringes, Tassels, and Narrow Textile Fabrics.......................................1,288,000
Furniture, (estimated).....................2,500,000
Furs 350,000
Gloves, Buckskin and Kid.................150,000
Glue, Curled Hair, &c......................775,000
Gold Leaf and Foil...........................325,000
Glassware.......................................1,600,000
Hardware, and Iron Manufactures not otherwise enumerated 1,169,000
Hats, Silk and Soft...........................800,000
Hose, Belting, &c..............................175,000
Hosiery 1,808,150
Hollow-ware, exclus'e of Stoves,&c...1,250,000
Iron, Bar, Sheet, and Forged...........1,517,650
Jewelry, and Manufactures of Gold..1,275,000
Lamps, Chandeliers, and Gas Fixtures 1,300,000
Lasts and Boot Trees 36,000
Lead Pipe, Sheet Lead, Shot, &c.......235,000
Leather, exclusive of Morocco.........1,610,000
Machinery 1,912,000
Machine Tools...................................350,000
Mahogany and Sawed Lumber.........580,000
Maps and Charts..............................400,000
Marble Work....................................860,000
Mantillas and Corsets.......................330,000
Matches, Friction.............................125,000
Medicines, Patent and Prepared Remedies..1,300,000
Millinery Goods, including Bonnet Frames, Wire, &c., but excluding Straw Goods & Artificial Flowers...360,000
Mouldings, &c..................................300,000
Morocco and Fancy Leather............1,156,250
Musical Instruments.........................485,000
Mineral Waters.................................350,000
Newspapers, Daily and Weekly, (estimated).......................................1,370,000
Oil Cloths...................................... 289,000
Oils, Linseed, Lard and Tallow, Rosin, and R. R. Greases.................2,131,230
Paints, Zinc, and Products of Paint Mills..770,000
Paper 1,250,000
Paper Hangings.................................800,000
Paper Boxes.....................................175,000
Patterns, Stove and Machinery.........115,000
Perfumery and Fancy Soaps.............850,000
Picture and Looking-Glass Frames, (estimated)....................................750,000
Preserved Fruits, &c., (estimated).....350,000
Printing, Book and Job...................1,183,000
Printing Inks....................... 160,000

Provisions — Cured Meats, Packed Beef, &c...................................$4,000,000
Rifles and Pistols.............................120,000
Saddles, Harness, &c......................1,500,000
Safes..150,000
Sails 135,000
Sash, Blinds, Doors, &c....................250,000
Saws..510,000
Scales and Balances.........................145,000
Shirts, Collars, Bosoms, and Gentlemen's Furnishing Goods 1,187,500
Shovels, Spades, Hoes, &c.................397,000
Show Cases..55,000
Sewing Silks.....................................312,000
Silver-ware......................................450,000
Soap and Candles, exclusive of Adamantine Candles.......................1,487,600
Springs, Rail-road and Coach...........238,000
Spices, Condiments, Essence of Coffee, &c., &c..................................350,000
Starch...155,000
Steel, Spring and Cast......................283,500
Stoves and Grates..........................1,250,000
Sand-stone, Granite, Slate, &c..........300,000
Straw Goods, including Hats 600,000
Surgical and Dental Instruments, Trusses, and Artificial Limbs........350,000
Sugar, Refined, and Molasses........6,500,000
Teeth, Porcelain..............................500,000
Tin, Zinc, and Sheet-Iron Ware.......1,200,000
Tobacco Manufactures, Cigars, Snuff, &c...3,256,500
Trunks and Portmanteaus................313,000
Turnings in Wood 550,000
Type and Stereotype.........................650,000
Umbrellas and Parasols, including Umbrella Furniture, Ivory & Bone Turning, Whalebone Cutting.......1,750,000
Upholstery, (estimated) 500,000
Varnishes 230,000
Vessels, Masts and Spars, Blocks and Pumps, &c...............................1,760,000
Vinegar and Cider............................300,000
Wagons, Carts, and Drays.................815,000
Watch Cases.....................................942,000
Whips..175,000
Whisky, Distilled 630,000
  "   Rectified...........................2,524,500
White Lead......................................960,000
Willow-ware, Baskets, &c., (estm'd)..120,000
Wire-work, (estimated,)....................250,000
Wooden and Cedar-ware 150,000
Works in Wood not otherwise enumerated...100,000
Miscellaneous Articles, not otherwise enumerated. For particulars see INDEX, (estimated).................3,000,000

Total Annual Product of Manufacturing Industry in Philadelphia **145,348,738**
Add for Leading Branches in the vicinity of Philadelphia, as before given......................**26,500,000**

Total for Philadelphia and vicinity..................................... **$171,848,738**

The Capital employed in Manufactures in Philadelphia, is..........................$72,500,000
Hands Employed,.. 132,000

Philadelphia and its Manufactures:

A HAND-BOOK

EXHIBITING THE

DEVELOPMENT, VARIETY, AND STATISTICS

OF THE

MANUFACTURING INDUSTRY OF PHILADELPHIA

IN 1857.

TOGETHER WITH

Sketches of Remarkable Manufactories;

AND A

LIST OF ARTICLES NOW MADE IN PHILADELPHIA.

BY

EDWIN T. FREEDLEY,

AUTHOR OF A "PRACTICAL TREATISE ON BUSINESS," "THE LEGAL ADVISER," ETC.

PHILADELPHIA:
EDWARD YOUNG, 333 WALNUT STREET.
1858.

PHILADELPHIA:
STEREOTYPED BY GEORGE CHARLES.
PRINTED BY KING & BAIRD.

PREFACE.

THE Title of this Volume defines its subject; and the subject, it is presumed, explains its object. The Author, however, desires to advert briefly to the circumstances that impelled him to undergo the vast amount of hard, thankless, profitless labor which the preparation of a volume like this, however imperfectly executed, necessarily involves, and to assume the responsibility of an undertaking, which, as an individual enterprise, unaided by municipal or corporate favor, is, it is believed, wholly unprecedented.

For many years it has been a source of mortification to the active friends of Philadelphia, that mainly through the misrepresentations of rivals, and the misapprehension of her resources, she has gradually receded from her former glorious position in the commercial firmament, until now she is regarded by many in Europe, and in some portions of our own country, as a mere speck on the horizon. Her enemies have industriously circulated, far and wide, reports which, if unexplained, must prove detrimental to her interests ; and the declension of her foreign commerce has been to them a harp of a thousand strings. The friends of Philadelphia, on the other hand, either at home or abroad, have not been furnished with facts to counteract these prejudicial statements ; and are themselves scarcely aware what a beautiful fabric she has erected,—more important in every truly national point of view than Foreign Trade,—dedicated to Home Industry and American Manufactures. The leading organs of the enlightened sentiment

(17)

of the city have earnestly and repeatedly called upon the corporate authorities to collect and publish statistics of its Productive Industry; and various attempts have been made by individuals, and by Committees of Commercial Associations, to effect this object, but the difficulties in the way of its accomplishment seemed insurmountable. More than a year ago, Mr. EDWARD YOUNG, the publisher of this volume, solicited the writer to undertake the preparation of a work on the Manufactures of Philadelphia, promising his assistance in the collection of materials. The Author will be pardoned a digression from the narrative to bear testimony how nobly his associate has redeemed his promise—how faithfully he has persevered. Without his co-operation, it is probable this volume would be far less complete than it now is. Foreseeing, perhaps, only a portion of the difficulties to be encountered and surmounted, and presuming that with certainty he could readily supply his own deficiencies by obtaining able assistance—a hope in which he has been grievously disappointed—the Author acceded to the request, and originated a plan of treating the subject, an outline of which was submitted to the Board of Trade, who honored it with an approbatory resolution, and to the Press who generally commended it, and invited the co-operation of Manufacturers. Other encouraging inducements were offered. BARTON H. JENKS, Esq., one of the "alive" men of this city, volunteered a liberal subscription; JOHN GRIGG, Esq., tendered his name and assurances of future influence in behalf of the enterprise; JOHN BIDDLE, of the firm of E. C. & J. Biddle, and WILLIAM L. REHN, of the firm of Brooke, Tyson & Rehn—men of well-known public spirit—contributed suggestions and valuable information from their stores of accumulated knowledge; and under these circumstances, and with this encouragement, the labor was commenced. Here, however, the historical record must close. Beyond this point there lies a dreary, panic winter; and a recital of facts attending the collection of materials and statistics would involve a revelation of too many difficulties,—interposed in part by the widely extended field over which we were compelled to travel in search of desired information, and in part by the indifference manifested by many of the very persons whose interests would be promoted by the publication,—to be pleasant in retrospection, though perhaps profitable for instruction to other adventurous persons.

Passing over the circumstances which rendered the task more arduous than it ought to have been, the Author desires to state, that in the prosecution of the undertaking he claims to have acted with strict impartiality, both as respects persons and facts. If injustice has been done in any instance to individuals—a fact of which he is at present not conscious—he would deeply regret it; but the omission to notice or mention a manufacturing establishment he cannot consent to consider an act of injustice. It was not deemed essential to the completeness of the narrative to notice individual establishments at all; and where such are introduced it has been done parenthetically, or for the purpose of illustration, or because they creditably represent the other establishments in the same branch. The insertion of names has been avoided as much as possible, for they are perpetually changing though the establishment or business which they represent survive unchanged. A Philadelphia *Business Directory*, for a full list of names, and *Ure's Dictionary of Manufactures*, for a description of processes, are a natural complement of this volume. In the selection of matter for insertion as facts, the Author has regarded first, accuracy; secondly, novelty or interest. It would have been easy by less exactness as to the accuracy of matters of fact to multiply details, and to increase the interest or "spiciness" of the volume; but even in a dearth of real facts "doubtful facts" have not been

resorted to.   But the circumstance which more than any other
caused the rejection of much interesting matter was—the *want of
space*.   It was the opinion of those who are conversant with the
secret springs of publishing success, that a volume on this subject,
to be useful, must not exceed in size or price certain prescribed
limits.   This principle—which, like the laws of the Medes and
Persians, was unalterable—was ever present to the Author's eye,
rendering him apprehensive lest by treating one subject at too
much length, he would be prevented from giving due consideration to another equally important.   The theme is so comprehensive, that what may be deemed essential preoccupied the
entire space to the exclusion of novelties, patented improvements,
undeveloped manufactures, and thus matters of great interest
have unavoidably and of necessity been omitted.

   The Author desires to render acknowledgments of indebtedness
to all who in any way aided him, and special acknowledgments
to WILLIAM C. KENT of James, Kent, Santee & Co., to Dr. J.
L. BISHOP, O. W. KIBBIE, THOMAS SHRIVER, WM. Y. CARVER,
JOHN D. STOCKTON, and Dr. JAMES MOORE.   For the mechanical execution of the volume he is indebted to the following persons : GEORGE CHARLES, Stereotyper ; Jos. W. RAYNER,
Proof Reader ; CHARLES MAGARGE & Co., Paper-makers ; KING
& BAIRD, Printers ; MILLER & BURLOCK, Binders.

   Philadelphia, July, 1858.   E. T. F.

PHILADELPHIA AND ITS MANUFACTURES.

MANUFACTURES—CAUSES OF EMINENCE IN.

"It requires a great deal of philosophy to observe what is seen every day."—ROUSSEAU.

THE term *Manufacture*, in its derivative sense, signifies —making by hand. Its modern acceptation, however, is directly the reverse of its original meaning; and it is now applied particularly to those products which are made extensively by machinery, without much aid from manual labor. The word therefore is an extremely flexible one; and as Political Economists do not agree in opinion, whether millers and bakers are properly manufacturers or not, we shall, if need be, take advantage of the uncertainty, and consider as Manufactures what strictly may belong to other classifications of productive industry.

The end of every Manufacture is to increase the utility of objects by modifying their external form or changing their internal constitution. In some instances, substances that would otherwise be utterly worthless, are converted into the most valuable products—as the hoofs of certain animals into Prussiate of Potash; the offal into Gold-beater's Skin ; and especially rags into Paper. Thus beneficent in their general object, it is scarcely remarkable that modern Manufactures are principally distinguished for their ameliorating influence upon man's social condition. By cheapening manufactured products they put

1   (21)

within the reach of the poorest classes what in former times was accessible only to the wealthy and noble. The servant, the artisan, and the husbandman of England, at the present time have more palatable food, better clothing and better furniture, than were possessed by " the gentilitie" in the " golden days" of Queen Bess. In no other equally extensive districts of the world are the people generally so well off as to physical comforts, or so intellectually progressive, as in England and Massachusetts, and in none have Manufactures as yet attained equal prominence as branches of industry.   In 1850 there were employed in *Textile Manufactures* alone, the following :

MILLS IN THE UNITED KINGDOM.

	England and Wales.	Scotland.	Ireland.	Total.
Mills,	3,699	550	91	4,340
Spindles,	22,859,010	2,256,408	532,303	25,647,721
Power Looms,	272,586	28,811	2,517	303,914
Moving Power, Steam, (horses)	91,610	13,857	2,646	108,113
" " Water, "	18,214	6,004	1,886	26,104

The persons employed in these mills numbered 596,082, of whom 40,775 were children under thirteen years of age, and 329,577 were females above thirteen.   In the United States, the most important of the Textile Manufactures are those of Cotton and Wool.   In 1850 there was employed in the Cotton Manufacture a capital of $74,500,931; consuming 641,240 bales of cotton annually, and producing about 763,000,000 yards of sheetings, shirtings, calicoes, &c., and 27,000,000 lbs. of yarn, valued for the entire product at $61,869,184.   The number of persons employed was 92,286, of whom 33,150 were males, and 59,136 were females.   Massachusetts contained about one-third of the whole number of spindles in the United States, and about one-half the capital invested in the Cotton Manufacture was owned in Massachusetts.   The Woolen Manufacture of the United States employed a capital of about $28,000,000; consuming 71,000,000 lbs. of Wool, worth $25,000,000; and the product was valued at $43,207,545. It is more generally distributed throughout the United States than the Cotton Manufacture, yet Massachusetts employs in it one third of the whole capital and consumes one third of the Wool.

But the future of manufacturing enterprise in the United States, except in its effects upon society, must not be judged from its present development in Massachusetts. In 1810, according to the census, Virginia, the two Carolinas and Georgia manufactured greatly more in quantity and value of Cotton and Woolen fabrics than the whole of New England; and North Carolina produced double as many yards as Massachusetts. We doubt not the superior intellectual energy of the people of Massachusetts has attracted much that, with equality in this particular, combined with superior physical advantages, will again be attracted elsewhere. Manufacturing enterprise in the United States is yet in its experimental stage. The people have but recently recovered from the delusion that Manufactures are injurious to national prosperity. They have not had time to study the conditions upon which success in Manufactures depends, or to comprehend the lines that naturally and properly separate human pursuits. In future times, a manufacturer will no more think of consulting merely his personal inclinations, or one favorable circumstance, in the location of his manufactory, than the agriculturist, for a similar reason, would choose for the field of his operations the Pilot Knob of Missouri, or the gold-seeker the sands of New Jersey. As yet manufacturers are working independently not only of each other, but of the general laws that underlie economical production. Being in doubt as to the proper locality, they have not concentrated or combined their efforts; and the buyers of manufactured goods being in doubt as to the Home

Market, give their confidence to European manufacturers.
It is the object of the present volume to submit, with due
deference, to the consideration of both these classes, some
suggestions based ón the experience of the past and of
other countries ; and to endeavor to aid them—First : *by
considering what are the requisites to prosperity or the causes
of economical production in Manufactures ;* Secondly : *by indicating a locality possessing the advantages for manufacturing in the highest degree of perfection ;* Thirdly : *by showing
the progress already made in Manufactures in that locality.*

I. Political Economists divide the essential requisites
of production into two—Labor, and appropriate natural
objects. To these, in Manufactures, we must certainly add
Capital. But the productive efficacy of all productive
agents, as every one has observed, varies greatly at various
times and places, and depends upon a variety and due combination of circumstances, partly *moral* and partly *physical.*
Foremost among the moral circumstances conducive and
essential to prosperity, especially in Manufactures, are
freedom of industry and security of property. We need
but glance at the history of any European nation, France
in particular, to discover that governmental interference
with industry is baneful in its effects, and that monopolies and corporation privileges retard progress. "I have
frequently seen," says Roland de la Platière, a minister
of state during the French Revolution, "manufacturers
visited by a band of satellites, who put all in confusion in
their establishments, spread terror in their families, cut
the stuff from the frames, tore off the warp from the
looms, and carried them away as proofs of infringement ;
the manufacturers were summoned, tried and condemned ;
their goods confiscated ; copies of their judgment of con-
fiscation posted up in every public place ; future reputation, credit, all was lost and destroyed. And for what

offense ? Because they had made of worsted a kind of cloth called *shay*, such as the English used to manufacture, and even sell in France, while the French regulations stated that that kind of cloth should be made with *mohair*. I have seen other manufacturers treated in the same way, because they had made camlets of a particular width, used in England and Germany, for which there was a great demand from Spain, Portugal, and other countries, and from several parts of France, while the French regulations prescribed other widths for camlets. There was no free town where mechanical invention could find a refuge from the tyranny of the monopolists —no trade but what was clearly and explicitly described by the statutes could be exercised—none but what was included in the privileges of some corporation."

In England freedom of industry dates from the abolition of monopolies in 1624 ; and there can be no question, as McCulloch observes, that "Freedom and security—freedom to engage in every employment, and to pursue our own interest in our own way, coupled with an intimate conviction that acquisitions, when made, might be securely enjoyed or disposed of—have been the most copious sources of our wealth and power.   There have been only two countries, Holland and the United States, which have, in these respects, been placed under nearly similar circumstances as England ; and notwithstanding the disadvantages of their situation, the Dutch have long been, and still continue to be, the most industrious and opulent people of the Continent—while the Americans, whose situation is more favorable, are rapidly advancing in the career of improvement with a rapidity hitherto unknown."

In the United States, industry, it is true, is generally free, and property in most places adequately protected by public opinion against both legislative and mob violence ;

1*

but our advantages in these respects for the development of enterprise in Manufactures have been modified and limited by fluctuating legislation on the subject of foreign competition. Very early in our constitutional history the question was agitated—Shall Government, in adjusting its taxes for revenue, so discriminate as to protect and encourage Home Manufacturers, or in other words, to diminish, if not exclude, foreign competition in our markets? This question was submitted to the people, but proved too vast for popular solution. Their opinions changed with the current of argument, like the judgment of the Dutch Justice; and the decision which they had made promptly in accordance with the wish of the attorneys on the one side, was as promptly reversed upon the suggestion of the attorneys on the other side. Finally, not knowing what to do, the majority seem to have concluded that, as posterity had done nothing for them they were under no obligations to do any thing for posterity. In the mean time legislation upon the question fluctuated with the vacillation in public sentiment; and capitalists being unable to calculate with certainty the risks involved, were timid in embarking in manufacturing enterprises. It would seem therefore that, in addition to security of property and freedom of industry, success in Manufactures implies a certain and stable, if not wise policy in governmental action upon questions affecting manufacturing interests.

2. Another moral cause contributing, and in fact essential to eminence in manufacturing industry, is *the general diffusion of intelligence among the people.* By intelligence, in this connection, we do not mean merely the understanding necessary to enable an individual to become the creator or the lord of a machine. The capacity to contrive and invent seems so much a part of the original

constitution of man, that we believe there is in every civilized community sufficient ingenuity and mental power to have originated all in physical science that has yet been devised by any.   The mind is God's machine, with powers seemingly unlimited, and capable of producing any thing from a bad pun to the lever of Archimedes, the flying pigeon of Archytas or the calculating machine of Babbage.   But the exercise of this faculty, the application of the best intellect in a community in the direction of practical improvements, depends largely upon the approbation and rewards bestowed upon successful enterprise in invention or mechanical labor.   It is in vain to hope that ambition will spur intellect to achieve mechanical triumphs, where an inventor is respected less than a tinseled soldier or a ragged lawyer.   It is in vain to expect that mechanics will strive to acquire any extraordinary skill where mechanical labor is degraded to serfdom, or even is not appreciated.   In the histories of nations, whose rise and fall are classical studies, we learn that the application of mind to invention as well as handicraft operations, was regarded as unworthy of freemen.   " In my time," says Seneca, " there have been inventions of this sort — transparent windows, tubes for diffusing warmth equally through all parts of a building; shorthand, which has been carried to such perfection that a writer can keep pace with the most rapid speaker.   *But the inventing of such things is drudgery for the lowest slaves.*   Philosophy lies deeper.   It is not her office to teach men how to use their hands."   Another ancient and eminent teacher, who can boast of a disciple here and there in our country, considered the true object of all education and philosophy to be—to fit men for war.   Need we wonder there have been dark ages in the world's history.   Need we say that in an atmosphere tainted with such a sterile philosophy, the arts which improve man's material condition cannot flourish. The proud position of New England—a position so enviable that her light reflects lustre on States with which she is allied—is due rather to her sound, intelligent, practical philosophy, than to any physical advantages or original intellectual superiority. A Yankee lad inhales from the surrounding atmosphere, if he do not hear from his father's lips, that it is an important part of his duty to aid in extending man's empire over the material world, and every available addition to human force for accomplishing that end, that he may originate, will be a sure passport to the respect of his neighbors, if not to fortune. The women and children are educated to regard ignorance and idleness as vices; and all, deeming it honorable to add something to the aggregate product of their country's wealth, co-operate and lighten the original curse, for

" All are needed by each one,
    Nothing is fair or good alone."

3. A third cause of eminence in Manufacturing, and essential to economical production, *is an abundant supply of the most effective laborers, and of those qualified to direct labor.* In view of the improvements already made, it would be rash to assert that a time will never come when automatic machines will dispense entirely with manual labor in manufacturing. So far, the introduction of machinery has stimulated the pressing demand for educated labor; and if we can at all judge of the future, success will depend more and more upon the quality of the labor employed. Labor is effective according as it is dexterous or as it is skillful. In purely routine processes, dexterity may be the quality of chief value, but laborers differ in dexterity almost as much as in mechanical skill. Englishmen say that a laborer in Essex is cheaper at 2*s.* 6*d.* per

day than a laborer in Tipperary at 5d.; and as operatives in cotton factories, our manufacturers assert that one American girl can accomplish as much in a given time as two English girls. "In England," said Mr. Kempton, before the Committee upon Manufactures of the House of Commons, " the girls tend two power-looms. In America our girls tend generally four power-looms ; some for years tended five power-looms, and some tended six for some time, and each of those power-looms turned off more cloth than I have found any power-looms turn off in this country."   Mr. Cowell, in illustrating the comparative efficiency of operatives, remarks—" At Mulhausen, which is styled the Manchester of France, one adult and two children are requisite for the management of 200 *coarse* threads, and they gain *among them* about 2s. (48 cents) at coarse work. At Manchester or Bolton one adult and two children can manage 758 threads, and gain among them 5s. 6d. per day. Thus, although wages are so much lower in France, the difference of product is so great that the cost, in money, of the commodity produced, is greater than in England. In the former, four men and two children are required to manage 800 threads, for which they receive 8s., while in the latter one man and two children are capable, with the best machinery, of doing the same, and their wages are 5s. 6d."

If then there be such difference in the productive efficacy of laborers, in operations calling for mere manual dexterity, it is obvious that the higher we ascend in those departments of mechanics and manufactures, in which the mind has a considerable part, the greater must be the advantage in favor of intelligence and skill. And such is the fact. The only standard by which to estimate the cost of labor, is the *amount of work done for the money paid*—the *per diem* earnings of the workmen being in itself no criterion by which to judge of the cost of labor.

That workman is the cheapest who can produce the most of a given quality for a given sum of money; whether he earn one dollar or five dollars per day, and that manufacturer can produce with the most efficiency, and the least expense, other things being equal, who can *at all times* command the requisite supply of such workmen.

As ingenious mechanics and rapid workmen, the Anglo-Americans have no superiors. As skillful workmen in departments for which they have been specially educated, the English are celebrated. Regular and habitual energy in labor, however, is a characteristic of both. They have no life but in their work—no enjoyment but in the shop. What other races consider amusement, is no amusement to them. But in England and America there is a marked difference between the quality of the labor that can be obtained in the country and in the towns. In fact, in or near large cities only can labor of the first quality be obtained. "As iron sharpeneth iron, so a man sharpeneth the countenance of his friend;" and away from the centres of population and competition, the face loseth its sharpness, and the hand its cunning. Cities are in nothing more remarkable than in their attractive, magnetic influence upon talent of every description. "The man who desires to employ his pen," observes Carey, "and who possesses only the ability to conduct a country newspaper, removes to the interior, while the man of talent leaves his country paper to take charge of one in the city. The dauber of portraits leaves the city to travel the country in search of employment, while the painter removes to Philadelphia, New York or London. The inferior lawyer, physician, surgeon, dentist or merchant removes to the West, while the superior one leaves the West and settles in those places in which population is dense; where the means of production are great; where talent is appreciated and best paid; and where reputation, when acquired, is worth possessing."

Superior mechanics and dexterous workmen manifest a similar preference for cities and an abhorrence of isolation ; hence, if for no other reason, extensive mechanical or manufacturing operations must be conducted at a great disadvantage in isolated localities.   In a limited experience, I have known of several establishments that have failed apparently from no other cause than the impossibility of filling orders promptly, in consequence of difficulty in procuring and retaining an adequate supply of good mechanics in an unattractive locality; and to the disposition to select such situations because of water power or some other circumstance, we ascribe much of the past embarrassments of our manufacturers.   In some of the secluded manufacturing villages of New England, it is the custom of the proprietors to fasten such superior workmen as they may have seduced thither, by aiding them to invest their earnings in a house and lot, which they cannot afterward dispose of except at a great sacrifice ; but the practice, it would seem, is rather to be commended for its shrewdness than its wisdom.   A dependent or dissatisfied workman can hardly be an efficient one.

As respects those who are well *qualified to direct labor*, the supply is, in all places, especially in isolated localities, far short of the demand.   Foremost in this class it can be no disparagement to place *scientific men*.   As agents of economical production, none are more effective.   The progress of Manufactures, in many of its departments, is intimately connected with and dependent upon the progress made in the exact sciences ; and to the experiments and investigations of scientific men—the men who peer into the secrets of Nature, whether concealed in plants, in animals or minerals, and who

" Find tongues in trees, books in running brooks,
Sermons in stones, and good in every thing,"—

that our Manufactures are largely indebted for their present development, and upon such men, we must rely principally, as we may do with confidence, for the discovery of new sources of wealth, that at a future day will give employment and wealth to millions of human beings. But scientific men are not abundant even in the centres where Libraries, Galleries and Academies are numerous; those best qualified to direct labor prefer the theatres offering the widest scope for the exhibition of their abilities; and even inventors have discovered, that in isolated localities, they may exhaust their efforts in attempting what has been better executed before.

II. Passing to the *physical causes* of eminence in manufacturing industry, we remark they are more obvious than the moral causes, but not more important. To produce manufactured goods of a given quality with the least expense being the great desideratum, it follows that whatever contributes to economy in production, whatever saves labor, or transportation, or raw materials, cannot safely be overlooked or despised. But to investigate carefully all the circumstances that have an influence upon economical production, would require a considerable volume, and be foreign to our main inquiry. Desiring merely to discover a locality within our extended country, that, by the use of the proper means, will certainly become the centre and chief seat of American manufactures, it is necessary to know what *circumstances have more influence than any others in facilitating manufacturing enterprise,* and thus sooner or later lead to superiority; but it is not necessary to exhaust the subject.

England, it is acknowledged, is pre-eminent in Manufactures over all other countries—but why? Her colonial system, her shrewd legislation, the simplicity of other nations and other accidental circumstances, have no doubt widened the market for her manufactured goods to an extraordinary extent, but her superiority nevertheless is the result of solid, substantial, not accidental circumstances. The physical advantages which have contributed more than any others to her eminence, as we think all must agree, are epitomized by the Edinburgh Review, in the following summary: 1st. Possession of supplies of the raw materials used in Manufactures; 2d. The command of the natural means and agents best fitted to produce power; 3d. The position of the country as respects others; and 4th. The nature of the soil and climate.

"1. As respects the first of these circumstances," the writer says, "every one who reflects on the nature, value, and importance of our manufactures of Wool, of the useful Metals,—such as Iron, Lead, Tin, Copper,—and of Leather, Flax, and so forth, must at once admit, that our success in them has been materially promoted by our having abundant supplies of the raw material. It is of less consequence whence the material of a manufacture possessing great value in small bulk is derived, whether it be furnished from native sources, or imported from abroad, though even in that case the advantage of possessing an internal supply, of which it is impossible to be deprived by the jealousy or hostility of foreigners, must not be overlooked. But no nation can make any considerable progress in the manufacture of bulky and heavy articles, the conveyance of which to a distance unavoidably occasions a large expense, unless she have supplies of the raw material within herself. Our superiority in manufactures depends more at this moment on our *superior machines* than on any thing else; and had we been obliged to import the iron, brass, and steel, of which they are principally made, it is exceedingly doubtful whether we should have succeeded in bringing them to any thing like their present pitch of improvement.

"2. But of all the physical circumstances that have contributed to our wonderful progress in manufacturing industry, none has had nearly so much influence as our possession of the most valuable coal mines. These have conferred advantages on us not enjoyed in an equal degree by any other people. Even though we had possessed the most abundant supply of the ores of iron and other useful metals, they would have been of little or no use, but for our almost inexhaustible coal mines. Our country is of too limited extent to produce wood sufficient to smelt and prepare any considerable quantity of iron, or other metal; and though

2

no duty were laid on timber when imported, its cost abroad, and the heavy expense attending the conveyance of so bulky an article, would have been insuperable obstacles to our making any considerable progress in the working of metals, had we been forced to depend on home or foreign timber. We, therefore, are disposed to regard Lord Dudley's discovery of the mode of smelting and manufacturing iron by means of coal only, without the aid of wood, as one of the most important ever made in the arts. We do not know that it is surpassed even by the steam engine or spinning-frame. At all events, we are quite sure that *we* owe as much to it as to either of these great inventions. But for it, we should have always been importers of iron; in other words, of the materials of machinery. The elements, if we may so speak, out of which steam-engines and spinning-mills are made, would have been dearer here than in most other other countries. The fair presumption consequently is, that the machines themselves would have been dearer; and such a circumstance would have counteracted, to a certain extent, even if it did not neutralize or overbalance, the other circumstances favorable to our ascendancy. But now we have the ores and the means of working them in greater abundance than any other people; so that our superiority in the most important of all departments —that of machine-making—seems to rest on a pretty sure foundation.

" It is further clear, that without a cheap and abundant supply of fuel, the steam-engine, as now constructed, would be of comparatively little use. It is, as it were, the hands; but coal is the muscles by which they are set in motion, and without which their dexterity cannot be called into action, and they would be idle and powerless. Our coal mines may be regarded as vast magazines of *hoarded* or *warehoused* power; and unless some such radical change be made on the steam-engine as should very decidedly lessen the quantity of fuel required to keep it in motion, or some equally powerful machine, but moved by different means, be introduced, it is not at all likely that any nation should come into successful competition with us, in those departments in which steam-engines, or machinery moved by steam, may be most advantageously employed.

" Since the introduction of steam-engines, Water-falls, unless under very peculiar circumstances, have lost almost all their value. Steam may be supplied with greater regularity, and being more under command than water, is therefore a more desirable agent. This, however, is but a small part of its superiority. Any number of steam-engines may be constructed in the immediate vicinity of each other, so that all the departments of manufacturing industry may be brought together and carried on in the same town, and almost in the same factory. A com-

bination and adaptation of employments to each other, and a consequent saving of labor, is thus effected, that would have been quite im-
practicable, had it been necessary to construct factories in different
parts of the country, and often in inconvenient situations, merely for
the sake of waterfalls.

"It may be supposed, perhaps, that a difficulty of this sort might
have been obviated by the employment of horse-power instead of steam;
but the following statement, which we extract from Dr. Ure's work,
shows conclusively that this would not have been the case :—

"'The value of steam-impelled labor may be inferred from the following facts, communicated to me by an eminent engineer, educated in the
school of Boulton and Watt :—A manufacturer in Manchester works a
sixty-horse Boulton and Watt's steam-engine, at a power of one hundred and twenty horses during the day, and sixty horses during the
night; thus extorting from it an impelling force three times greater
than he contracted or paid for.   One *steam* horse-power is equivalent
to 33,000 pounds avoirdupois, raised one foot high per minute; but an
animal horse-power is equivalent to only 22,000 pounds raised one
foot high per minute, or, in other terms, to drag a canal boat two
hundred and twenty feet per minute, with a force of one hundred
pounds acting on a spring; therefore, a steam-horse power is equivalent in working efficiency to one living horse, and one-half the labor
of another.   But a horse can work at its full efficiency only eight
hours out of the twenty-four, whereas a steam-engine needs no period
of repose; and, therefore, to make the animal power equal to the
physical power, a relay of one and a half fresh horses must be found
three times in the twenty-four hours, which amounts to four and a half
horses daily.   Hence, a common sixty-horse steam-engine does the
work of four and a half times sixty horses, or of two hundred and
seventy horses.   But the above sixty-horse steam-engine does one-half
more work in twenty-four hours, or that of *four hundred and five* living
horses !   The keep of a horse cannot be estimated at less than 1s. 2d.
per day; and, therefore, that of four hundred and five horses would be
24l. daily, or 7,500l. sterling, in a year of three hundred and thirteen
days.   As eighty pounds of coals, or one bushel, will produce steam
equivalent to the power of one horse in a steam-engine during eight
hours' work, sixty bushels, worth about 30s. at Manchester, will maintain a sixty-horse engine in fuel during eight effective hours,—and two
hundred bushels, worth 100s., the above hard-worked engine during
twenty-four hours.   Hence, the expense per annum is 1,565l. sterling,
being little more than one-fifth of that of living horses.   As to prime
cost and superintendence, the animal power would be greatly more ex-

pensive than the steam power. There are many engines made by Boulton and Watt, forty years ago, which have continued in constant work all that time with very slight repairs. What a multitude of valuable horses would have been worn out in doing the service of these machines! and what a vast quantity of grain would they have consumed! Had British industry not been aided by Watt's invention, it must have gone on with a retarding pace, in consequence of the increasing cost of locomotive power, and would, long ere now, have experienced, in the price of horses and scarcity of water-falls, an insurmountable barrier to further advancement: could horses, even at the low prices to which their rival, steam, has kept them, be employed to drive a cotton-mill at the present day, they would devour all the profits of the manufacturer.'"

Water power has heretofore been considered cheaper, especially for small manufacturing establishments, than steam power; but eminent engineers have carefully investigated the subject, and are of opinion that in any position where coal can be had "at ten cents per bushel," steam is as cheap as water power at its minimum cost. Even for cotton factories, the manufacturers of New England, according to Montgomery, consider the *advantages of a good location as fully equal to the extra expense of steam power, even when coal must be transported from Pennsylvania to Massachusetts,* and the largest mills now being erected are to have steam as a motive power. Steam, therefore, until superceded by some more effective agent, will be the power principally relied upon to propel Machinery; and as wood for the generation of steam upon an extensive scale is out of the question, we may safely conclude that at no very distant day, the centre of our Manufactures will certainly be in or near a district possessing inexhaustible supplies of cheap coal.

The importance of coal as a useful agent in the Arts, is not, however, limited to its capacity to produce power. It lies at the base of all manufacturing and mining operations, and surpasses all other natural products in the power of attracting to the vicinity where it can be obtained abundantly and cheaply—industry and population. In England, the Woolen Manufacturers were once scattered over Sussex, Kent, and other southern counties, but they have been attracted, principally by the wonderful magnetism of coal, to the North. In the coal districts of England we find all her great manufacturing cities and towns; Birmingham, with its population of perhaps 300,000; Leeds, with a population of 200,000; Sheffield, whose hardware manufactures are known all over the world, are located in districts abounding with coal, and its usual accompaniment—Iron. Manchester, the great seat of the Cotton Manufactures of Great Britain, whose population now exceeds 600,000, is situated on the edge of an immense and seemingly inexhaustible coal-bed. A like proximity may be noticed in the location of Bolton, Bradford, Carlisle, Huddersfield, Oldham and Wolverhampton in England; Merthyr Tydvil in Wales; Glasgow in Scotland; and Charleroy in Belgium.

The principal manufacturing cities of Europe, in this respect, present a striking contrast to those of the United States. In New England, the sites of the chief manufacturing towns seem to have been chosen solely with reference to abundant water power; and herein we have one reason for believing that their present pre-eminence is destined soon to be overshadowed, and finally obscured by that of other cities possessing all their other advantages, and having, in addition, a convenient proximity to our immense coal-beds. In spite of our warm regard for New England, and sincere wishes for her continued prosperity in Manufactures, we think the sceptre will eventually, and ere long, depart from Judah. But New England will be New England still. The virtues which make a great people are indigenous to her soil, and will continue to animate and ennoble the population when her capitalists

2*

and ingenious men have sought other localities, possessing greater physical advantages for the fulfillment of their " manifest destiny."

3. With regard to the third point, viz. *favorable situation as respects commerce with other countries,* its importance is second only to that which we have just considered. It is in the nature of Manufactures to be regardful of distant and foreign markets. The accelerated production which results from the application of machinery, enables one manufacturer to supply the wants of many hundreds of consumers, and a country or part of a country possessing superior facilities for Manufactures, can supply other countries with manufactured goods cheaper than they can produce them. Great Britain, it is well known, exports the bulk of her manufactured commodities. The writer whom we previously quoted, remarks :

" Owing to the facilities afforded by our insular situation for maintaining an intercourse with all parts of the world, our manufacturers have been able to obtain supplies of the raw materials on the easiest terms, and to forward their own products wherever there was a demand for them. Had we occupied a central situation, in any quarter of the world, our facilities for dealing with foreigners being so much the less, our progress, though our condition had been otherwise in all respects the same, would have been comparatively slow. But being surrounded on all sides by the sea, that is, by the great highway of nations, we have been able to deal with the most distant as well as with the nearest people, and to profit by all the peculiar capacities of production enjoyed by each."

In the United States, the consumption of manufactured goods is so vast, that we are apt to regard any foreign demand as unimportant. But for the year ending June 30, 1855, we exported manufactured commodities to the amount of $30,609,518. The list of articles exported embraced nearly all our prominent Manufactures—Cotton piece Goods being the most valuable item, amounting to

$5,857,181; Manufactures of Iron the next, $3,753,472; and Artificial Flowers and Billiard Tables the smallest, of which, however, the exports amounted to about $8000. The Canadas, the West Indies, the South American Republics, Spain and her dependencies, Russia, China, are all ready and willing to exchange their natural products for our manufactured goods, if we can compete with other manufacturing countries in their markets. Even English consumers have no objection to take our Manufactures, not excepting Cotton goods, if the price can be arranged satisfactorily. As early as 1826 we exported $664 cotton goods to England; in 1837, $11,889; and ever since, we believe, there have been small shipments annually. Hence, though it be true that, in the United States, the Home market is the one at present of chief importance, and though the consumption of manufactured goods is so immense that there is undoubtedly room for the establishment of many important local manufactories, if such can exist, at a variety of points; yet to supply a foreign demand, as well as to obtain the raw materials on the easiest terms, a situation on or near the sea-coast is desirable; and as large establishments can produce more cheaply than small ones, as we shall subsequently show, it is highly important for such to choose a locality possessing, in addition to the other moral and physical advantages, a complete communication, by railroads and canals, with all parts of our own country, and an established commerce or facilities for commerce with foreign countries.

4. A suitable *Climate* is also a consideration of very great importance. The influence of climate upon the productiveness of industry, especially in Manufactures, is very marked. A warm climate not only enervates the body, but enfeebles the mind. It diminishes the utility of money; and by rendering houses and clothing less

necessary to existence, relieves the inhabitants of one great spur to industry and invention. In very cold climates, on the other hand, the powers of Nature are benumbed, and the difficulty of preserving life overrides all considerations for making existence comfortable. The climate which seems most favorable to the development of manufacturing industry, is that which is also most conducive to health and longevity, imparting vigor to the frame and force to the intellect, and if we may judge from the past, it is found especially, if not exclusively, in that part of the Eastern Hemisphere which lies between the parallels of 45° and 55°, and in the Western between 39° and 45° North Latitude. Climate has also a direct influence upon the durability of buildings, the working of machinery, and the dyeing of fabrics—points that we may subsequently consider—and thus becomes an element of important consideration in many kind of Manufactures.

The *Soil* of a country or district well adapted for Manufactures, need not be naturally very fertile. In fact a soil naturally so rich that Agriculture is an easy art, will not afford sustenance to many kinds of Manufactures. In Southern Europe, for instance, where, according to one authority, the only art which the farmers know is to leave their ground fallow for a year, so soon as it is exhausted, and the warmth of the sun alone and temperature of the climate enrich it and restore its fertility, we look in vain for those enterprises which are the product of qualities and virtues that are nourished by difficulties, not facilities. In England, the soil is naturally coarse and stubborn, but capable of being made highly productive by labor, expense, and good husbandry ; and such a soil, with the habits of careful cultivation induced thereby, is the safest reliance for supplying the markets of a manufacturing district with the necessaries of life, at moderate prices.

III. But the one thing essential for the cheap production of manufactured commodities, and without which all the other moral and physical advantages are ineffectual, remains to be noticed. It is ASSOCIATION or COMBINATION OF LABOR. It is unnecessary to show that man, unaided by his fellow men, is a helpless being. If it were, we might refer to the savages of New Holland, who, they say, never help each other even in the most simple operations; and their condition, as may be supposed, is hardly superior, in some respects it is inferior, to that of the wild animals which they now and then catch. The first step in social improvement, is association for mutual security and mutual assistance; and every advance in civilization is directly the result of some new combination of efforts. All the marvels of past times, produced by human agency —the Temples, Pyramids and Catacombs—and all the wonders of the present—its Railroads, Telegraphs, Mines and Manufactures—have a common origin in association of numbers for a common purpose. All industrial pursuits depend more or less upon this principle for development, but in none are its advantages more strikingly manifest than in manufacturing operations.

To combine Labor effectually, it is necessary first to *separate employments into parts—that is, to assign to each co-worker a special occupation.* The Division of Labor, as Wakefield, it is said, was the first to point out, is only a single department of a more comprehensive Law, which he denominated Co-operation, or combined action of numbers. Its efficiency, however, as an aid to production, is none the less important, and has been abundantly illustrated by all who have written on Political Economy. Adam Smith illustrated it from pin-making; and mentioned that ten men, in a small manufactory, but indifferently accommodated with the necessary machinery, could make, by confining themselves as much as possible to distinct operations, upward of 48,000 pins in a day, or 4,800 for each individual, whereas if they all wrought separately and independently, they certainly could not, each of them, make twenty, perhaps not one pin in a day.

M. Say illustrates the principle by reference to the manufacture of playing-cards, and says that each card, before being ready for sale, undergoes no fewer

"Than seventy operations, and if there are not seventy clasess of workpeople in each card manufactory, it is because the division of labor is not carried so far as it might be ; because the same workman is charged with two, three, or four distinct operations. The influence of this distribution is immense. I have seen a card manufactory where thirty workmen produced daily 15,500 cards, being above 500 cards for each laborer ; and it may be presumed that if each of these workmen were obliged to perform all the operations himself, even supposing him a practiced hand, he would not perhaps complete two cards in a day, and the thirty workmen, instead of 15,500 cards, would make only sixty."

Henry C. Carey refers to weaving in India, and says :

"In India each weaver works by himself. He purchases at a high price, on credit, the materials with which he is to work, and the provisions required for his support, and he sells the product at a price not exceeding one-third of its market value. Here is no combination of action—no division of labor. The whole work is to be performed by the single individual ; and the time that might be employed in finishing the finest muslins, is wasted upon various processes requiring inferior ability, from the purchase of the cotton to its final sale."

Further illustrations are therefore superfluous. The principle is settled : quantity and economy of production are immeasurably aided by the division of employments into parts for the sake of combination of Labor.

Secondly, to combine Labor to the best advantage, it is essential to conduct operations *on a sufficientlg large scale to have a separate workman, or a separate machine, for each process into which it is convenient to subdivide the manufacture, and to afford each workman or machine full employ-*

ment in that special occupation. This we regard to be the natural limit of a manufacturing establishment. Any extension beyond this may be said to comprise two establishments in one; and any establishment of less size cannot realize the full benefits of a Division of Labor, and consequently cannot produce with the utmost efficiency and economy. The application of the principle, however, would, in most kinds of Manufactures, lead to moderately large establishments; and that such establishments can produce more economically, or in other words afford to work for a less percentage of profit, is simply a well-established fact. A Philadelphia miller is content with the bran alone as his toll for grinding his customer's corn; but a country miller, in a sparsedly populated district, must take a considerable portion of the grain for converting the balance into flour. The *expenses* of a business do not by any means increase proportionally to the quantity of business. A merchant, for instance, who, by advertising, has attracted trade to the amount of $1,000,000 per annum, is not required to pay ten times as much rent, nor does he need ten times more clerks, fuel, lights, &c., than the man who "never advertises," and perchance, does a business of $100,000 a year. In a large manufacturing establishment, the expenses of superintendence, repairs, etc., form but a trifling percentage on the aggregate product, while the time consumed in making a large purchase is very little more than in making a small one. Producers on a large scale can also afford to procure the best and most expensive machinery; and in some kinds of Manufactures, those who produce largely are content with "savings" as their profit, and are enabled to save what would be "waste" in a small establishment. Mr. Whitney, at his car-wheel establishment in Philadelphia, can save from the cinders, we are informed, enough iron to content a gentleman of his moderate views as to profit, but a manufacturer of car-wheels on a small scale, would not find it profitable to provide the machinery requisite for that purpose. From these and other considerations, which want of space forbids us to allude to, we infer that in future the manufacture of leading articles of consumption will be more and more conducted by large establishments, in a locality possessing in the highest degree of perfection, the moral and physical advantages that are essential to manufacturing prosperity. But it does not follow that large establishments will swallow up all smaller ones, unless it be those of a precisely similar kind, situated outside of the centres of combination. The economy which results from producing on a large scale, induces an increased demand for the manufactured goods; and an increased demand leads to a more minute subdivision of a manufacture into parts. When thousands of machines composed of Iron and Wood are required, we find establishments springing up, devoted exclusively to making parts—one, the nuts and washers; another the screws; another the bolts; another the nails; and others tools and machines to facilitate making parts, and so on, each extensive in its way, and thus large establishments in the leading branches of Manufactures are the parents of other extensive concerns in minor branches. A man who has not the requisite capital to conduct a leading Manufacture where large establishments abound, permit us to suggest, will not benefit himself by moving away from them. His policy is, we submit—to remain at all events, in their immediate vicinity, and then to accommodate his business to their operations and to his capital—that is, he will find it more profitable to be an extensive manufacturer of eyes for children's dolls in the centre of Manufactures, than a small manufacturer of machinery anywhere.

Lastly, to produce with the utmost efficiency and economy, *manufacturing establishments must be together.* The area of England and Wales is only about one-fourth more than that of Pennsylvania. In England all the large manufacturing establishments are situated, as we have stated, in close proximity to the coal beds. Manufacturers—one after another—have abandoned their factories in the Agricultural counties and moved their machinery to the district of which Manchester may be called the central point. Babbage has referred to one of the advantages resulting from this aggregation:

" The accumulation of many large manufacturing establishments in one district," he says, " has a tendency to bring together purchasers or their agents from great distances, and thus to cause the institution of a public mart or exchange. This contributes to increase the information relative to the supply of raw material and the state of demand for their produce, with which it is necessary manufacturers should be well acquainted. The very circumstance of collecting periodically, at one place, as large a number as possible, both of those who supply the market and those who require its produce, tends strongly to check those accidental fluctuations to which a small market is ever subject, as well as to render the average of the prices paid much more uniform in its course."

The accumulation of many large and excellent manufacturing establishments in one district, also gives a character and stamp to the Manufactures, which others who centre there receive the benefit of. There is also a mutuality of interest between manufacturers of even essentially different products, that renders aggregation highly desirable. The finished products of one class of manufacturers are often the raw materials of another. The power-looms of Mr. Jenks are but the instruments of production for the Manufacturer of Cotton and Woolen goods ; and the finished commodities of the latter, are the raw materials of those who manufacture ready-made Clothing. Pig iron—the finished commodity

3

of the smelter, is the raw material of him who rolls the bar; and the bar is again the raw material of sheet iron; which, in its turn, is the raw material of the nail and the spike. A sugar-refiner consumes the hogsheads, boxes and barrels of the cooper, paper of the paper-maker, and the finished products of coppersmiths, nail-manufacturers, twine-spinners, printers and various others. In fact, the largest and in many instances the sole consumers of certain manufactured articles, are the Manufacturers of other products; and finished commodities being, as a general rule, cheapest at the place of their production, without commissions or charges for transportation, it is certainly for the interest of those who buy to produce, and those who produce to sell, to be together. Aggregation, in fact, is the only effectual means of accumulating and combining all economies.

In Combination there is mystery like that of Trinity in Unity. Like the philosopher's stone, it turns all to gold —like the lever or the screw, it adds to man's power many hundred fold. Protective tariffs are useful as swaddling clothes to the infant; banks facilitate exchanges; but the perfection of combination cannot be attained except by aggregation in a suitable locality. If the Manufacturers of the United States ever hope to attain an independent position—independent of Foreign competition and of Home legislation, independent of commission merchants and of each other—they must centralize, so far as centralization is at all practicable. They must come out from sylvan retreats, deny themselves the advantages of mill-races and the harmonies of frog-ponds. They must tear down the miserable shingles " No admittance on any pretext whatever"—abandon their petty jealousies, enlarge their views, and co-operate like men and brethren. Blacksmiths, Cobblers and Wheelwrights may eke out an existence " in the neighborhood of the plow and the harrow," but in a Democratic country, whose people believe in buying where they can buy the cheapest, whether wisely or not we do not say, Manufacturers, in the true sense of the term, who attempt isolation, will inevitably find themselves, sooner or later, undersold, first, by those who operate in the centres of Combination, and finally, undersold by the Sheriff.

From *all* these considerations, which in substance we believe to be thoroughly sound, and to which we invite the closest scrutiny, we are led irresistibly to the conviction—that but few countries in the world, and but few places in any country, are well adapted for general Manufactures. Secondly: That the *best possible locality in the United States for general manufacturing is an attractive and suitable centre of Wealth, Population and Intelligence, situated in a populous district, abounding in well developed mines of Coal and Iron, and possessing established and superior facilities of intercommunication with all parts of our own country, and for commerce with foreign countries.* And Thirdly: *If there be two or more such localities, the one possessing desirable, in addition to the essential advantages in the highest degree of perfection, and the one already having the greatest number of large and well-managed manufacturing establishments, must be the best market in which to buy the commodities manufactured there, and eventually will be the chief seat of Manufactures in the United States.*

Now, have we such a locality? The centres of Wealth, Population and Intelligence in the United States are not numerous. Suitable centres for manufacturing, situated in close proximity to well-developed mines of Coal and Iron, and possessing established facilities for procuring raw materials on the easiest terms, and sending away manufactured produce, are very few; and of centres of Wealth, Population and Intelligence, we know of *but one* that possesses all the essential and most of the desirable advantages for manufacturing every variety of products, and which already contains many large and well-managed manufacturing establishments. To that one we invite the attention of all who produce, and deal in or consume manufactured commodities. The subject is one in which all these have a deep interest. If it be true that the highest degree of economy in production depends upon a *combination* of certain circumstances, rarely found, but which exist in the highest degree of perfection in a certain place, all who desire to produce cheaply, and all who desire to buy cheaply, have a direct pecuniary interest in knowing the facts, and in aiding to develop its capabilities. The place to which we invite earnest and sagacious attention, as the best manufacturing centre at present in the United States, is PHILADELPHIA, in the State of Pennsylvania.

PHILADELPHIA AS A MANUFACTURING CENTRE.

"I know thy *works.*"—Rev. iii. 8.

Philadelphia is a scriptural name, composed of two Greek words, which signify, as usually interpreted, brotherly love. St. John, as we are informed in the Revelations, was instructed to indite a consolatory epistle to "the church in Philadelphia," a city of Asia Minor, about seventy-two miles from Smyrna. The Philadelphia of which we write is a namesake of the biblical city; and though not very ancient, is yet a cotemporary with most of the important events in American history. It was founded in 1682–3, by William Penn, who with a colony of English Friends or Quakers, had come to America to settle a province or tract of land granted to him by Charles II., in payment of a debt due by the government to his father. Before attempting any overt acts of sovereignty, however, Penn was wisely "moved" to acknowledge and purchase the rights of the aborigines, and thus, as Raynal has remarked, signalized his arrival by an act of equity, which made his person and his principles equally beloved. He also promulgated a series of laws, in which *Liberty of Conscience* was the first in order and importance. "A plantation reared on such a seed-plot," says Chalmers, "could not fail to grow with rapidity, to advance to maturity, to attract notice of the world."

The site chosen for the proposed city was a nearly level

3*

(49)

plain between the Delaware and Schuylkill rivers, about six miles above their junction, and sixty miles from the ocean, by a direct line, though nearly a hundred miles by the course of the river. The influences that determined Penn in his choice of the spot are said to have been "the approach of the two rivers; the short distance above the mouth of the Schuylkill; the depth of the Delaware; the land heavily timbered; the existence of a stratum of brick clay on the spot, and immense quarries of building stone in the vicinity." In drafting the plan of his American city, Penn is supposed to have had in view the celebrated city of Babylon, which he certainly imitated in the regularity of the streets, and which he seemed desirous to emulate in size, for he gave orders to his commissioners to lay out a town that would have covered an area of 8000 acres. It was found, however, that "hundred-acre lots," which some of the squatter-sovereigns secured, would never answer the end of a city in a new country, and the plan was subsequently reduced. In 1701 it was again contracted, when the city was declared to be bounded by the "two rivers Delaware and Schuylkill, and Vine and Cedar streets as north and south boundaries." These continued to be the corporate limits of the city until 1854—the suburbs, as population extended, being divided into districts, as Spring Garden, Northern Liberties, Kensington, Southwark, Moyamensing and West Philadelphia, which in 1850 contained nearly twice as many inhabitants as the city proper.

The events in the early history of the town, prior to the Revolution, are not very striking. We subjoin a summary of the most important, as far as possible, in their chronological order. In 1687 a printing-press, the second in America, was set up; in 1689 Penn established a public High School with a charter. In 1742 Franklin projected an Academy and Free School, which became presently a College, and finally the "University of Pennsylvania." In 1765, the merchants of Philadelphia, in consequence of various restrictive and ill-advised Acts, particularly the Stamp Act, passed by the Parliament of Great Britain, pledged their word of honor not to order nor sell on commission any goods from Great Britain, except certain articles, more particularly those necessary for carrying on Manufactures, "unless the Stamp Act be repealed." In 1774 the first Congress in America assembled in Carpenters' Hall, (a building still standing in a court back of Chestnut street, between Third and Fourth streets,) to take into consideration the state of our relations with the mother country. In this city was adopted the Declaration of Independence, which was read from a stand in the State House yard, by Captain John Hopkins, July 4, 1776. From September, 1777 to June, 1778, in consequence of the disastrous battles of Brandywine and Germantown, the British army had possession of the city. The Convention that framed the present Constitution of the United States, met in Philadelphia, May, 1787. Here George Washington, when President of the United States, resided, in a building on the south side of Market street, one door east of Sixth, the lot being now occupied by a palatial business edifice, widely known as "Bennett's Tower Hall Clothing Store."

The first bank established in the United States was the Bank of Pennsylvania, opened at Philadelphia on the 17th of July, 1780, with a capital of £300,000, its special object being to supply the American army with provisions. In 1782 the Bank of North America went into operation; and in 1791 the United States Bank. In 1792 Congress passed an act establishing "a Mint for the purpose of a National coinage," to be situate and carried on at the Seat of Government of the United States for the time being, which was then at Philadelphia. In 1793, coinage was commenced in a building on Seventh street, opposite Zane, still known as the "Old Mint," and continued there until 1833, when the present noble edifice at the north-west corner of Chestnut and Juniper streets was completed.*

In the autumn of 1793 the yellow fever visited Philadelphia, and carried off more than 4000 persons, out of a population of a little over 40,000, of whom half, it was thought, had fled the city. The pestilence visited the city again in 1798, but was not so fatal as in 1793. The wars commenced by France in 1792 with other European powers, and which were continued until the abdication of Napoleon in 1814, had an immense influence in developing American Commerce, and Pennsylvania shared largely in this prosperity. Large importations were made from China and India into Philadelphia, for re-exportation to European markets. Our ships then enjoyed the carrying trade of the world, and numbers of our citizens accumulated large fortunes.

In January, 1801, Philadelphia was supplied for the first time with water from Water Works erected according to a plan proposed by Mr. Latrobe, viz. " to make a reservoir upon the banks of the Schuylkill, to throw up a sufficient quantity of water into a tunnel, and to carry it thence to a reservoir in Centre Square ; and after being raised there, to distribute it throughout the city by pipes." These works were superceded by the

* Since its establishment in 1793, to the close of the year 1856, the Mint at Philadelphia coined 525,536,141 pieces, of the value of $391,730,571 86; the gold coinage being $306,445,970 78, the silver coinage $83,685,297 99, and copper coinage $1,599,303 09. The entire coinage of the United States to the same period was $563,433,708 12.

The present officers of the Mint at Philadelphia, are:—*Director*, James Ross Snowden ; *Treasurer*, Daniel Sturgeon ; *Chief Coiner*, George K. Childs ; *Melter and Refiner*, James C. Booth ; *Engraver*, James B. Longacre ; *Assayer*, Jacob R. Eckfeldt ; *Assistant Assayer*, William E. DuBois.

present works erected at Fairmount, which we will subsequently notice.

In 1811, Dr. James Mease published a book which he entitled "A Picture of Philadelphia." At that time Philadelphia was the most populous city in the Union. From an enumeration made the previous year, it appears that the number of dwelling-houses in the city and districts, was 15,814, and the population of the city and county amounted to 111,210. The population of the whole of Manhattan Island, at the same period, embracing the city of New York, was 96,372. Philadelphia then, as now, was the most healthy city in the Union. The average of deaths per day, in Philadelphia, was $5\frac{2}{3}$, whereas in New York, with a smaller population, it was $6\frac{1}{3}$. We subjoin Dr. Mease's remarks on the Manufactures, from which it will be perceived that Philadelphia was already celebrated in various departments of Manufacturing industry.

"The various coarser metallic articles, which enter so largely into the wants and business of mankind, are manufactured to a great extent, in a variety of forms, and in a substantial manner. All the various edged tools for mechanics are extensively made : and it may be mentioned as a fact calculated to excite surprise, that our common screw auger, an old and extensively used instrument, has been recently announced in the British publications, as a capital improvement in mechanics, as it certainly is, and that all attempts by foreign artists to make this instrument durable, have failed.

"The finer kinds of metals are wrought with neatness and taste. The numerous varieties of tin ware in particular, may be mentioned as worthy of attention. But above all, the working of the precious metals has reached a degree of perfection highly creditable to the artists. Silver plate fully equal to sterling, as to quality and execution, is now made, and the plated wares are superior to those commonly imported in the way of trade. Floor-cloths of great variety of patterns, without seams, and the colors bright, hard and durable ; various printed cotton stuffs, warranted fast colors ; earthenware, yellow and red, and stone ware are extensively made ; experiments show that ware equal to that

of Staffordshire might be manufactured, if workmen could be procured.

" The supply of excellent patent shot is greater than the demand. All
the chemical drugs, and mineral acids of superior quality, are made by
several persons : also cards, carding and spinning-machines for Cotton,
Flax, and Wool. Woolen, worsted, and thread hosiery have long
given employment to our German citizens : and recently, cotton
stockings have been extensively made.

" Paints of twenty-two different colors, brilliant and durable, are in
common use, from native materials ; the supply of which is inexhaustible. The chromate of lead, that superb yellow color, is scarcely equaled
by any foreign paint. There are fifteen rope-walks in our vicinity.
We no longer depend upon Europe for excellent and handsome paper
hangings, or pasteboard, or paper of any kind. The innumerable articles into which leather enters are neatly and substantially made : the
article saddlery forms an immense item in the list. The leather has
greatly improved in quality ; the exportation of boots and shoes to the
Southern States is great ; and to the West Indies, before the interruption to trade, was immense. Morocco leather is extensively manufactured. The superiority of the carriages, either as respects excellence
of workmanship, fashion, or finish, has long been acknowledged. The
type-foundry of Binney & Ronaldson supplies nearly all the numerous
printing-offices in the United States. There are one hundred and two
hatters in the City and Liberties. Tobacco, in every form, gives employ
to an immense capital. The refined sugar of Philadelphia has long
been celebrated : ten refineries are constantly at work. Excellent
japanned and pewter ware : muskets, rifles, fowling-pieces and pistols
are made with great neatness. The cabinet-ware is elegant, and with
the manufacture of wood generally, is very extensive. The houses are
ornamented with marbles of various hues and qualities, from the quarries near Philadelphia.

" Mars Works, at the corner of Ninth and Vine streets, and on the
Ridge road, the property of Oliver Evans, consists of an iron foundry,
mould-maker's shop, steam-engine manufactory, blacksmith's shop,
and mill-stone manufactory, and a steam-engine used for grinding sun-
dry materials for the use of the works, and for turning and boring
heavy cast and wrought iron work. The buildings occupy one hundred
and eighty-eight feet front, and about thirty-five workmen are daily employed. They manufacture all cast or wrought-iron work for machinery
for mills, for grinding grain or sawing timber ; for forges, rolling and
slitting-mills, sugar-mills, apple-mills, bark-mills, &c. Pans of all
dimensions used by sugar-boilers, soap-boilers, &c. Screws of all sizes

or cotton-presses, tobacco-presses, paper-presses, cast iron gudgeons, and boxes for mills and wagons, carriage-boxes, &c., and all kinds of small wheels and machinery for Cotton and Wool spinning, &c. Mr. Evans also makes steam-engines on improved principles, invented and patented by the proprietor, which are more powerful and less complicated, and cheaper than others ; requiring less fuel, and not more than one-fiftieth part of the coals commonly used. The small one in use at the works is on this improved principle, and is of great use in facilitating the manufacture of others. The proprietor has erected one of his improved steam-engines in the town of Pittsburg, and employed to drive three pair of large millstones with all the machinery for cleaning the grain, elevating, spreading, and stirring and cooling the meal, gathering and bolting, &c., &c. The power is equal to twenty-four horses, and will do as much work as seventy-two horses in twenty-four hours : it would drive five pair of six-feet millstones, and grind five hundred bushels of wheat in twenty-four hours.

" All kinds of castings are also made at the Eagle Works, on Schuyl-kill, belonging to S. & W. Richards."

In 1812, *Steam* works for supplying the city with water were commenced at Fairmount, and in 1815 the use of the Centre Square Works was discontinued. In 1819, Councils resolved to erect the present *Water-power* Works, which for a long time were the only works of the kind in the United States, and which are yet unsurpassed by any in the whole country. The water from the Schuylkill is turned into a forebay 419 feet long and 90 feet wide ; whence it falls upon and turns eight wheels, from sixteen to eighteen feet in diameter, and one turbine wheel, each having its separate pump, and which elevate the water ninety-two feet to the top of a partly natural elevation, immediately at the works, and which give them their name. The reservoirs of these works, including the new one on Corinthian Avenue, furnish storage to the amount of 57,642,787 gallons. This is about equal to five days' supply in July and August. The total cost, including laying pipes, &c., to the present time, is about $3,500,000.*

* In addition to the Works at Fairmount, Philadelphia has, at the present

In 1829 the Pennsylvania Canal was completed, which, with the Schuylkill and Union Canals, previously constructed, formed a connection with the Ohio River, *via* Reading and Middletown.

In December, 1831, Stephen Girard, "Mariner and Merchant," died worth about $10,000,000, and bequeathed by his Will large sums to public uses, among others the sum of $2,000,000 for the erection of a College, now known as the Girard College. In 1835 Philadelphia was first supplied with gas from Works erected on Market street, near the Schuylkill.* In the same year a part of the Reading Railroad, to connect Philadelphia with the Schuylkill coal region, was put under contract, and in 1842 the first train passed over the whole line between Pottsville and Philadelphia. In 1837 the Philadelphia, Wilmington and Baltimore Railroad was completed. In 1838 the city was disgraced for the first time by a mob

time, three other Water Works, viz. Schuylkill Works, Delaware Works, and Twenty-fourth Ward Works. The total amount of water supplied by all these Works, in 1856, was 5,735,938,966 wine gallons. The Duplicates of the Water rents for the same year amounted to $359,906 08.

* Since that period the Northern Districts and Germantown erected Gas Works; and in 1854 the city completed new and additional works near Gray's Ferry Bridge, having the largest gasholder, it is believed, in the United States, being 160 feet in diameter and 90 feet high, and capable of holding 1,800,000 cubic feet of gas. The cost of the whole now belonging to the City Gas Trust, is about $2,500,000. We are furnished by John C. Cresson, Esq., who has been Engineer of the City Gas Works since 1836, with the following statistics:—

Street mains laid to January 1, 1858, - - -	214½ miles.
Service pipes, " " - - - - -	74 "
No. of Services and Meters in use, - - - -	25,180
" Lights in use, (private,) - - - -	332,487
" " " (public,) - - - - -	3,810
Gas made in 1857, - - - - -	469,067,000 cubic feet.
Total, made in 21 years, - - - -	3,198,088,000 " "

Present manufacturing capacity, 1857, 2½ million feet per diem.

that burned the Pennsylvania Hall, fired the Shelter for Colored Orphans, and attacked the negro quarters. In 1844 the city was again disquieted by riots incited by the presumed interference of Catholics with the elective franchise, and several Catholic churches were burned. In 1847, the Pennsylvania Railroad, to connect Philadelphia with the Ohio River, was commenced, and finally completed in 1854. In 1850 a census was taken, which showed that Philadelphia contained 23,601 more dwellings than the city of New York, and a population of 408,762, being an increase of $58\frac{1}{2}$ per cent. in the ten years preceding the census of 1850, and $953\frac{1}{2}$ per cent. in the sixty years since the first National census. Of the population of 1850, 17,500 were born in England; 72,312 in Ireland; 22,750 in Germany; 3,291 in Scotland; and 1,981 in France. Total foreign, 121,699.

In 1854 the corporate limits of the city were made coextensive with those of the county of Philadelphia, covering an area of 120 square miles, and placing the villages and towns of Bridesburg, Frankford, Holmesburg, Byberry, Nicetown, Flourtown, Andalusia, Bustleton, Rising Sun, Milestown, Jenkintown, Germantown, Chestnut Hill, Falls, Manayunk, Roxborough, West Philadelphia, Mantua, Haddington, Hamilton, and Darby under the wise guardianship of a Metropolitan Mayor and City Councils, *sans peur et sans reproche*.

These, we believe, may be called the most important events in the Annals of Philadelphia. In the history of a place whose "birth and spring-time" carry us back nearly a century anterior to the American Revolution, there are necessarily many events of greater or less importance that deserve to be commemorated. No city of equal age can present a fairer or more interesting record of the past than Philadelphia; none has been more prolific in

4

men who have been eminent in their day and generation; and not one has been so fortunate in inspiring that species of affection which manifests itself in culling and preserving, as a labor of love, the features and memorials of a time gone by. John F. Watson, in his "Annals," has done all that can be desired to preserve the lineaments and characteristics of what may be called the "olden time" of Philadelphia; and our Historical and Philosophical Societies have accumulated papers and disquisitions upon every conceivable subject pertaining thereto. Truly, if the prosperity of a city be promoted in proportion to the affectionate attachment of its inhabitants—a feeling, as Everett has observed, entitled to respect, and productive of good, even if it may sometimes seem to strangers over-partial in its manifestations—the citizens of Philadelphia may repeat, with confidence, the poetical prediction of Taylor, the astrological Hague of the eighteenth century:—

> "A city built 'neath such propitious rays
> Will stand to see old walls and happy days."

The Past of this city, therefore, has been well cared for; its historical incidents are preserved in its own and in the records of our country; the fame of its great men will survive "fresh in eternal youth"; and neophytes in Archæology may well despair unless they devote attention to its Present, which, with its material progress, its advance, especially in Manufactures, its Railroads and its Fire and Police Telegraphs, would at any time form a theme sufficiently comprehensive in itself to exclude any minute reference to the events of the past.

I. Philadelphia as it is.

PHILADELPHIA is usually described as the second city in the United States; and, if we except Paris, nearly equals the largest capitals on the continent of Europe in population. No census has been taken since 1850; but assuming that the increase has been in the same ratio as that which distinguished the ten years preceding the last national census, its present population cannot be far short of 600,000. Its entire length, as per Ellet's Survey, is twenty-three miles, and average breadth five and a half miles ; area, one hundred and twenty-nine and one eighth square miles, or 82,700 acres. The densely inhabited portion of Philadelphia extends about four miles on the Delaware, from Southwark north to Richmond, formerly Port Richmond, and two and a half miles on the Schuylkill, having a breadth between the two rivers, assuming South street formerly the Southern boundary of the city to be the standard, of 12,098 feet 3 inches. The plan of regularity in the streets,—originally adopted by Penn, and which, though condemned by some travelers accustomed to the crooked and narrow streets of European capitals, has been unqualifiedly approved by mathematical and scientific minds,—is adhered to ; and in the northern as well as the central parts of the city, there are avenues and streets which, for spaciousness and elegance, are unsurpassed by any. The elegance of the *public* buildings has long been a subject of remark, even in primary geographies; but, within the last few years, the architectural beauties of the city have been vastly enhanced by the erection of numerous costly *private* buildings: banks, stores, churches, dwellings—of granite, iron, sandstone, and marble ; and its upward growth, by the addition of stories upon stories, is not less remarkable. Beyond the compact or densely built-up portions, in the northerly direction, there is a wide expanding district between the two rivers, occupied in part by beautiful suburban residences, and by numerous Manufactories, surrounded by the habitations of industrious and contented artisans. The vicinity of Germantown is especially noted for the number of elegant cottages and villas, surrounded by handsomely laid out grounds, delightfully shaded; while the beauties of the Wissahickon, have they not inspired poets? But the citizens of Philadelphia, though appreciating her elegance in architecture, and scenes of natural beauty, cherish them less fondly, and point to them with less pride, than to the number and superiority of her charitable institutions, the excellence of her schools, the refinements of her society, her eminence in the Fine and the Mechanical Arts, the multiplied conveniences of life, promoting domestic comfort, and the celebrity of her Forum and Medical Schools, which, like the works of the Athenian orators, are regarded with veneration and respect by every polished nation.

Upon the minds of strangers and tourists, however, the external aspect of a city seems to leave the most permanent impressions; and if we may judge from their written opinions, that of Philadelphia has charmed those who charm the world. The learned and philosophical author of Mademoiselle Rachel's tour in America, was sagacious enough to remark—and in one so courteous, a trifling geographical inaccuracy can readily be pardoned—that "the capital of Pennsylvania, the Quaker city as it is called, is one of the richest, handsomest, and most flourishing cities in the United States of America." This is much from a gentleman who thanked God that he had visited North America, "because it is a duty disposed of," and he would never have to return there; but he proceeds to add: " Fortunately, it is superb weather here, and we can see this elegant capital at our ease. All the houses have a flaunting, coquettish air, which is pleasant to see. The streets are broad and clean. The shops are generally very large and very rich. There are superb goods in them. In fact, this city has a happy physiognomy, which is very agreeable." The ladies, especially the

Fannies, it is consoling to reflect, have also found much to delight them. Fanny Kemble was enraptured, we believe enchanted by the appearance of Fairmount, by moonlight; and Fanny Fern went off like an alarum clock at the beauties, and particularly the butter of Philadelphia. None, however, have expressed their admiration more gravely, deliberately, and ornately, than the writer of the following:

"Few great cities present such attractions for the stranger, as the city of 'Brotherly Love.' The American is proud that here the Declaration of Independence was signed; and his patriotic heart swells with a nobler emotion, while he looks upon the bell that pealed forth the joy of a nation's deliverance; and his heroic spirit will be stirred within him as he sits on the chair on which once sat the Father of his Country, yet, with many a relic of the past, preserved in Independence Hall. The philanthropist feels his heart throb with pleasure as he views the many noble institutions that a munificent charity has erected to ameliorate the condition of suffering humanity, supply the wants of the poor, minister to minds diseased, and alleviate the sufferings of the sick and wounded. The lover of science rejoices to see the city of Franklin abounding in Institutes whose object is the cultivation of all the arts that adorn, and all the sciences that tend to the progress of mankind. The philosopher will find kindred spirits in the great centre from which the rays of intellect emanate, whose brightness appears as a star of glory to the nation and the world. Medical students resort to Philadelphia for their professional training; the young aspirant to forensic honors seeks her classic shades; and while the admirer of the beautiful in architecture, and the architect, may exult in the stately proportions of her solemn temples, her gorgeous palaces, and the genius that adorned her with edifices whose beauty might vie with the Grecian models, the true Christian will find that the piety that erected the ancient church of Gloria Dei in the city's infancy, has diffused itself, and kept pace with its rapid increase. The merchant from other cities may look with wonder upon the commercial facilities of Philadelphia, her double port, the rich mineral treasures poured into her lap from the exhaustless resources of the Commonwealth, and the resources that put the numerous wheels of manufacturing industry in motion, and send the products of her skill, the results of her commerce, and the proceeds of her inland trade, to the furthest regions of the West, and almost all points of the

4*

compass. Her great Railway system, the most complete in the country, makes her pre-eminent for all the facilities of business, giving her a great advantage over all other cities in the Union. The exceeding beauty of her location, and the lovely scenery of the surrounding country, make her the resort of many who delight in beholding the fair face of Nature, seldom so full of beauty as in some portions of her enchanting rural scenery."

Such is Philadelphia as it appears to the optics of intelligent strangers. Such may it ever appear. If, however, a statistical description were wanted to convey a clearer idea of the magnitude of the city, we might say that Philadelphia is a collection of nearly 100,000 dwellings, Shops, and Manufactories, 7,404 Stores, 299 Churches,* 304 School-houses, 18 Banks, 11 Market-houses, 8 Medical Schools, 1 High School, 1 Girard College, 1

* The Directory assigns these churches to the different denominations, as follows: to Protestant Episcopal, 53; Methodist Episcopal, 42; Methodist Protestant, 4; Baptist, 30; Presbyterian, 44; Associate Presbyterian, 6; Associate Reformed Presbyterian, 3; Reformed Presbyterian, 9; Catholic, 28; Lutheran, 15; Friends, 13; Dutch Reformed, 4; German Reformed, 5; Jews' Synagogues, 6; Mariners', 2; Evangelical Association, 2; Universalist, 3; Independent, 2; New Jerusalem, 3; Unitarian, Second Advent, Moravian, Disciples of Christ, Christian, and Bible Christian, each 1. The colored churches are as follows: Methodist, 11; Presbyterian, 3; Baptist, 4; and Protestant Episcopal, 1. Several of the church buildings are beautiful specimens of architecture. The ST. MARK'S, (Episcopal), on Locust, above Sixteenth, cost, we believe, $120,000. The CALVARY CHURCH, (Presbyterian), Locust, above Fifteenth, and another at Seventeenth and Spruce streets, are also elegant structures. The Baptist Church, at Broad and Arch streets, has a steeple that cost about $16,000. The Catholics are now erecting, on Eighteenth street, opposite Logan Square, the CATHEDRAL of St. Peter and St. Paul, which, when completed, will cost more than half a million of dollars, and will be one of the most magnificent church-edifices in the country. The St. John's Church, (Catholic), Thirteenth near Market, is a fine Gothic structure, with a square tower on each of its front corners. The interior has some handsome paintings, and the windows are ornamented. St. Stephen's Church, (Episcopal), Tenth, between Market and Chestnut sts., is a "fine Gothic edifice, 102 feet long, 50 wide, having two towers at the front corners, octagonal, and 86 feet in height." This church contains the celebrated Monument to the BURD family, an object of considerable interest. Christ Church, in Second street, below Arch, is one of the oldest in the city, having been built in 1691, and enlarged in 1710. The spire was begun in

Polytechnic College, 1 State House, 1 Custom House, 1 Exchange, 1 Mint, 1 Navy Yard, 1 Naval Asylum, 3 Arsenals, 1 Blockley Almshouse,* 2 Insane Asylums, 1 Pennsylvania Institute for Deaf and Dumb,† 1 Blind Asylum, 1 Pennsylvania Hospital,‡ 1 Academy of Music, 1 Academy of Fine Arts, 1 Academy of Natural Sciences, 1 Athenæum, 1 Club House, numerous Libraries, 3 Theatres, 1 Masonic Hall, 15 Public Halls, 7 Gas Works, 5 Water Works, 1 County Prison, to which 15,809 persons were committed during 1857; 2 Houses of Refuge, containing 451 hopefuls; 1 Penitentiary, where 376 persons now chew the cud of reflection in silence; about 350 miles of cobble Pavements, 500 miles of Foot Pavements, 5631 Gas and Fluid Lamps, 9 Public Squares, 14 Cemeteries, 9 Railroad Depots, 90 Fire Engine-houses, 17 Station-houses, 3 Race Courses, besides Hotels, Restaurants, Savings Institutions, Insurance Companies, Charitable Institutions, Bridges, Vessels at wharves, Truck and other Farms, inclusive, " too numerous to mention." A statist, prosecu-

1753; its height is 196 feet. The money toward its completion was raised by lottery. This church has a chime of bells brought from England. The oldest church in the city, however, is the *Gloria Dei*, commonly called "Swedes Church," on Swanson street, near the Navy Yard.

* The *Almshouse* is an immense structure, situated on the west side of the Schuylkill, opposite South street. It consists of four main buildings, fronting on the Schuylkill, covering and enclosing ten acres of ground. The accommodations are excellent; and besides an almshouse capable of containing 3,000 persons, there is an Insane Asylum, in which there are over 300 patients of both sexes. Visitors admitted. Well worth seeing.

† The *Pennsylvania Institution for the Deaf and Dumb* occupies a building having a front of 200 feet on Broad street, and running back on Pine street 235 feet. Number in the institution, of both sexes, about 100. The Blind Asylum, situated at Race and Twentieth streets, is also a very useful and interesting institution.

‡ The *Pennsylvania Hospital*, in Pine street, from Eighth to Ninth streets, admits patients of all ages and sexes, who have received injury within twenty-four hours, provided they belong to the county. It possesses an Anatomical Museum, a valuable Medical Library of 10,000 volumes, and a Painting, by West, of Christ Healing the Sick, presented by the author.

ting his researches with due diligence, might ascertain that this wilderness of brick and mortar is inhabited by about 600,000 persons, white, black, mixed, and millionaires, including, as per the Directory for 1858, 1160 Smiths, 540 Browns, 480 Johnsons, 440 Joneses, 330 Thompsons, their heirs and assigns, and 1 George Munday. Further, if he be curious in such matters, he may probably discover that this people, collectively, "are well to do," owning real and personal property of a value of about $450,000,000, though assessed for much less; that, in 1857, they paid into the City Treasury $4,072,267, besides supporting about 600 lawyers, 1,000 physicians, over 900 teachers, and half as many preachers; that their city officials comprise, 1 Mayor, 1 City Solicitor, 1 City Controler, 1 Receiver of Taxes, 3 City Commissioners, 1 City Treasurer, 1 Chief Engineer of Water Department, 1 Chief Engineer of Fire Department, 1 Chief Engineer of Gas Works, 1 Chief Commissioner of Highways, 1 Commissioner of City Property, 1 Commissioner of Market-houses, 1 Chief Surveyor, and 12 Regulators; 24 Select Councilmen, with 3 Officers, and 89 Common Councilmen, with 5 Officers; 24 Members of a Board of Health, with 7 Officers, and 10 Executive Officers; 24 Guardians of the Poor, with 7 Officers, and 13 Out-door Visitors; numerous Assistants and Clerks in each Department; that their Police force, consists of

1 Mayor,	whose salary is	-	$3,500 per annum.
1 Mayor's Clerk,	- - - - -	1,000	" "
1 Chief of Police,	- - - -	1,500	" "
8 High Constables,	each - -	700	" "
4 Special Officers,	" - -	600	" "
1 Supt. of Fire and Police Alarm Telegraph,	1,200	" "
1 Assistant,	" " " -	600	" "
16 Lieutenants,	each - - - -	650	" "
32 Sergeants,	" - - - -	600	" . "
650 Policemen,	" - - - -	500	" "

Who made, in 1857, 21,537 arrests, and restored 3,430 lost children.

And notwithstanding the vast expenditure required for public purposes, the people had money enough left to contribute vast sums in charity, build 75 four-story dwellings, 991 three-story dwellings, 9 churches, 12 factories, and 4 school-houses; support 493 omnibuses, pay $500 per day to one Passenger Railway, spend about $5,000 per night in public amusements, lager-beer concerts, &c.; smoke about a million dollars worth of cigars, and purchase $2,000,000 worth of oysters; and they consumed, among other things, 60,425 beeves, 11,930 cows, 100,479 swine, 303,900 sheep, *exclusive of meat brought in market wagons;* and drank and wasted 6,318,880,116 gallons of water.

It is thus evident that Philadelphia, regarded from every point of view, is a centre of Wealth and Population; and, if the social characteristics of its inhabitants correspond with its external allurements, it must be an *attractive* centre. *What, then, are their characteristics, particularly with reference to the social position of the Mechanic and the Artisan?* What facilities are provided for their physical comfort and intellectual advancement? In the first place, the citizens of Philadelphia, who now give tone and direction to its popular sentiment, it may be relied upon, are far too clear-headed and practical in their views to do any thing tending to degrade labor and check useful enterprise. Even among the numerous sets of exclusives into which the descendants of great people sometimes divide themselves, there are none that I have heard of in this city who make idleness the "open sesame" to the enjoyments of their society. Nearly every citizen has some regular occupation; and prides himself upon diligence in the transaction of business and punctuality in fulfilling his engagements. The circle of those, at least among the male population, who aspire to distinction because of their uselessness, is like a wart on a man's nose, more looked at than important. The mass of the inhabitants believe in the Baconian philosophy, and illustrate its wisdom and beneficence by multiplying human enjoyments and mitigating human sufferings. The Press is emphatically a People's Press. The Quakers, whose influence, though diluted of late, continues to be felt in modifying the characteristics of our society, are true Benthamites in their views on individual and general happiness. They hold that the greatest happiness of the individual is, in the long run, to be obtained by pursuing the greatest happiness of the aggregate. They excel especially in the substantials of character, are fruitful in good works, zealous in education, and liberal in encouraging and rewarding decided mechanical and artistic triumphs. Constitutionally deliberate and prudent, the want of cordiality in their manners, which some strangers complain of, may be, and probably is, an unfortunate manifestation of these excellent qualities : or, in other words, of thinking twice before speaking once. Their city has been so prolific in great men, that the arrival of another does not create a sensation ; and being quite inexperienced in the art of giving entertainments at the *subsequent expense of their guests*,they prefer to conciliate mercantile visitors by giving them mercantile advantages. With respect to the want of enterprise—a standing accusation, which our fellow-citizens are accustomed to make against each other in tempestuous weather—we acknowledge the charge is seemingly reasonable and well founded, especially if it mean a total inability to comprehend the morality, or realize the pecuniary value of claptrappery, slap-dashery, or eclat. Adverse to puffing, they even refrain from scattering broadcast, as they ought to do, information relative to the mercantile and manufacturing advantages of their city; practical in their views, they sometimes forget that man does not live by bread alone; and straightforward in their own dealings, and governed exclusively in their own transactions by economical or commercial reasons, they do not suppose it possible that such trifles as " ancient and fish-like smells" in market-houses, can keep one customer away from where he ought to go ; or that such vanities. as popular preachers, big hotels, capacious theatres, palaces of mirrors, can possibly attract one customer where it is not his interest to go. The late panic, however, has dispelled many illusions; and if, moreover,—disabusing every mind of the feeling of entire security, and of the conviction that perfection is already attained,—it awaken a more active spirit, the anniversary of its advent may hereafter be celebrated as a civic holiday; and this beautiful city, having taken a new lease of Prosperity, will perpetuate the glory, as well as the memory of its Founders.

Secondly, the social and practical characteristics of the citizens of Philadelphia are in nothing more clearly and favorably manifested than in their zealous support of *free education*. According to the Controlers' Report of 1856, there were 304 Public Schools in the city, viz. : 1 High School, 1 Normal School, 55 Grammar Schools, 48 Secondaries, 156 Primaries, and 43 unclassified schools. The whole number of teachers was 940, of whom 78 were males, and 862 females ; the expense $456,089 14, and the number of scholars who enjoyed the benefits of gratuitous tuition was 55,099. But Public Schools are only a moiety of the educational establishments of Philadelphia. The city abounds in private schools and institutions of a semi-public character. Yet the quantity of the instruction given in the schools is perhaps less noteworthy than its quality. Public teachers must compete with private teachers ; while the latter are incited to emulation by the example of numerous eminent professors. From a mechanical point of view, however, the crowning distinction in this respect is the abundance of facilities provided for those who desire to increase their stock of *practical* and *scientific knowledge.* Books are at the command of such, rare in character and unlimited in quantity. The *Philadelphia Library,* one of the largest and best in the country, containing some seventy thousand volumes, is open to all, and access is thus given to works that probably are inaccessible to mechanics elsewhere. The work on British Patents, recently donated to the library, is valued at $3,000; the binding of the volumes alone having cost, we are informed, seven hundred dollars. For three dollars a year, any respectable person may enjoy the advantages of the *Mercantile Library,* whose members now number, we believe, 1,500. In various parts of the city there are Institutes with Reading-rooms and Libraries attached, where gratuitous lectures are given, especially adapted to the wants of mechanics. At the *Wagner Free Institute of Science,* twelve lectures are delivered weekly, during the Winter season, on Geology, Mineralogy, Mining, Astronomy, Botany, Anatomy, Physiology, Natural Philosophy, Chemistry, Chemical Agriculture, Ethnology, Comparative Anatomy, Zoology, Meteorology, and Civil Engineering. The apparatus is superior, and the lectures are well attended. The Spring Garden Institute gives instruction in the Mechanic Arts and Architecture, and has lectures on Literary and Scientific subjects. The Mechanics' Institute of Southwark, the Moyamensing Literary Institute, the Philadelphia City Institute, have reading-rooms and lectures, and the last has a School of Design. The Kensington Literary Institute, and the West Philadelphia Institute, are of the same character as the others; the latter having a School of Design. The Board of Trustees, in their report to contributors for 1856, state that the results of these Institutes show "that there is an aggregate of more than 11,000 volumes in the libraries; that during the past year more than 32,000 volumes have been loaned for home-reading; that more than 48,000 visits were paid to the reading-rooms by parties who partook of the intellectual food there dispensed; that one hundred pupils availed themselves of the valuable privileges afforded, for the culture of the eye and the hand in designing and drawing, by the schools of the Institutes; that sixty-seven lectures on literary, scientific, and artistic subjects, many of them replete with useful information, were listened to by thousands; and that, stimulated by your own generous contribution of more than $30,000, more than $50,000 additional have been contributed by our fellow-citizens to help onward the noble work commenced by you."

The *Franklin Institute* provides lectures at cheap rates every Winter, on Mechanical, Literary, and Scientific subjects, publishes a Scientific Journal, the oldest of its kind in the country, possesses a valuable Cabinet of Models and Minerals, and gives an Annual Exhibition that does much to promote progress in the Useful Arts. The *Academy of Natural Sciences* has a fine collection of objects in Natural History, embracing 25,000 specimens in Ornithology, and 30,000 in Botany; a library of over 26,000 volumes; and Mineralogical and Geological Cabinets, noted for their completeness. Professor Agassiz pronounced this institution the best out of Europe for its collections in the department of Natural History. At the *Polytechnic College*, opposite Penn Square, an engineer may obtain instruction in Physics that, before its establishment, he could not have obtained on this side of the Atlantic. In addition to the regular course, which embraces instruction in Civil Engineering, Mechanical Drawing, Mining, &c., the Managers have recently established a department designed to give instruc-

5

tion in "certain branches of knowledge that are demanded in common by every business pursuit, and are alike indispensable to the merchant, the farmer, the manufacturer, mechanic, and the manager of mining and other property." At the Girard College, drawing is taught from models of geometrical solids, and also in the High School, by competent teachers. The science of Accounts Book-keeping, Penmanship, and Commercial Law, are taught at a Commercial College, recently incorporated by the Legislature, and presided over by competent professors; and for the instruction of females in many departments of design, as applicable to manufactures, there is a school known as the "Philadelphia School of Design for Women," established a few years ago, by Mrs. Peter, the lady of the late British Consul at Philadelphia.

Among the educators of the people, too, the *newspapers* of this city are fairly entitled to rank. There are now twelve newspapers published daily—eight in the morning, and four in the afternoon; forty weeklies, and more than fifty publications properly designated as periodicals. The aggregate of those distinguished as newspapers, does not embrace any of a strictly scientific description; but the deficiency is in great part compensated for by many of the dailies, which never fail to advise their readers of whatever is important in the progress of the Mechanic Arts. The complement, also, lacks one or more of a metropolitan character, or those which can be said to possess universal interest; but as a faithful local Press, the newspapers of this city are models for those of the Union. The working-man here, for one cent, may enjoy a better morning newspaper than he can, for the same trifling sum, in any other place on the globe; while, for a larger expenditure, he may suit his taste from "grave to gay—from lively to severe." The sources then, it will be perceived, for acquiring that sort of knowledge which makes superior, efficient, intelligent mechanics, are very abundant in Philadelphia. It would be well, indeed, for affluent munificence to endow more completely one or two colleges, and establish an institution resembling, for instance, the British Museum; but in view of her present advantages, this city deserves now to be the resort of students in Art-education from all sections of the Union, as she long has been of students in Medical science. Here, there is an amount of scientific intelligence and professional skill concentrated, in part by the demands of the various institutions, seemingly sufficient to solve any thing in Mechanics but impossibilities; and which, conjoined with favorable physical circumstances, must enable manufacturers located near this city, to triumph over difficulties under which, in less favored localities, they would be compelled to succumb. Here, an educated hand-craftsman, or an inventor, may be said to stand at one of the great centres of intellectual life, with the world of mechanism in its practical forms on exhibition and in operation before him; Mentors on every side to enlighten him as to the recorded failures and triumphs of the ingenious men 'of all countries; and with the resources of the most scientific men of the present age, possessing the most perfect apparatus, at his command, to aid him in his experiments, or sustain him in his discoveries.

As a *place of residence*, Philadelphia enjoys the rare distinction of being desirable alike to the capitalist and to the artisan. In this respect, it is generally acknowledged, no other American city can compare with it. To the former, it offers all the attractions that can delight a cultivated mind, and all the luxuries that can please a fastidious palate; while an artisan, if industrious and intelligent, may command probably every thing essential to his present comfort, prospective independence, with constant participation in many of the chief pleasures of the capitalist.   In the important particulars of general cleanliness, healthfulness, wholesomeness of water, and the excellence of its markets, Philadelphia is unapproached by any of the other great cities ; and, as respects domestic accommodations, its superiority, at least over New York, is strikingly revealed by the census of 1850, which showed that, with a smaller population, this city contained about 23,601 more dwelling-houses : there being an average of 13½ persons to a house in the former city, and only 6½ in Philadelphia.   The custom, too, that prevails of selling lots on *ground-rent*, gives to the man of small means facilities that he cannot ordinarily obtain in other cities. For instance, if he have but money enough to erect a house, he can procure a lot on an indefinite credit; and so long as he pays the interest of the purchase-money, he will not be disturbed, nor can the principal be called for. By this means, it is quite common for mechanics, small tradesmen, and even laborers, to become owners of homesteads in the suburbs, which, by Passenger Railways that are being introduced, will be brought nearer to the centre than ever before.

A city, then, so attractive as Philadelphia, and possessing such superior educational advantages, can hardly fail, it would seem probable, to command, at all times, one of the first and most important requisites for success in Manufactures, viz. : *an abundant supply of skilled labor, and of those qualified to direct it.*   Experience demonstrates, that not only is the supply of labor generally abundant, but the surplus sometimes troublesome.   Here is congregated, at all times, an army of artisans from every civilized nationality—the majority employed, others seeking employment; and should the supply at any time fall short, an advertisement would bring a regiment from every place where it had been seen.   Men who would not go to "Raw Cheney," in Georgia, for $1,000 a year, nor to Pittsburg for $900, nor to Lowell for $850, eagerly come to Philadelphia for $800. Philadelphia has thus the pick and choice, at less wages, of the mechanics of the Union. Hence, too, the name, PHILADELPHIA MECHANIC, has become synonymous with skill and superiority in workmanship. We simply state a well tested fact, when we assert that a mechanic, traveling with favorable credentials from reputable workshops in this city, will be preferred to fill the first vacancy in any similar establishment, not merely in most places throughout the United States, but in portions of Europe.

So much for Philadelphia as it is. Its *status* establishes the fact, that it possesses the *moral* circumstances that are essential to success in manufacturing operations, and we might proceed immediately to consider those that are properly denominated *physical*. Before doing so, however, it may be proper to glance at the present,

II. Commercial Relations of Philadelphia.

A Review of Commercial Transactions, for a year financially so disastrous as 1857, can hardly be expected to be very imposing, or even favorable. We notice, however, that the statistics of the commerce of the Port of Philadelphia, for 1857, show in several particulars an increase over the previous few years; indicating that the city is just beginning to realize fully the benefits long expected from an immense expenditure incurred to develop the trade, and especially the mineral wealth of the interior.

The number of vessels that arrived during the year was 505 foreign, and 32,241 coastwise: being an increase of 5,702 over 1856, and 2,523 over the arrivals in

5*

1855. The value of merchandise entered for consumption was $11,845,205; and the value entered for warehousing was $6,706,017. Total Imports for the last three years, being :—

		Imports.		Withdrawals.
1857,	-	$18,551,222	-	$5,421,092
1856,	-	18,303,288	-	3,050,400
1855,	-	15,104,478	-	. . . .

The cash duties received at the Port, for the last three years, were,

1857.		1856.		1855.
$3,096,324 24	-	$4,301,123 80	-	$3,358,517 41

The Exports for 1857, included, of breadstuffs, 198,867 barrels Flour, 48,572 barrels Corn Meal, 8,254 barrels of Rye Flour, 191,400 bushels Wheat, 625,556 bushels Corn ; and a great variety of manufactured articles.

The construction of vessels at all places was quite limited during the year, but at Philadelphia 147 new vessels, having an aggregate tonnage of 17,917 tons, were admeasured by the United States officers.

Of COAL, the following statement shows the comparative shipment by the four principal lines to Philadelphia, for 1857 and 1856 :—

	1857.		1856.	
	Tons.	Cwt.	Tons.	Cwt.
Philadelphia and Reading R. R.	1,709,692	19	2,088,903	03
Schuylkill Navigation	1,275,988	00	1,169,453	08
Lehigh Navigation	900,314	06	1,186,294	00
Lehigh Valley Rail-road	418,235	11	165,740	00
Total	4,304,230	16	4,610,390	11
Decrease from Schuylkill and Lehigh regions, in 1857,			306,159	95

The following Table Exhibits the Tonnage of the Pennsylvania Railroad for 1857.

ARTICLES.	From Phil'a to Pittsburg.	From Pitts'g to Philad'a.	From Phil'a to Way Stations.	From Way Stations to Philad'a.
Agricultural Implements and Productions....	1,391,797	620,846	115,962	....
Boots, Shoes, Hats, &c................................	4,481,376	....	709,267	....
Books and Stationery	2,360,675	183,700	173,441	....
Butter and Eggs....................................	....	4,428,779	....	2,665,382
Brown Sheetings and Bagging....................	5,374,835	....	322,735	....
Bark and Sumac....................................	....	....	....	980,988
Cedarware...	212,326	236,082	245,377	....
Confectionery and Foreign Fruits................	2,138,853	....	335,566	....
Coffee...	5,729,353	....	1,591,561	....
Cotton	161,100	733,651	34,245	....
Coal	....	11,011	7,655	255,382,087
Copper, Tin and Lead.............................	2,373,751	96,974	348,961	....
Camphene..	....	....	....	....
Dry Goods..	48,442,442	723,735	3,311,476	136,663
Drugs, Medicines and Dye Stuffs................	7,064,227	626,100	988,423	....
Earthenware.......................................	....	152,692	....	....
Flour	48,816	65,163,024	....	2,842,806
Fresh Meats, Poultry and Fish..	....	276,785	....	455,097
Feathers, Furs and Skins........................	....	51,970	....	....
Furniture and Oil Cloth...........................	2,504,485	311,990	444,033	36,667
Glass and Glassware	868,914	1,555,943	296,103	27,140
Green and Dried Fruits...........................	....	2,630,054	....	298,415
Grass and other Seeds............................	....	81,971	....	873,123
Grain, of all kinds................................	....	11,296,545	....	8,377,902
Groceries, (except Coffee)........................	18,755,092	149,426	6,586,644	104,724
Ginseng	....	101,322	....	....
Guano...	12,250	....	46,913	....
Hardware	10,008,923	1,113,370	1,966,422	492,370
Hides and Hair....................................	....	2,724,863	1,222,211	....
Hemp and Cordage................................	1,334,638	1,153,344	149,992	....
Iron, rolled, hammered, &c......................	217,720	....	39,796	1,710,957
Iron, rail-road.....................................	389,097	....	1,638,862	....
Iron Ore..	171,450	....	263,900	....
Iron, Blooms and Pigs........	....	....	....	2,574,356
Live Stock..	122,075	40,056,014	46,505	9,955,769
Leather...	2,428,264	844,025	134,708	2,285,018
Lead and Shot.....................................	....	....	....	....
Lard, Lard Oil and Tallow........................	....	7,155,977	....	166,212
Lumber and Timber...............................	27,304	714,730	55,541	25,728,000
Machinery and Castings..........................	6,796,518	775,171	2,518,666	....
Marble and Cement	2,577,776	....	991,885	....
Malt and Malt Liquors............................	43,945	405,446	5,245	....
Nails and Spikes..................................	....	....	92,424	1,680,182
Oils...	2,454,893	....	722,329	....
Oysters...	240,825	....	1,065	....
Paper and Rags....................................	1,702,745	614,837	4,205	330,062
Plaster..	....	....	....	....
Potatoes, Turnips, &c............................	....	....	....	....
Pot, Pearl and Soda Ash.........................	8,332,527	107,495	38,315	....
Queensware...	4,928,353	....	672,934	....
Salt	73,665	....	355,398	....
Salt Meats and Fish..............................	3,552,423	24,704,577	2,061,291	25,829
Soap and Candles.................................	....	796,674	....	....
Tobacco...	1,830,837	1,086,581	405,166	....
Tar, Pitch and Resin..............................	430,656	....	79,380	....
Wines and Liquors, Domestic....................	....	....	....	....
Do. do. Foreign....................	2,974,461	5,580	939,007	....
Wall Paper	....	....	....	....
Whisky and Alcohol...............................	....	12,571,537	....	1,219,631
Wool and Woolen Yarn......	37,669	4,978,191	....	171,486
Miscellaneous......................................	1,739,600	568,113	500,730	123,726
Total First Class	62,058,365	6,840,642	7,034,142	3,373,186
Total Second Class...................	53,563,860	17,611,766	11,940,214	4,582,337
Total Third Class	8,160,805	42,060,772	3,036,988	10,053,200
Total Fourth Class...................	30,553,626	123,295,945	8,452,995	300,635,869
Total for the Year....................	154,336,656	189,809,125	30,464,339	318,644,592

During the past year, a movement was made, in various departments of trade, toward a union of energies and aims, and effected, as we are informed by the official report of the Corn Exchange Association, amongst the Dry-goods merchants, the Workers in metals, the Dealers in Queensware, and others. This favorable result was coincident with an effort to represent all these branches in a Commercial Congress, by means of a reconstruction of the Philadelphia Board of Trade. This board, heretofore exclusive in its character, was, during the year, completely reorganized. Its floor, and its offices, are now open to all who have a right, by force of character, to the attention and respect of their fellow-members. Since its reorganization, many of its meetings have been attended by unusual numbers; and its future action is looked to with interest, as likely to exert a favorable and important influence upon our trade and commerce.

In Foreign Commerce, the Port of Philadelphia, it is undeniably obvious, has not maintained its original supremacy. In 1796, the exports amounted to $17,513,866; and from 1795 to 1826, the aggregate exports more than trebled those of the last thirty years. For a long period of time, nearly a century, Philadelphia was regarded throughout Europe as, commercially and numerically, the great city of the Western Continent. Vessels of the largest tonnage then known, and laden with the richest merchandise of Europe and the Indies, sailed up the Delaware, and found accommodations at her wharves. Large fortunes, besides that of Girard, were accumulated by her citizens from well-planned adventures to foreign countries. The names of her principal merchants were known and respected at every Exchange in Europe. A decline, then, from a position so commanding in the world's markets, would naturally cause a pang of regret in the breasts of every one not indifferent to the city's future, were the causes that induced it, less creditable, patriotic, and honorable than they are. It is especially consoling to know that they do not impinge in the slightest the commercial capabilities of the Port. No one, if called upon to correct the assertions of the willfully ignorant, need trouble himself to controvert the assertion, that the inland situation of Philadelphia is an effectual barrier to her commercial supremacy. The position of the chief commercial cities of the Old World,—London on the Thames, Liverpool on the Mersey, and Paris on the Seine,—proves that immediate proximity to the ocean is not essential to constitute a great shipping port. Besides, the channel of the Delaware is known to be abundantly wide and deep to float, as it has floated, the largest vessels in the Naval service. According to the official chart of the Coast Survey, it is seldom less than a quarter of a mile in width, and ranges in depth, *at low water*, from 4 to $9\frac{1}{2}$ fathoms, excepting at the bar below Fort Mifflin, where, for a few rods, it varies from 18 feet to 25 feet 8 inches, according to the state of the tide. Moreover, the fact that the " Cathedral," a vessel of too large tonnage to obtain entrance into the port of her destination, New York, was therefore sent hither, where she was amply accommodated ; the length of wharves on the Eastern front, extending as they do, for about three miles ; and especially her former eminent success in Foreign Commerce, all establish and fortify the assertion, that Philadelphia possesses all essential, in fact ample facilities for shipping. To ascertain the true and principal causes of the decline in this particular, then, we must direct our attention to other channels that have absorbed capital ; and we may possibly discover that the chief sources of the present prosperity of Philadelphia have their origin in a comparative neglect of Foreign Commerce.

About thirty years ago, the citizens of Philadelphia may be said to have become thoroughly aware of the immensity of the riches concealed in the mountains and ravines of their native State. They then, for the first time, comprehended the value and the vastness of the deposits of Coal and Iron near their metropolis, and realized, as vividly perhaps as subsequent experience has justified, how great a boon would be conferred upon the whole country by their development. It is true, that for many years previously, it had been known that a peculiar species of coal abounded in the counties of Lehigh and Schuylkill, but it was regarded as worthless. Even as late as 1817, when Col. George Shoemaker forwarded ten wagon loads of a coal which he had discovered about one mile from Pottsville, to Philadelphia for sale, he could with difficulty find purchasers; and some of those who did purchase it, were wholly unsuccessful in their attempts to use it. "Nearly every one considered it a sort of *stone*, and, saving that it was a 'peculiar stone'—a stone-coal—they would as soon have thought of making a fire with any other kind of *stone!* Among all those who examined the coals, but few persons could be prevailed upon to purchase, and they only a small quantity,—'to try it.' But, alas! the trials were unsuccessful. The purchasers denounced Colonel Shoemaker as a vile impostor and an arrant cheat! Their denunciations went forth throughout the city; and Colonel Shoemaker, to escape an arrest for swindling and imposture, with which he was threatened, drove thirty miles out of his way, in *a circuitous route, to avoid the officers of the law!* He returned home, heart-sick with his adventure. But, fortunately, among the few purchasers of his coal, were a firm of iron factors in Delaware County, who, having used it successfully, proclaimed the astounding fact in the newspapers of the day. The current of prejudice thereafter began to waver somewhat; and new experiments were made at iron-works on the Schuylkill, with like success, the result of which was also announced by the Press. From this time, Anthracite began gradually to put down its enemies; and among the more intelligent people, its future value was predicted." But it was not until 1825 that the first successful experiment to *generate steam*, with Anthracite coal, was made at iron-works at Phœnixville. From that year, too, the existence of the Schuylkill trade may be said to date, though some coal had been shipped previously. The speculative mania in the coal regions, however, did not commence until a few years later.

About 1829, the news of fortunes accumulated by piercing the bowels of the earth, and bringing forth from the caverns of mountains " metals which shall give strength to our hands," became generally current, and aroused an enthusiasm less wide-spread than that which fevered Europe upon the discovery of silver in Mexico, or recently America upon the discovery of gold in California, but certainly not less intense. " Capitalists awoke as if from a dream, and wondered that they had never before realized the importance of the anthracite trade. What appeared yesterday but as a fly, now assumed the gigantic proportions of an elephant! The capitalist who, but a few years previously, laughed at the *infatuation* of the daring pioneers of the coal trade, now coolly ransacked his papers, and ciphered out his available means; and whenever met on the street, his hands and pockets would be filled with plans of towns, of surveys of coal lands, and calculations and specifications of railways, canals, and divers other improvements until now unheard of. The land which yesterday would not have commanded the taxes levied upon it, was now looked upon as 'dearer than Plutus's mine, richer than gold.' Sales were made to a large amount; and in an incredibly short space of time, it is estimated that upward of *five millions of dollars* had been invested in lands in the Schuylkill coal-field alone! Laborers and mechanics of all kinds, and from all quarters and nations, flocked to the coal region, and found ready and constant employment at the most exorbitant wages. Capitalists, arm-in-arm with confidential advisers, civil engineers, and grave scientific gentlemen, explored every recess, and solemnly contemplated the present and future value and importance of each particular spot. Houses could not be built fast enough ; for where nought but bushes and rubbish were seen one day, a smiling village would be discovered on the morrow. Enterprising carpenters in Philadelphia, and elsewhere along the line of canal, prepared the timber and frame-work of houses, and then placing the *materiel* on board a canal boat, would hasten on to the enchanted spot to dedicate it to its future purposes. Thus *whole towns* were arriving in the returning canal boats; and as ' they were forced to play the owl,' a moonlight night was a godsend to the impatient proprietors, for with the dawning of the morning would be reflected the future glory of the new town, and the restless visages of scores of anxious lessees."*

---

* The late Joseph C. Neal, who was one of the motley mass, some years afterward wrote the following humorous description of the speculating scenes :

In the memorable year to which we allude, rumors of fortunes made at a blow, and competency secured by a turn of the fingers, came whispering down the Schuylkill and penetrating the city. The ball gathered strength by rolling—young and old were smitten with the desire to march upon the new Peru, rout the aborigines, and sate themselves with wealth. They had merely to go, and play the game boldly, to secure their utmost desire. Rumor declared that Pipkins was worth millions, made in a few months, although he had not a sixpence to begin with, or to keep grim want from dancing in his pocket. Fortune kept her court in the mountains of Schuylkill County, and all who paid their respects to her in person found her as kind as their wildest hopes could imagine.

The Ridge road was well traveled. Reading stared to see the lengthened

A reaction attended this, as all other speculative ma-
nias. But disastrous as it was, and involving hundreds
in ruin, it did not prevent the continued investment

columns of emigration; and her astonished inhabitants looked with wonder
upon the groaning stage-coaches, the hundreds of horsemen, and the thousands of footmen, who streamed through that ancient and respectable
borough; and as for *Ultima Thule*, Orwigsburg, it *has not recovered from its
fright to this day!*

Eight miles further brought the army to the land of milk and honey, and
then the sport began—the town was far from large enough to accommodate
the new accessions; but they did not come for comfort—they did not come
to stay. They were to be among the mountains, like Sinbad in the valley
of diamonds, just long enough to transform themselves from the likeness
of Peter the Moneyless into that of a *Millionaire;* and then they intended
to wing their flight to the perfumed saloons of metropolitan wealth and
fashion. What though they slept in layers on the sanded floors of Trout-
man's and Shoemaker's bar-rooms, and learned to regard it as a favor that
they were allowed the accommodation of a roof by paying roundly for it,
a few months would pass, and then Aladdin, with the Genii of the Lamp,
could not raise a palace or a banquet with more speed than they!

One branch of the adventurers betook themselves to land speculations,
and another to the slower process of mining. With the first, mountains,
rocks, and valleys changed hands with astonishing rapidity. That which
was worth only hundreds in the morning, sold for thousands in the evening,
and would command tens of thousands by sunrise, in paper money of that
description known among the facetious as slow notes. Days and nights
were consumed in surveys and chaffering. There was not a man who did
not speak like a Crœsus—even your ragged rascal could talk of his hundreds
of thousands.

The tracts of land, in passing through so many hands, became subdivided, and that brought on another act in the drama of speculation: the
manufacture of towns, and the selling of town lots. Every speculator had
his town laid out, and many of them had scores of towns. They were, to be
sure, located in the pathless forests; but the future Broadways and Pall
Malls were marked upon the trees; and it was anticipated that the time was
not far distant when the deers, bears, and wild-cats would be obliged to give
place, and take the gutter side of the belles and beaux of the new cities.
How beautifully the towns yet unborn looked upon paper! the embryo
squares, flaunting in pink and yellow, like a tulip show at Amsterdam; and
the broad streets intersecting each other at right angles, in imitation of the
common parent, Philadelphia. The skill of the artist was exerted to render
them attractive; and the more German text, and the more pink and yellow,

6

of capital in the coal regions.   Every year more mines were opened, more iron-works erected, more improvements of a stupendous character planned, more tons of coal

the more valuable became the town! The value of a lot, bedaubed with vermilion, was incalculable; and even a sky-parlor location, one edge of which rested upon the side of a perpendicular mountain, the lot running back into the air a hundred feet or so from the level of the earth, by the aid of the paint-box was no despicable bargain: and the corners of Chestnut and Chatham streets, in the town of Caledonia, situated in the centre of an almost impervious laurel swamp, brought a high price in market, for it was illustrated by a patch of yellow ochre!

The bar-rooms were hung round with these brilliant fancy sketches; every man had a roll of inchoate towns in the side pocket of his fustian jacket. The most populous country in the world is not so thickly studded with settlements as the coal region was to be; but they remain, unluckily, in *statu quo ante bellum.*

At some points a few buildings were erected to give an appearance of realizing promises. There was one town with a fine name, which had a great barn of a frame hotel. The building was let for *nothing;* but after a trial of a few weeks, customers were so scarce at the "Red Cow," that the tenant swore roundly he must have it on better terms, or he would give up the lease.

The other branch of our adventurers lent their attention to mining; and they could show you, by the aid of a pencil and piece of paper, the manner in which they must make fortunes, one and all, in a given space of time—expenses, so much; transportation, so much; will sell for so much; leaving a clear profit of 000,000! There was no mistake about the matter. To it they went, boring the mountains, swamping their money and themselves. The hills swarmed with them; they clustered like bees about a hive; but not a hope was realized. Calculations, like towns, are one thing on paper, and quite another when brought to the test.

At last the members of the expedition began to look haggard and careworn. The justices did a fine business; and Natty M., Blue Breeches, Pewter-Legs, and other worthies of the catchpole profession, toiled at their vocation with ceaseless activity. When the game could not be run down at view, it was taken by ambuscade. Several bold navigators discovered that the county had accommodations at Orwigsburg (at that time the seat of justice, now located at Pottsville) for gentlemen in trouble. Capiases, securities, and bail-pieces, became as familiar as your garter. The play was over, and the farce of "*The Devil to Pay*" was the after-piece. There was but one step from the sublime to the ridiculous, and Pottsville saw it taken!

sent to market. Canals which had been projected but suffered to languish were speedily completed ; rail-roads were built, not only above ground but under ground; and in a comparatively brief period of time, it has been estimated by competent authority, *a hundred millions of dollars* were withdrawn from commercial activity, and invested in productive and unproductive improvements and partially abortive schemes. Many of the works, however, constructed in Lehigh and Schuylkill counties, are imperishable monuments of solidity and beauty, and will be objects of admiration in after ages.

At the present time there are, within the borders of Pennsylvania, upward of 800 miles of Canal, and 1,600 miles of Rail-road, of which the revenues are mainly derived from freight on Anthracite and semi-Anthracite coal. Many of those, projected with other views, have become large transporters of coal; and certainly the amount of capital expended in Pennsylvania for one object, viz., *for constructing avenues to convey Anthracite coal to market*, is now at least SEVEN TIMES GREATER than *the whole amount invested in all the manufactories at Lowell.* (See annexed TABLE.)

Gay gallants, who had but a few months before rolled up the turnpike, swelling with hope, and flushed with expectation, now betook themselves, in the gray of the morn, and then the haze of the evening, with bundle on back—the wardrobe of the Honorable Dick Dowles tied up in a little blue-and-white pocket-handkerchief—to the tow-path, making, in court phrase, "mortal escapes"; and, in the end, a general rush was effected—the army was disbanded—*sauve qui peut!*

The following Table, which we have prepared principally from official information, exhibits the

Names, Length, and Cost of the Canals and Rail-roads in and leading to the Anthracite Coal Regions of Pennsylvania.

	Canals. Length.	Railroads Length.	Cost. Jan. 1, 1858.
Philadelphia and Reading R. R. (including City Br.).....		98	19,262,720
Catawissa, Williamsport and Erie R. R., including the Quakake Branch, with equipments...........................		79	4,200,000
Williamsport and Elmira, with real estate and basins at Williamsport and Elmira, and equipments complete...		78	3,850,000
Lebanon Valley R. R., (consolidated with Reading)......		53½	4,000,000
Schuylkill Navigation Co..	108		10,950,000
Lehigh Coal and Navigation Co., viz.:			
Canal and Improvements................................	72		4,455,000
Lehigh and Susquehanna R. R.		20	1,380,000
Summit and Branch Rail-roads.........................		25	1,400,000
Delaware Division of Pennsylvania (State) Canal	60		2,200,000
Eastern Division of Pennsylvania (State) Canal............	46		1,737,236
Susquehanna Canal.......................................	41		897,160
Lower North Branch Canal..	73		1,598,379
Upper North Branch Canal..	94		4,500,000
Union Canal..	99		5,000,000
North Pennsylvania R. R..		68	5,773,925
Lehigh Valley R. R...		46	3,276,523
Philadelphia and Sunbury R. R., (unfinished)..............		33	1,348,812
Sunbury and Erie R. R., whole amount expended on finished and unfinished..		40	3,693,492
Dauphin and Susquehanna R. R..................................		54	4,000,000
Lackawana and Bloomsburg R. R.		57	1,650,000
Delaware, Lackawana and Western R. R......................		113	8,701,888
Lackawana R. R., (unfinished).................................		9	300,000
Little Schuylkill Railroad, exclusive of land................		33	1,402,651
Mine Hill Railroad, Extension and Branches................		120	2,750,000
Schuylkill Valley Rail-road and Branches...................		28	568,000
Mill Creek Rail-road and Branches............................		13½	311,000
Mount Carbon and Port Carbon R. R., including land....		2½	282,000
Mount Carbon R. R..		7	200,000
Beaver Meadow Rail-road and Branches......................		20	1,000,000
Hazleton Coal Co.'s R. R...		14½	400,000
Lehigh and Luzerne R. R., (unfinished)......................		8	200,000
Buck Mountain Coal Co.'s R. R., (exclusive of land)......		7	430,000
Big Mountain Coal Co.'s R. R....................................		2½	35,000
Trevorton R. R.		14	700,000
Tioga R. R..		29¾	1,093,263
Barclay R. R. and Coal Co.'s R. R..............................		16¼	438,000
New York and Middle Coal Field Co.'s R. R., (unfin'd)..		5	150,000
Columbia Coal and Iron Co.'s R. R............................		7	150,000
Carbon Run Coal Co.'s R. R.....................................		3½	60,000
Lykens Valley R. R..		16	443,000
Union Canal Co.'s R. R..		4	150,000
Swatara R. R...		6	100,000
Wisconisco Canal..	12		381,836
Lorberry Creek R. R...		2½	25,000
Sundry Coal roads, private and underground, to the Mines, say..		300	6,000,000
	605	1,433½	111,444,885
Morris Canal, (present total cost)...............................	102		5,612,000
Pennsylvania Coal Co.'s R. R....................................		44	1,994,819
Delaware and Hudson Canal and R. R., (estimated)	108	24	3,250,000
Central Rail-road of New Jersey.................................		63	5,048,340
Total,	815	1,564⅓	$127,350,044

The total Capital invested in Manufactories, at Lowell, was, in 1846...$10,550,000
On January 1st, 1855, the latest date at hand, Hunt's Merchants' Magazine, (October No., 1855,) states that the capital invested in *all* the Manufactories, at Lowell, was..... ....$14,000,000

But the development of the mineral regions of Eastern Pennsylvania was not the only scheme that abstracted the attention and capital of the citizens of Philadelphia from the prosecution of Foreign Commerce. The West was becoming known as " The Great West." Regiment after regiment of hardy pioneers, armed with axes and plowshares, had entered the wilderness to subdue it; each successive year the frontiers of civilization were carried further westward; production outran consumption; and the people of Pennsylvania were called upon to furnish superior avenues and outlets for the produce of the West to the best markets on the Atlantic coast. A grand system of internal improvements was therefore resolved upon, and undertaken, to connect the metropolis of Pennsylvania with the Ohio River and the Lakes. The Erie Canal in New York was then near its completion, and herculean and partially successful efforts were being made to divert the trade of the West away from its natural and geographical channels by a circuitous route to New York. But the means adopted by Pennsylvania to establish superior connections with the West were less successful in execution than praiseworthy in conception. The Alleghanies defied the skill of the engineers, broke up the chain of communication into disjointed links; and the attempts made to unite them—constructing part rail-road, and part canal—instead of affording to shippers and producers the promised benefits, only fully succeeded in arousing the fears of foreign creditors, and provoking the sarcasm of the witty Dean of St. Paul's at the "the drab-coated gentry." No one acquainted with the physical characteristics of this State —its magnificent scenery, its rugged acclivities and impenetrable fastnesses—need be told that to construct railroads and canals within its limits, is and must be a serious and costly undertaking. The cost of the Commercial

6*

Marine of many recognized commercial nations is a mere bagatelle in comparison with the vast sums expended in Pennsylvania for internal improvements alone.

On the first of January, 1858, Pennsylvania had 2773¼ miles of rail-road, costing $135,166,609; or, estimating the population of the State at three millions, the amount expended was at the rate of $45 for each man, woman, and child in the Commonwealth. The cost of constructing the canals within its borders, exceeding as they do 1200 miles in length, has been stated at thirty millions of dollars. To these immense sums, if we add the amounts expended in seeking for minerals, sinking shafts, opening mines, disinterring iron ore, and erecting works to manufacture it, the vastness of expenditure incurred for the development of internal wealth may well astonish and appal even those to whom the theme has become familiar by daily contemplation. In all these enterprises, the capital and credit of Philadelphia are conspicuous. Owning property equal to one third the assessed value of the property in the entire State, the city has contributed more than one half of the cost of public and private improvements. To aid these, her merchants sold their ships: to sustain them, her capitalists declined the profits of Bottomry and Respondentia.

But the prodigies achieved within the limits of Pennsylvania, great as they are, did not exhaust the zeal of the citizens of Philadelphia in behalf of internal improvements. Their brethren in neighboring States, in the South and the West, have drawn largely for contributions to such projects; and, to the extent of our ability, their drafts have not been dishonored. The portfolios of our merchants are now plethoric with such obligations and bonds; and when presently available, will build an Armada of merchant ships. If it were practicable to ascertain how many thousands of merchants are now thriving, how many tens of thousands of farmers in the States of Ohio, Indiana, Illinois, Wisconsin, and the South, are now comparatively wealthy, because of their present facilities for reaching good markets—facilities encouraged and perfected through aid from Philadelphia—the revelation would so interweave the ties of friendship with those of mutual mercantile interests, as to form a bond indissoluble by any assaults.

The citizens of Philadelphia, it is then safe to aver, are eminently patriotic, even in their business predilections. They have withdrawn their capital largely from prosperous commerce, to invest it in Mines, Rail-roads, Iron-works, and Manufactories, preferring to aid the development of the resources of the interior even at the expense of commercial importance and reputation abroad. Without giving assent to the doctrines of Chinese economists, who hold that Foreign Commerce is generally prejudicial to a State, because, by diminishing the quantity of desirable products, it must raise their price to the home consumer, they nevertheless believe that a prosperous, active interchange of products between citizens of the respective States is more conducive to the permanence and well-being of the Republic than even a more profitable commerce with foreigners. Cherishing, then, as they have done, and as they do, what they presume to be the best interests of our whole country, and having proved, by abandoning their share in the rich commerce of the Indies, the sincerity of their desire to accelerate its industrial development, the Merchants and Capitalists of Philadelphia would seem to be entitled to praises rather than taunts for the decline of their city in direct Foreign Commerce; and certainly they have established a claim to the high place which they hold in the friendly regard of their intelligent fellow-merchants throughout the Union.

But while acknowledging a decline in the Foreign

Commerce of Philadelphia, it is but justice to state that the decline is more apparent than real. The number of foreign arrivals, and the amount of duties paid at the Custom House here, are no index to the imports of the merchants of this city. Many of the most extensive importing-houses—and there are some, we were about to say, quite too extensive for the country's welfare—import nearly all their goods *via* New York. The largesses given by Government to steamers connecting with that port, and the peculiar facilities and inducements said to be held out to shippers, not to mention the rumor recently current that duties are sometimes lower there than elsewhere, influence our merchants in directing their foreign correspondents to ship goods to Philadelphia *via* New York. The advantage that the New York importer has over his Philadelphia competitor is simply a saving in freight between the two cities—an item perhaps not exceeding $2 per ton, or at least so unimportant on imported light and costly fabrics as to add no appreciable per-centage to the cost. That this advantage is overbalanced by other circumstances—lower rents, less extravagant expenditures for personal gratification, etc.—is evidenced by the fact that scores of New York jobbers visit Philadelphia every season, to replenish their stocks from the shelves of the importers, knowing that they can do so, besides paying fare, freight both ways, and all other expenses, at a cheaper rate than they can purchase the same goods from any of their neighbors. One Fancy Goods importing-house in particular, whose operations came within the range of my personal observation, attracts New York and Boston jobbers as regularly and more extensively than Cincinnati and St. Louis buyers. This is explained in part by the fact that the house has more favorable connections in Europe than their competitors in other cities, and partly by their ability to sell at a lower per-centage of profit in consequence of diminished expenses. These two circumstances, and especially favorable connections with the foreign manufacturers, would seem to be of more importance, in a regular importing business, than any other; and these, Philadelphia merchants,—whose honorable character and mercantile probity have ever been understood and appreciated in Europe,—enjoy peculiar facilities for obtaining. But in all probability I would not misrepresent popular feeling if I were to say that Philadelphia does not covet the distinction of being a great importing mart. She would be content if other cities monopolized the doubtful honor of importing hither French gimcracks and German cloths in exchange for gold and silver—our commercial life-blood—provided her merchants were encouraged to devote their energies successfully and uninterruptedly, to building up Home Industry and American Manufactures.

III. Commercial Relations with the South and West.

Pennsylvania, it has been frequently observed, is the only State in the Union that has a navigable outlet to the ocean, a footing upon the Lakes, and a command of the Ohio and the Mississippi. This position necessarily gives the metropolis of the Commonwealth points of superiority over all the other great cities on the Atlantic coast, for the purpose of receiving and distributing merchandise to and from a great portion of the South and West. With the ocean, and the principal cities of the Southern seaboard, Philadelphia has regular and direct communication by way of the Delaware River; and in consequence of improvements in locomotion, the distance is now less than at any previous time. With the gate of the West, Philadelphia is connected by canal and a magnificent railway; and at Pittsburg, with all the cities and towns on the navigable waters east of the Rocky Mountains, by thousands of miles of river navigation, and also by rail-roads joining Cleveland and Chicago on the one side, Wheeling and Cincinnati on the other, continuing through Kentucky to Nashville, and prolonged with a continuous, unbroken gauge westwardly, beyond St. Louis, on the Mississippi. Philadelphia has also an advantage over New York and Boston, in being considerably nearer to all the prominent foci of the products of the Great West. The principal rail-road lines from New York—the Erie and Central—it has been aptly remarked, lie on the circumference line to the West; while the great rail-road of Pennsylvania—the Pennsylvania Central—is on a diameter line. Their direction is to the Lakes—ours to the West. But to exhibit more clearly the relative position of Philadelphia, New York, and Boston, with reference to proximity to the chief centres of trade in the West, we have prepared the following Table, from data furnished in "Dinsmore's Railway Guide," published in New York:

	Clevel'd, Ohio. Miles.	Cincinnati, Ohio. Miles.	Chicago, Ills. Miles.	Indianapolis, Ind. Miles.	St. Louis, Mo. Miles.
From Philadelphia, *via* Pennsylvania Rail-road, to Pittsburg; thence by shortest Rail-road route to..	501	703	851	746	1000
New York, *via* Hudson River to Piermont, and the Erie Rail-road to Dunkirk, 468 miles; thence by shortest Rail-road route.............	612	867	954	893	1154
New York, *via* Hudson River Rail-road to Albany; thence by Rail-road to Buffalo, 442 miles; thence as above.............................	625	880	967	906	1167
Boston, *via* Western Rail-road to Albany and Buffalo, 498 miles; thence as above..........	681	936	1023	962	1223

Hence, it is manifest that Philadelphia has considerable advantage over New York and Boston, in nearness to the principal centres of trade in the West. The saving in distance will be regarded as an important one by the weary traveler, while its effects in reducing the cost of transportation will be shown hereafter. It is true, New York has a shorter route to the places named than by the above-mentioned rail-roads; but that is, *via Philadelphia and Pittsburg.* Pennsylvania is truly the Key-stone State; and those who would pass and repass from the

West to the East, may congratulate themselves that their most direct route carries them over a rail-road so well managed as the Pennsylvania Central, and through a city so beautiful as Philadelphia.

To understand, more especially, *the relative cost of transportation by railway*, from Philadelphia, New York, and Boston, to the commercial centres of the West, we procured the *published Tariffs of Freight for* 1857, of the principal rail-road lines receipting through : and the following Tables will exhibit the results. Any one choosing to do so, may verify the statements made by procuring the freight tariffs for 1857, of the Boston and Worcester, New York and Erie, and Pennsylvania Central Rail-roads.

SUMMER RATES--1857.

	1st Class.	2d Class.	3d Class.	4th Class.
Philadelphia to Columbus, Ohio,	1.05	90	80	60
New York, " "	1.38	1.05	88	71
Boston, " "	1.43	1.08	91	84
Philadelphia to Dayton	1.10	95	85	60
New York, " "	1.43	1.05	93	73
Boston, " "	1.48	1.08	96	86
Philadelphia to Cincinnati	1.10	95	85	60
New York, " "	1.43	1.05	93	73
Boston, " "	1.48	1.08	96	86
Philadelphia to Indianapolis	1.20	1.00	85	65
New York, " "	1.48	1.10	93	75
Boston, " "	1.53	1.13	96	88
Philadelphia to Louisville (all Rail)	1.45	1.20	1.02	80
New York, " " "	1.63	1.25	1.08	88
Boston, " " ....(River from Cincinnati)...	1.68	1.28	1.11	1.01
Philadelphia to Terre Haute	1.45	1.20	1.00	80
New York, " "	1.73	1.30	1.08	90
Boston, " "	1.78	1.33	1.11	1.03
Philadelphia to Forte Wayne	1.30	1.10	90	75
New York, " "	1.48	1.13	96	77
Boston, " "	1.48	1.11	96	87
Philadelphia to Lafayette	1.46	1.20	1.00	80
New York, " "	1.62	1.25	1.08	86
Boston, " "	1.67	1.28	1.07	95
Philadelphia to St. Louis	1.70	1.40	1.20	98
New York, " "	1.80	1.35	1.20	98
Boston, " "	1.98	1.48	1.31	1.20
Philadelphia to Cairo	1.88	1.56	1.30	1.10
New York, " "	1.90	1.45	1.30	1.08
Boston, " "	2.08	1.58	1.42	1.32
Philadelphia to Cleveland	1.00	80	68	55
New York, " "	1.08	80	68	53
Boston, " "	1.13	83	71	66
Philadelphia to Chicago	1.50	1.10	1.00	80
New York, " "	1.58	1.18	1.06	87
Boston, " "	1.58	1.18	1.06	97

WINTER RATES--1857 and 1858.

	1st Class.	2d Class.	3d Class.	4th Class.
Philadelphia to Columbus,	1.20	1.05	85	70
New York, " "	1.50	1.23	95	82
Boston, " "	1.60	1.33	1.03	90
Philadelphia to Dayton,	1.28	1.10	95	80
New York, " "	1.58	1.28	1.08	85
Boston, " "	1.68	1.38	1.16	1.00
Philadelphia to Cincinnati.	1.30	1.15	1.00	80
New York, " "	1.60	1.30	1.10	85
Boston, " "	1.70	1.40	1.18	1.00
Philadelphia to Indianapolis.	1.45	1.20	1.08	85
New York, " "	1.75	1.40	1.20	90
Boston, " "	1.85	1.50	1.28	1.05
Philadelphia to Louisville.	1.60	1.35	1.23	1.00
New York, " "	1.90	1.55	1.35	1.00
Boston, " "	2.00	1.65	1.43	1.20
Philadelphia to Terre Haute.	1.70	1.40	1.15	1.00
New York, " "	2.00	1.60	1.35	1.05
Boston, " "	2.10	1.70	1.43	1.20
Philadelphia to Fort Wayne.	1.55	1.25	1.05	85
New York, " "	1.84	1.46	1.18	1.00
Boston, " "	1.94	1.56	1.26	1.05
Philadelphia to Lafayette.	1.71	1.35	1.14	96
New York, " "	1.98	1.58	1.30	1.09
Boston, " "	2.08	1.68	1.38	1.14
Philadelphia to St. Louis.	2.00	1.65	1.40	1.15
New York, " "	2.30	1.90	1.60	1.30
Boston, " "	2.40	2.00	1.68	1.35
Philadelphia to Cairo.	2.05	1.72	1.48	1.20
New York, " "	2.35	1.95	1.65	1.35
Boston, " "	2.45	2.05	1.73	1.40
Philadelphia to Cleveland.	1.20	95	85	70
New York, " "	1.35	1.05	85	75
Boston, " "	1.45	1.15	93	80
Philadelphia to Chicago.	1.90	1.45	1.25	1.00
New York, " "	2.10	1.60	1.40	1.10
Boston, " "	2.10	1.60	1.40	1.10

A Table showing the Saving on a Ton (2240 lbs.) of First Class Freight by Shipping, from Philadelphia instead of New York or Boston.

	NEW YORK.		BOSTON.	
	Summer.	Winter.	Summer.	Winter.
To Columbus, Ohio	$7.39	$6.72	$8.51	$8.96
Dayton, Ohio	7.39	6.72	8.51	8.96
Cincinnati, Ohio	7.39	6.72	8.51	8.96
Indianapolis, Indiana.	6.27	6.72	7.39	8.96
Louisville, Kentucky	4.03	6.72	5.15	8.96
Terre Haute, Indiana.	5.15	6.72	7.39	8.96
Fort Wayne, "	4.03	6.50	4.03	8.74
Lafayette, "	3.58	6.05	4.70	8.29
St. Louis, Missouri.	2.24	6.72	6.27	8.96
Cairo.	45	6.72	4.48	8.96
Cleveland, Ohio.	1.79	3.36	2.91	5.60
Chicago, Illinois.	1.79	4.48	1.79	4.48

These Tables "speak for themselves"—comment cannot add to their force. They demonstrate, conclusively, that every shipper who, during the past year, sent Western merchandise by rail-road from the points designated, to New York or Boston, that could have been sold as well in Philadelphia—every Western merchant who purchased goods in those cities on no better terms than he could have purchased them in Philadelphia, and sent them home by rail-road, expended unnecessarily, or in other words, lost from one dollar and seventy-nine cents to eight dollars and ninety-six cents on every ton usually classified as first-class freight. These are the facts, and the deductions from facts, with respect to shipments "all the way by rail-road."

Now, it may be said that New York and Boston have the advantage of Lake navigation to many prominent points in the West. We assert—and appeal to the managers of the New York and Erie and Boston and Worcester rail-roads, who receipt through both "all the way by rail-road or by steamer on the Lakes," as shippers prefer—that this is no advantage. The Lake freights are the regulators of the rail-road charges, which barely exceed them by the cost of insurance necessary to cover the great risks attending navigation on the Lakes. But Philadelphia, on the contrary, has a very important advantage, in addition to that stated in the Tables, by communicating at Pittsburg with thousands of miles of safe river navigation, extending southwardly to New Orleans and the ocean, and westwardly to St. Paul, on the Mississippi; and, in fact, to all the cities and towns on navigable waters east of the Rocky Mountains. The advantage in shipping from Philadelphia to Pittsburg, and thence by the Ohio River to Cincinnati and Louisville, over shipping to those points by the Northern rail-road lines, amounts, in addition to the saving stated above, to about $5 per ton on first-class goods, $4 on second, $3 on third-class, and $2 on very heavy goods; while to Nashville, Memphis, Cairo, St. Louis, and all points south of New Albany, Ind., the additional saving is nearly double this amount—that is, about $10 per ton on first-class goods, $8 on second, $6 on third, and about $3 per ton on fourth-class. It is thus evident, as experienced shippers know,

7

that freight from the West, bound for European markets, can be brought to Philadelphia, and shipped hence,
landing it at its destined port abroad, at cheaper paying
rates than by way of New York. Indeed, the leading
products of the West—for instance, flour, the products of
the hog, whisky, etc.—can be shipped to Philadelphia, and
hence at least *half the distance to Liverpool*, for the cost
of transporting them to New York. Further, in view
of the facts stated, it is also obvious that a Western merchant, purchasing goods in Philadelphia, may have his
preference rewarded by a saving in the cost of transporting them home. The only practical question, then, for him
to consider is, *whether it is probable he can make his purchases in the Philadelphia market as cheaply as in any other;*
for, supposing the terms to be the same, he will nevertheless, by doing so, obtain an advantage. We beg per-
mission to offer a suggestion or two upon this probability,
for the consideration of those who study and appreciate
economy.

We may remark, at the outset, that any one who has
taken time to examine, compare, and reflect upon the
characteristics of the respective markets, the development of Manufactures, and the comparative facilities for
manufacturing, will not need any arguments to convince
him that the probability of an advantage in price must be
altogether in favor of the Philadelphia market. Let those,
however, who have not already done so, examine the
subject in its details, and they will be astonished to discover how few classes of goods constituting a country
trader's usual assortment are not, to greater or less extent,
made in or near Philadelphia. For instance, with regard
to *Domestic Dry Goods:*—According to the census, Pennsylvania, in 1850, contained within her borders a larger
number of factories for the making of cotton and woolen
goods than any State in the Union; even more than the

great manufacturing State of Massachusetts, and considerably more than New York. The former had 213 cotton and 119 woolen factories, and the latter 86 cotton and 249 for wool; while, in Pennsylvania, there were 588 of these establishments in all, of which 208 were employed in the cotton and 380 in the woolen manufacture. The extent to which Philadelphia is engaged in the production of these goods, will be illustrated in another place (see DRY GOODS MANUFACTURE); but we may state here, that one firm, Messrs. ALFRED JENKS & SON, manufacturers of cotton and woolen machinery, supplied the mills of this city alone, during the past year, with 800 looms for weaving checks, and on which could be woven twenty thousand yards per diem. The *New York Tribune* of May 1, 1857, in an editorial, urging greater attention to manufactures in that locality, remarked, we suppose with truth, "*Philadelphia has at least twenty manufactories of textile fabrics where New York has one; and her superiority in the fabrication of metals, though less decided, is still undeniable.*" Cottonades, checks, carpetings, Germantown hosiery and woolen goods, ribbons, sewing-silks, military goods, &c., are manufactured here to an immense extent; and of these, New York and other jobbers are constant and acceptable customers to the amount of millions annually. But, besides the vast quantities of dry goods manufactured in and near this city, all the principal mills of New England, and elsewhere, consign their fabrics to agencies established here, with authority to sell them, frequently at an abatement from invoice prices. The first agency for the sale of domestic fabrics in the United States, was that of ELIJAH WARING, established in this city about the year 1805; and from that day to this, the domestic Dry Goods Commission-houses of Philadelphia have maintained a position alike honorable to themselves; advantageous to American manufactures; and with one exception, viz., too great liberality in giving credit to strangers,* beneficial to the city.

With respect to *Foreign Dry Goods*, the importing-houses of Philadelphia certainly possess the same facilities for procuring desirable selections on advantageous terms as any others do ; and in some instances enjoy unusually favorable connections in Europe, established long since, and by means of them secure perhaps more than their share of bargains.   The stocks are generally selected by resident partners, who know the wants and consult the interests of purchasers ; and therefore they consist, less than some others, of the unsaleable refuse of London warehouses.

Proposing, as we do, to make a minute and detailed examination of the manufacturing industry of Philadelphia, it would not be proper here to anticipate its results; but, for the benefit of anxious mercantile inquirers, we may state further, that more than four millions of dollars worth of *fine Boots and Shoes* are annually made in this city ; while of the common, cheap, pegged-work of New England, Philadelphia is also a large purchaser, consumer, and distributer.   The quality of our manufactures in this department is so generally and highly appreciated, that several of the manufacturers in Lynn, Mass., with a view of attracting additional custom, announce on their signs, " Philadelphia Shoes for sale."   Of *Educational and Medical Books*, the publishers of Philadelphia are generally recognized as leaders; and for the distribution of books of all kinds, Henry C. Carey, the distinguished political economist, has asserted that Philadelphia has the largest book distributing house in the world. As respects *Iron*, the last census showed that nearly one-half of the pig, cast, and wrought iron, made in the

---

* The late panic disclosed the fact, that a prominent dry goods jobbing-house that failed, in New York city, was indebted to a commission-house in Philadelphia, in a sum but little short of $100,000—a line of credit entirely beyond the limits of prudence.

United States, was the product of the furnaces and forges of Pennsylvania; and the latest statistics show, that of the 782,958 tons of iron produced in the United States, in 1856, Pennsylvania produced 448,515 tons. Of the *Manufactures of Iron*, as stoves, hollow-ware, and those articles, usually denominated *Hardware*,—nails, screws, saws, forks, shovels, enameled-ware, hinges, bolts, nuts and washers, Philadelphia is an immense producer; and, for the sale of their products, the hardware manufacturers of Old and New England have agencies established in this city, authorized to sell at factory prices. In short, the market of Philadelphia differs in many important respects from most others, resembling from one point of view a Leipsic Fair, and from another the Eastern Bazaars. Manufacturers' depots are often situated between a commission-house and a house importing the same class of goods; fabrics, fresh from the loom, may be found close to the gold-tipped embroideries of France, or the crasse dresses of Turkey; factories adjoin stores, and stores are surrounded by manufactories; while, diverging from the city, are numerous roadways, constantly traversed by iron horses, bringing fuel from Nature's vast magazines not far distant; and from the East, caravans of boats, propellers, cars, come laden with the products of distant workshops, seeking here a central point for redistribution throughout the South and the West. Hence it is obvious, that a purchaser of a miscellaneous stock, adapted to the wants of a rural, town or city population, must be, when in Philadelphia, as near the fountain head, where goods are as yet in first hands, as it is possible for him to get; while the merchant, who visits the city to replenish his mind as well as his stock, can hardly fail, in a world of machinery, literature and art, as this is, to note much that is to him novel, and carry back suggestions that will be useful to himself and his neighbors.

Is it not probable, then, that the merchants of Philadelphia, in view of their advantages, with manufactories all around them, consignments from abroad seeking their markets and supplying their auction-houses, with abundance of capital and good credit, can buy and sell on terms as favorable as any of their competitors? We have no doubt they do this; but we go further, and insist that those now doing business have mistaken their vocation, unless, to responsible buyers, they *actually do undersell all others.* One reason that we have for entertaining this opinion is, that expenses for conducting business are less here than in most other large cities. In the city of New York, the leading Dry Goods jobbing-house pay, or did recently pay, as we are informed, an annual rent of $22,000 for their store; and a prominent wholesale clothing-firm pay, or did pay, $28,000; while the greatest amount of rent paid by a leading firm, in a similar business in Philadelphia, that I have heard of, and for which equal, or at least all necessary accommodations are procured, is $8,000. It is true, the "Stewart" of Philadelphia deems $14,000 a moderate compensation for his magnificent store, but his customers are principally the wealthy of the city. A proportionate difference in favor of Philadelphia prevails in rents, generally, for dwelling-houses as well as stores. The room for expansion afforded by the plan and locality of the city multiplies the number of eligible sites, and consequently diminishes speculation and prevents monopoly. The demands of fashion and extravagance, also, though sufficiently exorbitant, are less onerous in Philadelphia; and, from these and other circumstances, it would seem evident, without ocular demonstration, that a merchant in Philadelphia can afford to sell at a per-centage of profit, which, on the same amount of business, would not pay the expenses of his less favorably situated competitor.

These are the deductions of reason and common sense. Their importance entitles them at least to consideration, reflection, and experiment; hence we beg those who are engaged in buying and selling, inasmuch as their mercantile success, and the prosperity of the mercantile class throughout the country, depend upon the wisdom of their action, to test the respective markets fairly,—disregarding " baits," which are quite too common in all, and extending their view beyond exceptional circumstances,— and if there be an atom of truth in that principle of political economy, which demonstrates that the nearer the place of production the cheaper the price, they will discover, as thousands of thriving merchants have already done, that Philadelphia is the CHEAPEST SELLER, and NATURAL DISTRIBUTER OF MERCHANDISE ADAPTED TO THE WANTS OF THE SOUTH AND THE WEST.

Returning from this digression to subjects more immediately connected with our inquiries, and having already adverted to the moral circumstances that have an effect upon economy of production in Manufactures, we now proceed to consider the position of Philadelphia with respect to

IV. Physical Advantages for Manufacturing.

In considering Philadelphia as a Manufacturing centre, it must be obvious, from previous remarks, and still more obvious from minute information respecting the topographical and geological features of Pennsylvania, and the intimacy of connection between the metropolis and the principal mineral sections of the State, that Philadelphia and its vicinity command, in the first place, the *most important raw materials used in Manufactures; and secondly, the agents best fitted to produce power.* But the celebrity of Pennsylvania for its vast deposits of IRON and COAL —those primary sources of England's manufacturing greatness—is so widely extended, that to dilate upon their abundance would hardly convey any additional information to any person of ordinary intelligence. The census of 1850, as we previously stated, showed that nearly one half of the pig, cast, and wrought Iron made in the United States was from her forges and furnaces; while her mines of " black diamonds," it is a proverb, are only equalled in national importance by the gold mines of California. The district in Pennsylvania that produces the most Iron and the cheapest Coal, viz., the Valleys of the Lehigh, the Schuylkill, and a part of the Delaware—is directly tributary to Philadelphia, procuring its supplies from this city, and selling its products here almost exclusively. We therefore record the latest statistics of these important products.

1. Iron. The statistics of the Iron production of Pennsylvania, for 1856, as furnished us by the Secretary of the *American Iron Association* in Philadelphia, are as follows :

Anthracite Iron.

	Tons.
Valley of the Delaware and Lehigh, - - - -	108,367
Valley of the Schuylkill, - - - - - - -	60,882
" " Susquehanna and Juniata, - - - -	76,971
" " North Branch of Susquehanna, - - -	56,411
" " West Branch of " - - - -	4,340

Charcoal Iron, including Five Coke Furnaces.

Eastern and Northern Pennsylvania, - - - -	53,160

Charcoal, Bituminous Coal and Coke Iron.

Western Pennsylvania, - - - - - - -	88,384
Total, - - - - - - - -	448,515

The statistics of the Iron production, for 1857, are not as yet prepared, but will probably exhibit an increase over that of the previous year. A comparison of results shows, that the production of Anthracite Iron in Eastern Pennsylvania is an increasing one,—that of 1854 having been 67.8 per cent., that of 1855 having been 74.4 per cent. ; while that of 1856, was 78.3 per cent. of the total product of the entire United States. The value in dollars of the product for 1856, assuming the average price of Anthracite Iron to be $27 per ton, and Charcoal Iron to be $33 per ton, is as follows:

Anthracite Iron, made near Philadelphia, 169,249 tons at $27,	$4,569,723
Charcoal Iron, made in Eastern Pennsylvania, 53,160 tons, at $33,	1,754,280
Product of Anthracite Iron, in Pennsylvania, 306,971 tons, at $27,	8,288,217
Value of Anthracite Iron in United States, 393,509 tons, viz., at $27,	10,624,743

It is thus manifest, that Philadelphia is situated in the district that is entitled to be called the *centre of the Iron production of the United States.* It is further manifest, that the centre of the Iron interest is likely to *remain* in the district tributary to Philadelphia, inasmuch as the production has been an increasing one ; and, the establishments situated within its limits have been able to survive disasters that have borne down those in other places, and consequently there must exist circumstances peculiarly favorable to economy of production.

2. COAL. The quantity of Coal sent to market from the district tributary to this city, was as follows :—

Product of the Anthracite Coal Fields of Pennsylvania, for 1857.

	Area in Acres.	Production in Tons.
1. Southern Coal District, comprising the Schuylkill, Pine Grove, and Lyken Valley regions,	75,950	3,256,891
2. Middle Coal District, comprising the Lehigh, Mahanoy, and Shamokin regions,	85,525	1,582,786
3. Northern Coal District, comprising the Wyoming and Lackawanna regions,	76,805	1,958,362
Total of the three fields,	238,280	6,798,039

The value of this product, at $2 per ton, the minimum

price at the mines, would be $13,596,078, while the market value certainly exceeds *thirty millions of dollars.* In addition to Anthracite, the mines of Eastern Pennsylvania produced, last year, 494,100 tons of semi-Anthracite and Bituminous coal, and those west of the Alleghanies, about thirty-four millions of bushels of Bituminous coal, of an estimated value exceeding three millions of dollars.

The qualities of different coals have necessarily been made the subject of careful analysis; and their relative value has been tested by frequent experiments. We believe it is conceded by both scientific authority and practical experience, that *Pennsylvania Anthracite is practically the cheapest and best fuel that the United States afford.** It contains about 90 per cent. of carbon, and,

---

* For the purposes of *steam navigation,* an impression formerly prevailed that the Pennsylvania Anthracite was inferior to the Cumberland coal, which, it is acknowledged, surpasses in strength the foreign bituminous coals of Newcastle, Liverpool, Scotland, Pictou, and Sydney. In January, 1852, a series of experiments were undertaken at the New York Navy Yard with the boilers of the United States Steamer "Fulton," to settle the question of relative value and superiority for this purpose. The result is given in the following extract from the Report of the Engineer-in-Chief, CHARLES B. STUART, to Commodore JOSEPH SMITH, Chief of Bureau of Yards and Docks.

COMPARISON.

The coals used in these experiments were the kinds furnished by the agents of the Government for the use of the United States Navy Yard and Steamers, and was taken indiscriminately from the piles in the yard, without assorting.

The bituminous was from the "Cumberland" mines. The anthracite was the kind known as "White Ash Schuylkill."

From the preceding data, it appears that, in regard to the rapidity of "getting-up" steam, the anthracite exceeds the bituminous thirty-six per cent.

That, in economical evaporation per unit of fuel, the anthracite exceeds the bituminous in the proportion of 7.478 to 4.483, or 66.8 per cent.

It will also be perceived, that the result of the third experiment on the boilers of the pumping-engine at the New York Dry Dock, which experiment was entirely differently made and calculated from the first and second

next to charcoal, gives out more heat than the same weight of any other fuel. So far as at present known, Pennsylvania is the only State where this valuable mineral can

experiments, gave an economical superiority to the anthracite over the bituminous of 62.3 per cent.—a remarkably close approximation to the result obtained by the experiments on the "Fulton's" boilers (66.8 per cent.), particularly when it is stated that the boilers and grates of the pumping-engine were made with a view to burning bituminous coal, which has been used since their completion, while those of the "Fulton" were constructed for the use of anthracite. The general characters of the boilers were similar, both having return drop-flues.

Thus it will be seen, from the experiments, that, without allowing for the difference of weight of coal that can be stowed in the same bulk, the engine using anthracite could steam about two-thirds longer than with bituminous.

These are important considerations in favor of anthracite coal for the uses of the Navy, without taking into account the additional amount of anthracite more than bituminous that can be placed on board a vessel in the same bunkers; or the advantages of being free from *smoke*, which in a *war* steamer may at times be of the utmost importance in concealing the movements of the vessel; and also the almost, if not altogether, entire freedom from spontaneous combustion.

The results of the experiments made last spring on the United States steamer "Vixen" were so favorable, that I recommended to the Bureau of Construction, &c., the use of anthracite for all naval steamers at that time having, or to be thereafter fitted with, *iron* boilers; particularly the steamers "Fulton," "Princeton," and "Alleghany," the boilers for all of which were designed with a special view to the use of *anthracite*, and with the approval of that Bureau.

The "Fulton's" bunkers are now filled with anthracite; and the consumptions referred to in the engineer's report on that steamer show, during the short time she has been at sea, that the anticipated *economy* has been fully realized.

In view of the results contained in this report, I would respectfully recommend to the Bureau of Yards and Docks, the use of anthracite in the several Navy Yards, and especially for the engine of the Dry Dock at the New York Navy Yard.

In conclusion, I desire the approval of the Bureau to make such investigations as my duties will permit, with regard to the *experience* of the durability of *copper* boilers, when used with bituminous or anthracite coal; which can be done without any specific expenditure.

The inquiry may prove highly important to the Navy Department, as the

be obtained cheaply and in unlimited quantities; but within her borders the supply is seemingly sufficient to satisfy the probable wants of this country for centuries to come.

The rapidity with which Anthracite coal has appreciated in popular estimation, is shown by the increase in the demand for it. In 1820, only 365 tons were sent to tide-water; in 1840, the product amounted to 867,000 tons; in 1852, it had reached five millions of tons: being an increase in 12 years, from 1840 to 1852, of 600 per cent. Supposing this rate of augmentation to continue up to 1870, Gov. Bigler once amused himself by calculating that the production would be forty-five millions of tons, worth, at the present prices of the Philadelphia market, the sum of $180,000,000. No wonder the worthy Governor was moved to pronounce this a gratifying picture, confirming his belief "that, before the close of the present century, Pennsylvania, in point of wealth and real greatness, would stand in advance of all her sister States."

In the cost of fuel, Philadelphia has an admitted advantage over New York of about twenty-five per cent.; over Providence, R. I., from $1.75 to $2.25 per ton; and over Boston, from $2 to $2.50 per ton. The advantage, moreover, which Philadelphia enjoys from controlling the production of the best fuel, in addition to proximity, is too evident to need illustration; and being also the central and chief market of the district producing the

use of anthracite under copper boilers has been heretofore generally considered as more injurious than bituminous coal, and is consequently not used by Government in vessels having copper boilers.

Respectfully submitted, by your obedient servant,

CHARLES B. STUART,
Engineer-in-Chief, U. S. Navy.

Commodore JOSEPH SMITH,
Chief of Bureau of Yards and Docks.

best and cheapest iron, it would seem almost superfluous to inquire further as to her capabilities for Manufactures.

But Iron and Coal, though the most important, are not the only useful mineral products that abound in Eastern Pennsylvania. *Copper* exists extensively in several counties; *Plumbago* is obtained in Bucks County, and *Zinc* in the vicinity of Bethlehem. *Marble*, well adapted and extensively used for building purposes, has long been obtained from quarries in Montgomery County, a few miles above Philadelphia. *Steatite*, or *Soapstone*, is quarried extensively on the Schuylkill, above Manayunk. Roofing and Ciphering *Slates* of the best quality are found in the counties of Lehigh, Monroe, and Northampton; there being in the county of Lehigh alone some thirty quarries open, with a capital of $60,000 invested, employing about 300 men, and producing at least 25,000 squares of roofing-slates per annum, valued at $3 per square on the quarry bank, and at $5 and $6 in the Philadelphia market. Nearly all the best school-slates in this country are from the Pennsylvania quarries; and many of them are manufactured at an establishment in this city. Of *Salt*, the census of 1850 states the produce of Pennsylvania at 184,370 barrels. *Kaolin*, or *Porcelain earth*, is abundant at several points within a radius of thirty miles from Philadelphia. About $2\frac{1}{2}$ miles north of Camden, N. J., there is an extensive bed of Fire Clay, of which specimens have been sent to England, and pronounced by competent judges superior to the German clay, which commands $25 per ton. Besides these, Barium, Chromium, Cobalt, Nickel, Magnesium, Titanium, Lead, Silver, Zirconium, and Fire and Potter's Clay, are scattered over the State, and in some instances of superior quality.

With all the points in Pennsylvania producing mineral and mining products, Philadelphia is directly connected by rail-roads and canals, and thus may be said to be situated in close proximity to the original sources of many of the most important articles that can be enumerated in a list of raw materials of Manufactures. And if we were to pass from the products of the mine to those of the forest and of Agriculture, we would find them equally abundant, cheap, and accessible. *Lumber*, in immense quantities, is obtained on the Susquehanna and the Delaware, and floated down those rivers every Spring and Fall. In 1852, it was estimated that 250,000,000 feet were sent down the former river; while the Lehigh region supplied the Philadelphia market, *via* canal, in the same year with 52,123,751 feet. At the present time, we are informed by persons intimately acquainted with the subject, Philadelphia has a larger stock of *seasoned* lumber than any other mart in the Union.*   *Wool*, of the very

---

* Many of the forest trees most useful in the Arts, Manufactures, and Medicine, are natives of Pennsylvania. We condense from *Trego's Geography of Pennsylvania* the following list, which may be of value to some of our readers:

OAKS. At least twelve varieties. The *White Oak*, the most esteemed of this noble family of trees, is found throughout the State; and in the Southeastern districts the wood is exceedingly compact and tough. The *Black Oak*, which is very abundant, and one of our largest trees, furnishes *Quercitron Bark*, which is exported in large quantities, and used in dyeing wool, silk, &c., a yellow color. When used by tanners, it imparts a yellow tinge to the leather. The *Spanish Oak*, of which the bark commands a high price, is less common in Pennsylvania than further South. The other species, valuable for their bark, which is highly esteemed by tanners, is the *Rock Chestnut Oak*, the *Scarlet Oak*, and the *Red Oak*. In addition to these, there are the *Iron Oak*, confined to the Eastern part of the State, and resembling the White Oak; the *Swamp White Oak*, the *Swamp Chestnut Oak*, *Laurel* or *Shingle Oak*, *Scrub Oak*, and *Pin Oak*.

WALNUTS. Two principal kinds, the *Black* and *White Walnut*. The former is much used for cabinet-work, and for the stocks of military muskets; also for the posts of fences, which, it is said, will last from twenty to twenty-five years. The bark of the *White Walnut*, or *Butternut*, yields an excellent cathartic medicine, said to be efficacious in cases of dysentery. It is also used in the country for giving a brown color to wool.

best American grades, is grown in the Western counties of the State ; and all, or nearly all, of which, as the woolen manufacturers of Rhode Island and Massachusetts, who

HICKORY. The most common species are the *White Heart Hickory, Pig Nut, Bitter Nut, Shell Bark*, and *Thick Shell Bark*—highly valued for axle-trees, handles, flails, &c., and also as a fuel, affording in the same bulk more combustible matter than any other wood.

MAPLE. The *Red Maple* is the most common, and probably the most valuable species. Its wood is much used by chairmakers, and for bedsteads, saddle-trees, &c. In many of the old trees, the fibres of the wood, instead of following a perpendicular direction, are undulated and waving. This is known as the *Curled Maple*, and when skillfully polished, produces the most beautiful effect of light and shade. The bark of the Red Maple yields a purplish color by boiling, which, by the addition of copperas, becomes dark-blue, approaching to black. It is used in the country for dyeing, and for making ink. The *true Sugar Maple* is abundant, particularly along the elevated range of the Alleghanies, and the *Black Sugar* tree along the Western rivers. Large quantities of maple sugar are made in the Northern and Western counties. The *Striped Maple* grows in the mountainous parts of the State, and the *Ash-leaved Maple*, or *Box Elder*, west of the mountains.

DOGWOOD. The most valuable species grows to the height of twenty or thirty feet. The wood is used for tool handles, and other purposes, and the inner bark has medicinal properties resembling those of the cinchona or Peruvian bark, from which quinine is made, and has been successfully used in intermitting fevers.

The POPLAR or TULIP tree is common in Pennsylvania, and surpasses most of our forest trees in height and the beauty of its flowers and foliage. Its wood is applied to many purposes where lightness and strength are desirable, as trunks, chairs, &c., and the bark is said to possess tonic and antiseptic qualities; and a decoction of it, combined with a few drops of laudanum, has been found efficacious in giving tone and vigor to the stomach after fevers and inflammatory diseases. It has been also used in dyspepsia and cholera infantum.

WHITE and RED BIRCH grow abundantly along the Delaware above Philadelphia, and *Black*, or *Sweet Birch*, in deep, loose, and cool soils. It is said that articles of furniture made from this acquire with time the appearance of mahogany.

Of woods remarkable for their durability, we have the *Locust*, which is abundant in limestone valleys ; and the *Red Mulberry*, frequently met with in fertile soils, when seasoned, is nearly equal to the Locust; also, the *Red Cedar*, exceedingly durable, and highly esteemed for making fence posts, is common in most parts of Pennsylvania.

come hither to purchase it, can testify, is secured to the Philadelphia market. About ten millions of pounds are sold annually. But a still wider range of raw materials is open to the manufacturers of Philadelphia. Those which are the product of other States or foreign countries are, by means of direct commerce, brought to her wharves, and concentre in her warehouses. The hides of Buenos Ayres, the woods of Guiana, the marble of Italy, the dye-stuffs of Calcutta, and the cotton of our Southern States, are delivered to the doors of our factories, in many instances as directly from the producers as the Minerals, Lumber, and Wool of Pennsylvania.

The CHESTNUT may also be ranked among very durable woods. It grows most abundantly in the hilly regions, and frequently attains an extraordinary size; one on Mount Etna being 53 feet in diameter, or 160 feet in circumference, but *hollow* to the bark. The wood is much used for posts and rails; and it is largely consumed in the manufacture of charcoal for the supply of the iron-works in the interior of the State. Its fruit is particularly appreciated by the boys.

Of PINES, there is every variety, though the *true Yellow Pine* is not very common in the State. The *Pitch Pine* is abundant, and in some places tar is manufactured from the more resinous parts of it. *White Pine*, so useful, and applied to such a variety of objects, is becoming comparatively scarce, in consequence of the enormous consumption for shingles, lumber, &c.; but nevertheless, it is still found in considerable quantities on the upper streams of the Lehigh, the head waters of the Susquehanna, and some of the tributaries of the Alleghany.

The *Hemlock Spruce*, however, which is more common, growing on the steep banks of streams, and in dark and shaded situations, is being substituted for White Pine, wherever it can well be done.

The other forest trees which are natives of Pennsylvania are the *White* and *Red Ash*, highly esteemed for strength and elasticity, several species of the *Aspen, White* and *Red Beech, Buttonwood* or *Sycamore, Catalpa* or *Bean* tree, *Crab-apple, Cucumber* tree, so called because the cones or fruit somewhat resemble a small cucumber, *Chincapin, White* and *Red* or *Slippery Elm, Sweet* and *Sour Gums, Hornbeam, June Berry* or *May Cherry, Linden, Lime* tree or *Basswood, Magnolia* or *Beaver* tree, *Papaw, Persimmon. Sassafras, Black* or *Double* Spruce, *Tamarack* or *American Larch, Willow,* and *Wild Cherry,* of which the wood is used as a substitute for mahogany, and the bark as a valuable tonic medicine.

But the term raw material, though ordinarily limited to natural or unmanufactured products, is more comprehensive in its scope, embracing *Chemicals, substances used as food, and such substances of vegetable and animal origin as are used in Manufactures.*

3. CHEMICALS. We before remarked that the finished product of one class of manufacturers is often the raw material of another class. If the proposition needed further illustration, we might advert to Chemicals, which are such important reagents in manufacturing operations, that without them it would be difficult, (in fact, by any known processes,) impossible, to produce several articles of daily and essential utility. Without Sulphuric Acid or Oil of Vitriol, for instance, we could not probably produce Alum, Ammonia, Sal-ammoniac; Iodine and Bromine, upon the existence of which the daguerreotype art is dependent; Bleaching powder or Chlorid of Lime; Corrosive Sublimate and Calomel; Bichromate of Potash, and consequently the pigments of chrome-red, chrome-green, and chrome-yellow; Phosphorus, and consequently friction matches; or lastly, Stearic acid candles. By means of this acid, more than 100,000 tons of Soda-ash are extracted from common salt in Great Britain yearly. Without Muriatic and Nitric acids, the art of refining gold and silver, the jeweler's art, the art of electrotyping, and numerous other branches of industry, could not flourish, and some of them could not exist. The useful Arts and Manufactures, it is thus evident, are largely dependent upon Chemicals; and, consequently, a locality possessing those of the best quality in abundance, has necessarily secured an important and undoubted advantage. The chemical factories of Philadelphia, every one acknowledges, rank among the first in extent and celebrity throughout the Union. About seventeen millions of pounds of Sulphuric Acid are made yearly, and other acids and

8*

alkaline salts in proportion. The products of the establishments of Powers & Weightman, Rosengarten & Sons, Lennig & Co., Harrison Brothers, Buck, Simonin & Co., and others, are recognized as of standard excellence in the markets of the world; and where such establishments exist, we can hardly err in presuming, that at least, those Manufactures which are dependent upon the Chemical Arts must certainly flourish.

4. Agricultural Products, Provisions, &c. Again, no one need be told that with *substances used as food*, the markets of Philadelphia are always abundantly supplied, at moderate prices. As a wheat-growing State, the census of 1850 shows that Pennsylvania excels all her sister States; the product for that year having been 15,367,691 bushels, which exceeded that of Ohio, and was two millions of bushels more than that of New York. Of Rye, the product was 4,805,160 bushels; of Indian corn, 19,835,214 bushels; of Oats, 21,538,156 bushels; and hay, grass seeds, wool, butter, maple sugar, &c., in proportionate quantities. The counties immediately surrounding Philadelphia vie with each other, and rival the best counties in any other State, both in the quality and quantity of their productions. In 1850, Montgomery produced greater quantities of *hay* and *butter* than any other one county in the State; Lancaster produced more oats than any other county in the United States, more wheat than any, excepting Monroe County, New York, and more corn than any other county in Pennsylvania. In Chester, the quantity of corn produced exceeded that of any other county of the State except Lancaster, and of hay, except Montgomery; while Delaware excels in dairy products, supplying the markets of Philadelphia with butter, cheese, milk, and ice-cream, and the Union with whetstones.

Fifty years ago it was remarked, and the remarks are as

true now as then, "Much of the land within five or six miles North and South of the city is devoted to the purpose of market-gardens, and is kept in the highest state of cultivation. Two crops are very commonly produced on the same ground in one season. The neighboring State of New Jersey contributes to the abundant supply of those species of fruit and vegetables to which its light soils are particularly adapted: such as the grateful musk-melon, the water-melon, sweet-potato, cucumbers, and peaches, immense quantities of which are brought in boats across the Delaware. The superiority of the butter of Philadelphia, and the great neatness with which it is prepared for market, are generally acknowledged. One fourth of a dollar may be said to be the average price of a pound of butter, throughout the year."*

* The abundance and superior quality of the Agricultural products, for which the markets of Philadelphia are distinguished, are probably the fruition, and certainly the just reward, of the interest that has always been manifested by her citizens in Agricultural improvement. As early as 1785, a number of gentlemen, among others Robert Morris, Dr. Rush, and Richard Peters, met together and established the first Agricultural Society on this continent, under the title of the "Philadelphia Society for Promoting Agriculture," which still survives, surrounded now, however, by almost innumerable sister associations, diffusing information on rural affairs throughout the entire Union. At a later period, in September, 1826, a company of Philadelphians, principally through the instrumentality of the late Dr. James Mease, founded the "Pennsylvania Horticultural Society," which, like its predecessor, has the proud distinction of having led the way in its own particular sphere, and induced the creation of many kindred associations, promoting refinement and kindling a taste for Horticulture, even at the verge of Western settlements. One of the means early adopted by both associations to stimulate improvement was holding Annual Exhibitions, at which live stock, implements, fruits, vegetables, and flowers, were brought into competition. The exhibitions of the Agricultural Society attracted, years before they were held elsewhere, throngs of intelligent observers and practical cultivators from neighboring States, as well as from Pennsylvania, and diffused a most salutary and beneficial influence. The development which it is possible for such societies to attain, was witnessed in October, 1856, when the exhibition of the "United States

Of *Fish*, the markets of Philadelphia are constantly supplied, from the river, the bay, and the sea, with almost every desirable variety. We can imagine the delight with which epicures, a half century ago, read, that " early in the spring large sun-fish are caught in the Bay, and are succeeded by herrings, shad, roach, four kinds of cat-fish, four kinds of perch, rock, lamprey eel, common eel, pike, sucker, sturgeon, gar-fish. These are river fish, and appear in the order mentioned. From the sea, come

Agricultural Society," held in Philadelphia, by invitation of the Philadelphia Society and the officers of the City Government, attracted the most imposing display probably ever witnessed in the United States, on any similar occasion. Upward of *forty thousand dollars* were received, and the entire sum expended in premiums and for the necessary preparations. Competitors from distant States carried off many well-earned and important premiums; but it would be only justice rewarding merit, to record the fact, that to a Philadelphia firm, that of DAVID LANDRETH & SON, was awarded the first and most important premium, viz., that *for the best display of Agricultural Implements manufactured by the exhibiter.*

In the importation of *Live Stock*, Philadelphians were among the first to embark, and they have had the satisfaction of introducing to agriculturalists some of the most valuable foreign breeds that are known. The first "short-horned" cow that probably ever crossed the Atlantic, was landed at the wharf in Philadelphia, in 1807. This importation was in advance of the appreciation of such stock, and the cow was returned to England; but a bull-calf, dropped by her whilst here, was fortunately retained, and impressed his stamp on the cattle of the country.

About the year 1828, Mr. JOHN HARE POWELL, imported *South Down Sheep;* and the same enterprising gentleman, some years subsequently, commenced his importations of Short Horns, (Durhams). Not long afterward, Mr. Whittaker, the noted English breeder, consigned similar animals to his care for sale. Other gentlemen in this vicinity followed the example of Mr. Powell; and shortly afterward further importations were made for Kentucky, and other Western States. Mr. Sarchet, of Philadelphia, has the credit of the first importation of "Alderneys"; afterward, in 1840, the late Mr. Nicholas Biddle, imported specimens of the "Jersey" or "Alderney" cattle. Their descendants are now spread into the adjoining counties, and have produced a sensible improvement in the quality of the cream and butter wherever the strain has been infused. It seems to us proper, that early enterprise in this direction should be recorded.

cod, sea-bass, black-fish, sheep's-head, Spanish mackerel, haddock, pollock, mullet, halibut, flounder, sole, plaice, skait, porgey, tom-cod, and others. Of Shell-fish, there are oysters, (several kinds,) clams, lobster, crab, snapping-turtle, and terrapin—all excellent. Oysters abound throughout the year, and are sold at a low price. The shad caught in the vicinity of Philadelphia are generally esteemed superior in flavor, and more delicate than those caught elsewhere. It is supposed that the situation of the fishing-places influences the size and the flavor of shad." But the abundance, cheapness, and excellence of provisions in Philadelphia are conceded. The *New York Tribune* of May 1, 1857, stated that "*Philadelphia has about twenty-five per cent. the advantage of us in fuel, and perhaps ten per cent. in the average cost of provisions.*"

III. SITUATION, RAIL-ROAD CONNECTIONS, &c. The third point that we have considered essential to success in Manufactures, is a *favorable situation*. Viewing Philadelphia with respect to situation, we remark, in the first place, that it is far enough from the ocean to be exempt from a salt atmosphere, which has been found decidedly injurious in several Manufacturing and Chemical processes;* yet it is near enough to the great highway of nations to partake of the advantages of a port on the seacoast, in receiving raw materials and sending away manufactured products. Secondly, Philadelphia now possesses unrivaled means of communication with the interior of our country, and directly or indirectly with all foreign countries. Shippers of freight, destined for other seaports, have a choice of routes to the ocean, viz., the Delaware River—the ordinary and natural channel— and the Camden and Amboy Rail-road, the Philadelphia and Trenton Rail-road, and the Delaware and Raritan

* In paints, a pure Carbonate of Lead cannot well be made near to the sea.

Canal. By way of the river and the ocean, merchandise may be forwarded and received cheaply and expeditiously from all parts of the world, regular lines being established to all principal cities of the United States— Boston, New York, Baltimore, Richmond, Savannah, New Orleans, California, and to Liverpool, &c. During the last year, (1857,) as we have stated elsewhere, there were 505 foreign, and 32,142 coastwise arrivals, principally at the Delaware wharves. Since 1845, the vessels annually employed in the coal trade alone, from Port Richmond, largely exceed in number and capacity the whole foreign tonnage of the city of New York. But, though the Delaware River be the natural channel for freight destined to distant sections, it is by no means the only one. Immense quantities of goods are daily sent and received by the Propeller lines, *via* the Delaware and Raritan Canal; and the Camden and Amboy, and Philadelphia and Trenton Rail-roads, which are far-famed thoroughfares. A shorter and more direct route to the ocean than any of these, may now be finished for a trifling expenditure, viz., by the extension of the Camden and Atlantic Railway; and when the contemplated project of building a magnificent roadway from Florence to Union, N. J., is carried into execution, another avenue for the conveyance of light freight, cheaply and expeditiously, between Philadelphia and the ocean, will be opened.

But the highways in which Philadelphia has invested the greatest amount of capital, and which probably will in future be of the most advantage to her industrial interests, are those which communicate with the interior. To the North, and connecting her with the coal regions, there are several canals and two principal rail-roads— the Reading, and the North Pennsylvania. The latter is a new and promising road, communicating with the populous towns of Lehigh and Northampton counties, and in connection with the Lehigh Valley Rail-road, affording another outlet for the Coal and Iron products of the Lehigh regions. The former was constructed primarily as an avenue for the transportation of coal from Schuylkill County; but by means of connections established with other roads, it now forms part of a great through route to the Falls of Niagara, the Lakes, the Canadas, and the West.

The READING RAIL-ROAD, being unquestionably one of the most magnificent freight roads in the world, is entitled to further notice. It was the first rail-road that revolutionized popular opinion with respect to the adaptation of railways for carrying heavy burdens. Having a slightly descending grade in the direction of the loaded trains, the entire distance from Schuylkill Haven to the Falls of Schuylkill, 84 miles, it is able to transport heavy freight at a cost which is insignificant, even in comparison with the usual tolls on canals. The cost of transporting a ton of coal, per round trip of 190 miles— that is, from the coal region to tide-water and back with empty cars, was, during the last year, only 36.3 cents; whereas, the tolls on a ton of merchandise on the Erie Canal were nearly double that amount. The average load of an engine, during the busy season, is nearly 500 tons of coal; and a single engine has conveyed a train of 166 cars, weighing 797 tons of 2240 lbs. each.

The original charter, passed in 1833, contemplated Reading as the northern terminus of the road—hence its name; but subsequently the charter was extended, and the road constructed to Pottsville. The first locomotive and train passed over the entire line on the first day of January, 1842. The event was celebrated with military display, and "an immense procession of seventy-five passenger cars, 2,225 feet in length, containing 2,150 persons, three bands of music, banners, &c., all drawn by a single engine. In the rear was a train of fifty-two burden cars, loaded with 180 tons of coal, part of which was mined the same morning, 412 feet below water level." The road now consists of a double track; rail of the H pattern; whole length 258 miles, of which 127¾ miles have been relaid during the last seven years, at a cost of $796,735 43. The rolling stock includes 142 locomotives, 58 passenger cars, 924 merchandise cars, and 4,831 iron and wooden coal cars, besides over 600 used by the company, but owned by other parties; and the whole, if placed in a line, *would extend for a distance of fifteen miles.* The equipments are ample for the transportation of 2,500,000 tons per annum; the tonnage, in 1855, being 2,213,292. The road has nearly ninety stone and iron bridges, and over forty wooden bridges; four tunnels, the largest of which at Phœnixville is 1,934 feet cut through solid rock; numerous depots, wharves, and workshops, (those at Reading furnishing employment to about 400 hands, including boys,) and a vast deal of valuable real estate.* The entire cost of the whole, on

* "At Richmond, the lower terminus of the road, at tide-water on the river Delaware, are constructed the most extensive and commodious wharves, in all probability, in the world, for the reception and shipping, not only of the present, but of the future vast coal tonnage of the railway; forty-nine acres are occupied with the company's wharves and works, extending along twenty-two hundred and seventy-two feet of river front, and accessible to vessels of six or seven hundred tons. The shipping arrangements consist of some twenty wharves or piers, extending from three hundred and forty-two to eleven hundred and thirty-two feet into the river, all built in the most substantial manner, and furnished with chutes at convenient distances, by which the coal flows into the vessel lying alongside, DIRECTLY FROM THE OPENED BOTTOM OF THE COAL CAR IN WHICH IT LEFT THE MINE. As some coal is piled or stacked in winter, or at times when its shipment is not required, the elevation of the tracks, by trestlings, above the solid surface or flooring of the piers, affords sufficient room for stowing upward of two hundred and fifty thousand tons of coal. Capacious docks extend in-shore, between each pair of wharves, thus

November 30, 1857, was $19,262,720 27. The officers are: President, R. D. CULLEN; Treasurer, SAMUEL BRADFORD; Sec'y, W. H. McILHENNEY, and Gen'l Supt., G. A. NICOLLS.

With Pittsburg, and the "Gate of the West," Philadelphia is connected by a magnificent Railway, to which we have more than once referred, and to which it seems proper to refer again, if for no other purpose than to aid in perpetuating the names of those who have been most active in contributing to the success of so great an undertaking. While the Reading cheapens fuel to the citizens of Philadelphia, the Pennsylvania Central cheapens food, and both are entitled to rank among the most important enterprises of modern times.

The act incorporating the Pennsylvania Central Railroad Company was passed April 13th, 1846. As soon as the news of its passage had reached Philadelphia, a large meeting was held, and a committee appointed to prepare an address inviting the co-operation of the citizens. This committee consisted of THOMAS P. COPE, (since dec'd,) Chairman; DAVID S. BROWN, JOHN GRIGG, THOMAS SPARKS, GEORGE N. BAKER, RICHARD D. WOOD, JAMES MAGEE, and J. R. TYSON. The address issued by these gentlemen met with a warm response, and public and private subscriptions were freely tendered. The city,

---

making the whole river front available for shipping purposes. Over one hundred vessels can be loading at the *same moment;* and few places present busier or more interesting scenes, than the wharves of the Reading Rail-road, at Richmond. A brig of one hundred and fifty-five tons has been loaded with that number of tons of coal in less *than three hours time,* at these wharves. The whole length of the lateral railways extending over the wharves at Richmond will probably exceed ten miles, affording a shipping capacity for upward of *three millions of tons!* and it will probably not be many years before this amount, extraordinary as it may seem, (as, indeed, it really is,) will be annually transported over this great thoroughfare. The company has laid the *foundation* for a trade as broad as the future destiny of the coal trade itself."

9

in its corporate capacity, subscribed two and a half millions of dollars, and this gave an impulse to the enterprise that left no longer any doubt of its success. The first Board of Directors consisted of the following gentlemen, most of whom had been active in promoting this great work, viz.: S. V. Merrick, Thomas P. Cope, Robert Toland, David S. Brown, James Magee, Richard D. Wood, Stephen Colwell, George W. Carpenter, Christian E. Spangler, Thomas T. Lea, William C. Patterson, John A. Wright, and Henry C. Corbit. First officers—S. V. Merrick, President; Oliver Fuller, Secretary; George V. Bacon, Treasurer; J. Edgar Thomson, Chief Engineer; William B. Foster, Jr., Associate Engineer, of the Eastern Division; Edward Miller, of the Western.

During the past year this Company made a most important and extensive negotiation, being no less than the purchase from the Commonwealth of 285 miles of Canal, between Philadelphia and Pittsburg; and 37 miles of Railway, between Johnstown and Hollidaysburg; and 80 miles of double track between Philadelphia and the Susquehanna River, with all the appurtenances, giving their bonds, bearing five per cent. interest, for the sum of $7,500,000, payable $100,000 on July 31st, 1858, and $100,000 annually thereafter, until July 31, 1890, when the payments will be at the rate of $1,000,000 per annum until the whole is paid. Present total cost of roads and canals belonging to Company, $27,266,981 58. The rolling stock consists of 216 locomotives, 99 passenger cars, 27 baggage cars, and 1,945 freight cars. The aggregate tonnage of the road, for 1857, was 530,420. The surplus earnings were $1,854,926 86. The present officers of the road are— President, J. EDGAR THOMSON; Vice-President, W. B. FOSTER, JR.; Treasurer, THOMAS T. FIRTH; Secretary, EDMUND SMITH; Gen'l Superintendent, THOS. A. SCOTT; Controler and Auditor, H. J. LOMBAERT; Superintendent

Philadelphia Division, G. C. FRANCISCUS; Superintendent Eastern Division, A. L. ROUMFORT; Superintendent Middle Division, THOMAS P. SARGENT; Superintendent Western Division, JOSEPH D. POTTS; General Freight Agent, H. H. HOUSTON.

The last link in the chain is now perfected, connecting Philadelphia and Chicago, *via* Pittsburg, Fort Wayne and Chicago Rail-road; other connections are constantly being made; and the Pennsylvania Central Railway, fortunate in its mode of construction, and fortunate in its officers, will hereafter still further reduce the cost of transportation between Philadelphia and the West, and perpetually prove an increasing source of benefit to both.

The other great trunk line diverging from Philadelphia, and increasing its rail-road connections with the South, is the PHILADELPHIA, WILMINGTON, AND BALTIMORE RAIL-ROAD. This road forms part of the great Southern mail route—and being one of the oldest, is consequently one of the best known rail-roads in the country. The low charges for water carriage between Philadelphia and the prominent points of the South, have heretofore deprived this road of any considerable revenue from freight; but, nevertheless, the Company is now free from floating debt, has paid all the demands that were made upon it, and its regular dividend, without borrowing a dollar. This Company is peculiarly fortunate in its President, S. M. FELTON, Esq., who is regarded as one of the ablest rail-road officers in the country.

A new road to Baltimore, entitled the *Baltimore Central*, connecting with the Westchester Rail-road, at Grubb's Bridge, is in course of construction, and thirty miles will probably soon be finished. This road, it is said, can bridge the Susquehanna. The minor rail-roads diverging from Philadelphia, are the *Philadelphia, Germantown, and Norristown*, which, in 1857, carried 1,378,228 passengers; and having as tributaries the Chester Valley and Chestnut Hill Rail-roads; and the *Westchester* Rail-road, of which seventeen miles are completed, and the balance (ten miles) graded and ballasted.

The following Table exhibits the

Names, Length, and Cost of the Rail-roads centering in Philadelphia, with their Receipts, Expenses, and Surplus Earnings, for 1857.

Names.	Length	Cost	Gross Receipts.	Expenses.	Surplus Earnings.
Pennsylvania................*	393	$19,766,981.58 / 7,500,000.00	$4,855,669.76	$3,000,742.90	$1,854,926.86
Reading............................	98	19,262,720.27	3,065,521.56	1,481,745.22	1,583,776.34
Philad, Wil'g'n and Balt....	98	8,568.369.32	1,143.852.69	764,917.10	378,935.59
Camden and Amboy.........†	91	5,563,580.11	1,598,124.91	880,131.17	717,993.74
Philadelphia and Trenton...	28	1,000,000.00	operated in	part by C. &	A. Co.
North Pennsylvania...........	68	6,106,280.57	248,783.80	112,186.65	136,597.15
Philad. Germt'n & Norris'n.	21	1.810,812.28	312,958.63	132,852.25	170,268,75
Westchester and Philad......	17	1,300,000.00	50,986.00	39,000.00	unfinished.
Camden and Atlantic..........	61	....	....	....	....

* This includes the Indiana Branch, 19 miles; Hollidaysburg Branch, 9 miles; and Johnstown Branch, 37 miles—but excludes the Harrisburg, Lancaster and Mountjoy Rail-road, (35 miles) which is leased, not owned by the Pennsylvania Rail-road.

† This includes Trenton and other Branches.

The rail-road system of Philadelphia, we may remark, in conclusion, adopting the language of one who has made it the subject of careful consideration, extends to all points of the compass, pushes out toward the ocean, pierces the coal regions of the North, reaches Eastward to the great seaports of the nation, drains the rich and fertile agricultural counties of our own State, and extends Westward toward the Rocky Mountains and the gold region beyond. It is a grand plan, and needs but one important line to make it perfect. The Sunbury and Erie road must be completed to Lake Erie, to develope the resources of that portion of the State through which it passes, while our legitimate portion of the trade of the Northwest runs along it into the lap of Philadelphia, which will be nearer to the great inland seas than either of her rivals—Baltimore or New York.

IV. The fourth and last subdivision of *essential* physical advantages is a *suitable climate*—a climate favorable to vigor of mind and health of body, and chemically

adapted for manufacturing processes. The climate of
Philadelphia, in common with other portions of the
State, we may say the country, has undergone important
changes within a half century. The winters are less
uniformly cold than formerly, and the summers less uniformly warm. Except during the winters of 1855-6,
and 1856-7, which were entirely exceptional, ice in the
Delaware has not presented any formidable obstruction
to navigation for many years, and sleighing has been a
sport of short duration. In the present winter (1857-8),
no snow, worth mentioning, has fallen up to the middle
of February; and the weather during January was as
genial as spring. In the summer, the thermometer sometimes rises for a few consecutive days above 93°; but
the temperature invariably diminishes sensibly after
sunset, and the nights are generally comfortable and refreshing. The most disagreeable feature of the climate
in summer is liability to sudden variations, amounting in
some rare instances to 30° in twenty-four hours. These
variations, however, it would seem, are more unpleasant
than permanently injurious in their effects.

The air of Philadelphia, compared with that of New
York, has less keenness; and being free from saline impregnation, it is less irritating to weak lungs. It was observed long since, and remarked by physicians, that persons did spit blood in New York who were entirely free
from any pulmonic affection in Philadelphia. Compared
with New England, generally, the winters in Philadelphia
are less severe, and consequently *less fuel is consumed;
while the days are of greater average length*, thereby diminishing the consumption of gas. Both of these items have
a bearing upon economy of production in Manufactures.
But the climate of Philadelphia has further some peculiar and remarkable properties, as is evidenced by its
effects upon certain chemical processes. It is conceded,
9*

even by Englishmen, that the woven fabrics of southern Europe are superior to those of England in the richness and clearness of their colors; and this superiority is accounted for by ascribing it to atmospheric qualities and peculiarities, for which neither the science of chemists nor the skill of dyers in England, has been able to provide a complete equivalent. So, experience demonstrates, that it is possible in Philadelphia to attain a degree of excellence in dyeing fabrics, unattainable by the same processes anywhere else except in Southern Europe. A celebrated French dyer, whose local partialities are distant from this city, experimented in various localities in France and the United States, and found the climate and water nowhere in either country so well adapted for his purposes as those of Philadelphia. Hence, every year the practice is becoming more common with the merchants of Philadelphia, New York, and elsewhere, to import silks, and woolen goods in an unfinished state, have them dyed in Philadelphia, and then they readily command prices equal to the best French or European finished fabrics.

In addition to these circumstances, which are considered essential to success in Manufactures, there are many others so desirable and important, that they can scarcely be ranked as secondary. Foremost in this class is—

1. PURITY OF WATER. Water, like climate, has a sanitary, and also a chemical bearing. The water principally used in Philadelphia proper is from the Schuylkill; while in Frankford, Bridesburg, and other important manufacturing adjuncts, there are springs possessing some remarkable properties. The Schuylkill water, as we learn from the report of Messrs. Booth and Garrett, who, in 1854, made it the subject of careful analysis, is

distinguished above almost all other waters for its purity and freedom from organic matter. Their very able report concludes with the following opinion :—

" We may further observe, that a comparison of our waters, with waters used elsewhere in the United States and in Europe, highly esteemed for their excellency, may be characterized by its greater purity, its slightly alkaline impregnation, and by being nearly free from organic matter. In conclusion, we infer that the Schuylkill water has deteriorated, in no important respects, from its former excellent quality ; is superior to most waters for domestic and manufacturing purposes; and lastly, a comparison of the past and present, leads to the inference, that no plan of improving the water will be required for many years to come."

By analysis, it has been ascertained that the water of the Cochituate, (used in Boston,) contains 1.16 grs. of solid organic substances in one gallon ; and the Croton, (used in New York,) contains 4.28 grs., and that, too, after it had passed through forty-one miles of aqueduct; while the Schuylkill water, taken directly from the river, before it had entered into the reservoir, and had time to deposit its solid particles, contained but a trace of organic matter. The chairman of the Philadelphia County Medical Society concludes, that we possess the advantage of a purer quality of water for drinking purposes than any other city in the United States, or perhaps the world over, a prerequisite as essential to the enjoyment of health, as it is necessary for the preservation of life itself.

The sanitary results of the climate and the water are manifested in—

2. THE STATISTICS OF HEALTH. The comparative healthfulness of various cities has been made a subject of careful observation by physicians and others, for more

than a half century, and the tables of mortality have uniformly shown that *Philadelphia is the most healthy of the great cities of the United States.* In 1806, when the city contained a larger population than New York, the deaths per day in the former were 5⅔, and in the latter 6⅛. In 1810, the proportion of deaths to population, in Philadelphia, was one to fifty. In 1855, WILSON JEWELL, M. D., as chairman of the Committee on Epidemics of the State Medical Society, presented a report full of valuable suggestions, and containing the following Table and remarks relative to the sanitary condition of our principal cities:—

1855	Population.	Total mortality.	Ratio of deaths to population.	Deaths to every 1000 inhabitants.	Per ct. of deaths under 5 years to total mortality.	Deaths under 5 years to every 1000.	Ratio of still-born to deaths.
New York	650,000	22,728	1 in 28.59	35.	53.40	18.67	1 in 13.70
Philadelphia	500,000	10,458	1 in 47.81	20.91	44.86	9.38	1 in 17.85
Baltimore	215,000	5,465	1 in 39.52	25.41	44.88	11.40	1 in 14.01
Boston	162,748	4,308	1 in 39.36	26.59	46.63	12.40	1 in 19.33

" The averages, deductions, and comparisons drawn in this Table, prove conclusively that the mortality in our own city is much less, compared with the total of deaths, with the deaths to population, or with every thousand, than in the other Atlantic cities.

" While in New York 1 in every 28 of the population dies annually, and in Baltimore and Boston 1 in every 39, in Philadelphia there is only 1 in every 47; more favorable by one half than the death rate of New York; and, by nearly one fourth, more favorable than that of Boston and Baltimore.

" Again, the health of Philadelphia, contrasted with that of the other cities named in the Table, is shown by estimating the deaths to every thousand of the population. While New York contributes 35, Boston 26, and Baltimore 25, Philadelphia gives only 20.

" Nor can it be overlooked, that the infantile population in New York suffers by death to a far greater extent than in either of the other cities. Those under five years of age (exclusive of still-born) make up 53 per cent. of the total mortality; Boston 46 per cent.; while Baltimore and Philadelphia are each 44 per cent.: less by 8 per cent. than the former, and 5 per cent. than those under five years in the latter city.

" The deaths under five years in every thousand of the population presents an equally favorable contrast; New York furnishing 18, Bos. ton 12, Baltimore 11, and Philadelphia only 9 in every thousand.

"It will be seen, too, while the population in New York was but 13 per cent. greater than that of Philadelphia, the deaths for the year 1855 were 35.90 per cent. more than in our own city. The ratio of still-born children to the mortality is less in Philadelphia than in either of the other places.

"The preceding estimates are sufficiently clear to maintain the position, that we are the healthiest of the large Atlantic cities, and that for salubrity, we should have the preference before the others named in the Table.

The aggregate mortality in the four cities, in 1856 and 1857, was as follows:—

	1856.		1857.			
Philadelphia,	12,090	-	10,950	-	Decrease,	1,140
New York,	21,496	-	23,370	-	Increase,	1,874
Baltimore,	5,677	-	5,524	-	Decrease,	153
Boston,	4,170	-	4,005	-	Decrease,	165
Total,	43,433		43,849		Increase,	416

The proportion of deaths to population, it will be perceived, is about the same as in 1855, and the result equally favorable to Philadelphia.

3. PROTECTION AGAINST FIRES, &c. Disastrous fires, it is well known, occur more frequently in American than European cities; and there was a period when Philadelphia enjoyed an unenviable distinction in this respect, even among her sister cities. Fortunately that period has gone by, and we now may proclaim confidently that in no American city is life more secure, or property better protected, than in Philadelphia. One of the causes, it was ascertained, of the former prevalence of fires, and the destruction of property, was the feuds which in course of time had sprung up between the various organizations originally established for the extinguishment of fires. The system of voluntary association for this purpose,—inaugurated, it is said, by Franklin, in 1732,—though manifestly calling forth a great deal of self-sacri-

fice and heroism, was regarded by many as a failure, or in other words, as better adapted for small towns than for large cities. But many of the evils developed from this source have been obviated by the *reorganization of the Fire Department*, recently effected: that is, by disbanding the most disorderly companies, dividing the city into districts, permitting only a prescribed number of companies to go into service except in case of a large fire, when the general alarm rung on the State-House bell calls the whole Department into requisition.*

In 1856, another very important improvement was made by the establishment of a *Police and Fire Alarm Telegraph*, by which information can be communicated, at a moment's notice, to and from any of the sixteen Police-stations that comprise the jurisdiction. During 1857, by this means, 34,207 messages were transmitted, 3,430 lost children restored to their parents, 884 strayed and stolen animals were restored to owners, 392 fire alarms given, the Coroner notified 387 times, and 1,361 Police-officers subpœnaed to testify before the courts. Still more recently, another safeguard was originated by the establishment of the *Fire Detective Police*—a department of the General Police—specially charged with the duty of investigating fires and detecting incendiaries. The inception of this wise measure is due, we believe, to our Mayor, the Hon. RICHARD VAUX, and its success and efficiency, largely to the signal ability of the chief officer, A. W. BLACKBURN. But the improvement that will probably be found the most effective of all, as a protection against serious loss by fire, is the introduction of *Steam Fire Engines.*

* The Fire Department now consists of 42 Engine Companies, 43 Hose Companies, 5 Hook and Ladder Companies, and 1 Steam Fire Engine. Members about 8,500. Officers—Chief Engineer, SAMUEL P. FEARON; Assistant Engineers: WM. E. STANCLIFF, DAVID M. LYLE, WM. M. LOUGHEAD, MICHAEL YOUNG, JOHN GIVEN; Secretary and Treasurer, EDWIN F. MILLER.

Within a few months past a Philadelphia firm, Messrs.
REANEY, NEAFIE & Co., have produced a machine in all
respects a striking contrast to its cumbersome and inefficient Western predecessor, and which has revolutionized
popular opinion with regard to the practicability of steam
for this purpose. One Steam Fire Engine is now in use
in this city, and three others in course of construction.
The law regulating the erection of buildings will, to
some extent, diminish fires; but the most efficient protection against serious loss which manufacturers and
owners of property in Philadelphia have, exists in the
reliable character of numerous Insurance Companies,
who are always prepared to take risks at low rates, and
to meet losses with creditable promptness.* In view of
all these circumstances, we are not surprised to learn,
that the losses by fire within seven months, from May,
1857, to January, 1858, were a quarter of a million of
dollars less than for the corresponding period of 1856;
and that the losses to the owners of property, that is,
over and above insurance, within said period, amounted
to only $54,780.

3. ABUNDANCE OF CAPITAL. Another matter that has
a bearing upon the adaptation of localities for manufactures, is *the quantity of floating or loanable capital—or, in
other words, the normal state of the money market.* The
success of the English manufacturer, compared with that
of the American, is probably due less to the low rate of
wages in England, or to any other one circumstance,
than to the low charges for the use of capital. In this
country the rates of interest, advanced by the competition engendered by the tempting opportunities for profit,

* We may probably insert in the Appendix a list of Insurance Companies
of undoubted solvency, as in some degree a protection for our distant
friends, and even our own citizens, against bogus Insurance Agencies, dat-
ing from this city.

are in most places too high for Manufactures yet in their infancy, and weighed down by an inhospitable political sentiment, to sustain. There is a marked difference, however, in this respect between different localities, and we think we do not err in saying, that in no city in the United States have the rates of interest on second-class paper—for an average of years—been so uniformly low as in Philadelphia; none in which there are so many small surplus capitals, say from ten to one hundred thousand dollars and upward, constantly seeking investment in temporary and permanent loans. Large fortunes are, it is probable, less numerous now, or at least less prominently conspicuous, than formerly;* the banking capital of the city, being about $12,000,000, is hardly one third of the amount twenty years ago; nevertheless, that unfailing barometer of money centres—the average rate of interest—has generally indicated an abundance of loanable capital. If our Manufacturers have not as yet derived their proper share of benefit from this circumstance; or, if bank officers, in distributing their loans, have not exercised a wise discrimination in their favor, we sincerely hope that the mistake originated solely

* A Book of Millionaires in Philadelphia, if published at this time, would be more imposing from its subject than its size, unless the author adopted the New York plan, and inserted the biographies of all who are worth a hundred thousand dollars, or so. There are, however, within our knowledge twenty-five individuals in this city accredited, by their intimate acquaintances, with the ownership of a million and more, viz.; JOHN GRIGG, (retired Publisher); JOS. HARRISON, Jr., (of Russian celebrity); GEO. W. CARPENTER, (druggist); RICHARD ASHURST, (private banker); JOHN B. MYERS, (auctioneer); ALEX. BENSON, (broker); F. A. DREXEL, (banker); JACOB STEINMETZ, (farmer); J. S. LOVERING, (sugar refiner); WM. H. STEWART, (planter); J. P. CROZER, (manufacturer); RICHARD WISTAR, JOHN WISTAR, JAMES DUNDAS, J. J. RIDGWAY, DR. J. RHEA BARTON, DR. JAMES RUSH, CHARLES HENRY FISHER, J. FRANCIS FISHER, MRS. LOGAN, EVANS ROGERS, H. MESSCHERT, JOHN A. BROWN, DAVID JAYNE, J. L. FLORANCE, and several Millionaire estates. But the author of the "Wealthy Men in New York," who included NICHOLAS LONGWORTH, of Cincinnati, among the number, would find no difficulty in extending this list in Philadelphia.

in misconception as to the predominant interest of the city; and that, with the aid of the late panic, in destroying the blinding fascination of "gilt-edged paper,"—and perhaps in some humble degree, with the aid of this volume—Manufacturers and Mechanics will hereafter approximate more closely to that position in the scale of mercantile credit, to which the advantages of the locality, and their own solvency and usefulness, unquestionably entitle them.

5. SUPERIOR MACHINES. The immense productive power of machinery, compared with mere manual operations, can require at this day no illustration. For instance, by the improvements effected in Spinning Machinery, one man can attend to a mule containing 1,088 spindles; each spinning three hanks, or 3,263 hanks a day; so that, as compared with the operations of the most expert spinner in Hindostan, an American operative can perform the work of *three thousand* men. The efficiency of machinery, however, like that of labor, depends upon its quality; and this, it would seem, depends upon the cheapness and abundance of the materials that enter into the composition of machines. In England, it has been supposed, that if Iron, Steel, and Brass, were less abundant, the machines would be in a less degree superior; and in the United States, though the mechanical appliances in use are almost everywhere deserving of admiration—none are probably more remarkable for power and efficiency than those in Philadelphia. A gentleman, who has quite recently made the Manufactures of Iron in this city the subject of investigation, publishes the following observations respecting the machines in use in the Iron establishments:

"In the course of our inquiries into the Manufactures of Iron in this city, the bearing of machinery upon production has been constantly brought to notice, and striking instances of its value have been observed.

10

In a leading establishment, where foundry work is the principal business, six thousand tons of iron being melted per year, the economical power of machinery in moving all the masses of iron is such that the production of each man exceeds three thousand dollars annually for the average of all the employed. This is three times the production of equally skilled workmen, without machinery. The lowest average for foundry work, as well as for artisans in wrought iron, is below a thousand dollars per annum, and this whether they handle a large weight of iron or not, if the processes are conducted by physical strength alone, and wholly without the use of machinery. In short, the economy of machinery applies alike to all forms of iron working, and to the processes which change its value least, equally with those which increase its value many times.

" The introduction of machinery has revolutionized the simple production of Iron from the ores also. It has been stated to us that the anthracite furnaces now make six thousand tons of iron more easily than six hundred tons were made fifteen or twenty years ago. In every thing that relates to the making or working of iron there is the greatest possible inducement to the employment and perfecting of machinery, intended to economize the force required, and the labor employed. In this direction investment is safe, and capital is certain of satisfactory returns. The leading departments of iron manufacture furnish articles of universal use and universal necessity, in which accumulation of stocks is not to be dreaded so much as the narrow margin between cost and sale prices. Reduce the cost of manufacture fifteen or twenty per cent., and the proprietor may proceed in the face of even a dull market, and indeed, under a total cessation of orders. The direction in which improvement lies is in perfecting and introducing powerful machinery, and every inducement concurs to urge attention to this point.

" It is noticeable that the machinery employed in American manufactures of iron is new and original in almost all cases. The most signal economies of power in the establishments of this city are not by the use of purchased machinery, but they are the creations of the proprietors who use them, suggested in the course of their work, and devised and applied by themselves. In all forms of machinist manufacture those inventive and constructive processes are making rapid progress. The great capital they represent when finished, is capital created by the establishment, and not an investment from the outside. This fact guarantees the permanent efficiency of these manufactures, since such capital is not easily withdrawn, and the establishment is not broken up by temporary depression of a business, or even by the dispersion of workmen for a considerable time.

" It has been recently stated that the machinery invented and applied in American armories, private, as well as those belonging to the government, is much sought in Europe, and will soon be in almost universal use abroad. This fact bears directly on the point we are stating. Machinist machinery is equally advanced here; and at two or three of our great establishments it is confessedly superior to that of the celebrated Lowell Machine Works, while constant improvements are being made. In the appliances for handling iron in heavy foundry work, the world may be challenged for comparison with the machinery of at least one great establishment here, and the most important items in that case are the absolute creation of the proprietors. It is obvious that such machinery differs widely, in its economical importance, from that which is purchased by direct expenditure, and particularly from any form of machinery imported from other quarters.

"The direction in which this city always will excel is in the handling of heavy masses of metals. Power is cheapest here, and necessity first impels to the economy of forge-work, iron rolling, foundry-work, ship building, and costly machine building. In the minor manufactures the application of improvements is more rapidly made at the North; but this is from want of attention here, instead of from want of the requisite field and facilities. No location in the Union can compare with this in natural advantages for the manufacture of arms of every sort, cutlery and tools, implements of every kind, and the multitude of minor manufactures in which inventive talent and machinery decide the whole question of profitable attention to the business. The market is the whole world. At this moment many superior instruments of steel and iron are actually made here for European sale; and the skill which does this now on a small scale, only requires the aid of more perfect machinery, and the capital necessary to work it, to make the business all that the most sanguine might wish."

In the production of MACHINE TOOLS, and fine as well as heavy machinery, very marked success has attended the efforts of our mechanical engineers. The Lathes, Planers, Drills, Borers, and the machinery for working metals generally, made in Philadelphia, are wonderful specimens of workmanship, and celebrated not only throughout the United States, but in portions of Europe. A few years ago, Commissioners were sent to this country to procure tools and machines for the government workshops in Russia. Discharging their duty faithfully, they visited, we believe, all the manufactories of these important articles in New England and the principal cities; and, though they found the prices in some instances nominally cheaper, their order was reserved until they again reached this city. The machines of New England, in consequence of the great cost of iron, are remarkable for their *lightness;* but in substantial excellence and quality of workmanship, none can compare with those of Philadelphia.

In reflecting upon the causes conducive to superiority in this particular, it has occurred to me as probable, that the establishment and continuance of the United States Mint, in this city, have tended in some degree, by creating a demand for a finer and higher class of workmanship, to centre here the best skill in this department of Mechanics. No expense being spared by the able managers of that Institution to procure the most perfect machines; and every reasonable facility being afforded for experiment, we need scarcely wonder at the degree of perfection that has been attained. Our Mint has probably originated a greater number of valuable improvements than any similar establishment in the world; and all persons familiar with its past history and present management, unite with the Committee of the Board of Assay Commissioners, in stating "that the Institution, in their opinion, is conducted and maintained in such a manner as to merit the highest confidence of the Government and the public."*

* The Director of the Mint has favored me with the following letter, in answer to a request for some information respecting the machines, and the curiosities to be seen in that establishment; and, as it will be read with interest, I trust he will pardon its publication.

"MINT OF THE UNITED STATES,
Philadelphia, Jan. 21, 1858.

"DEAR SIR:

"Without being able at present to go minutely into the subjects mentioned in your note of the 5th, I may state, that the establishment and

6. ESTABLISHED REPUTATION. Established reputation, though in its nature etherial, is an object of substantial value—a power in the money market. It is of two kinds—personal and local. The marketable value of

continuance of the Mint, in this city, have undoubtedly had their share in calling forth the various kinds of scientific and mechanical talent, which are requisite for the successful conduct of such an Institution.

"Within a period, now embracing more than sixty years, there has been a large amount of machinery manufactured for, and within the Mint establishment, from the more ordinary workmanship up to the most delicate and elaborate. A number of important mechanisms and processes have had their origin and invention here; and others, borrowed from other places, have been modified and improved. Some instruments, it is true, are still imported; but they are now of comparatively trivial account, being such as are of so limited demand, as not to be an object for the attention of our artists.

"The most important improvements introduced into the Department of the Chief Coiner, have been, the press for cutting out blanks or planchets; the draw-bench for equalizing the strips—afterward adopted in the London Mint; the old self-feeding lever-coining press, and after it the steam press; the milling machine; the counting machine; and the arrangements for cleaning; also, many fine balance beams, large and small; and an assorting machine, not as yet brought into use. The system of hardening dies was originated at the Mint, and is greatly superior to the methods heretofore practiced.

"In the Melter and Refiner's Department, we may specify the parting arrangement, for separating gold and silver; the hydraulic press, for condensing the powdered gold or silver; the sweep machine; and the various arrangements by which the melting has been made a neat and economical operation.

"In the Assayer's Department: the delicate balances; the gas-bath; and generally, the systematic arrangements for the assay of gold, silver, and copper.

"The Cabinet of coins, medals, and ores, which occupies a suite of apartments at the Mint, is an attractive feature in the Institution. The collection is not very large, if compared with similar Cabinets in Europe; but it is sufficiently so to furnish valuable information on the subject of Coinage, and useful monuments of history. Besides, an examination of the collection gratifies popular curiosity, as well as educated taste.

"I may add in conclusion, that the Mint has, within a year or two past, been rendered thoroughly fire-proof in all its departments, and the arrange-

10*

the products of mechanical industry, every one will concede, is affected not merely by the reputation of the maker, but also to a greater or less extent by the general reputation of the place of their manufacture. No illustration of the principle can be necessary; but if it were, we might refer to France, the stamp of whose city, "Paris," on articles of *vertu*, of itself commands a premium; or again, we might refer to New England, whose stamp unfortunately, in many instances, does not tend to elevate the price of articles to which it is attached. The value of a good name is appreciated, perhaps, by none so forcibly as by those who have lost it. The manufacturers and mechanics of New England would no doubt give millions to obliterate from human recollection the impressions produced, in part, by operations in wooden nutmegs, mahogany hams, oak-leaf cigars, and paper-soled shoes. Deceptions of this kind, and trickeries frequently practiced by Yankee operators, though we believe and insist only by a few, and the production of a vast quantity of cheap, fragile fabrics, have so impaired confidence in Yankee contrivances in general, that all, no matter how excellent in themselves, are prejudged unfavorably from the place of their origin. To avoid this prejudice, or to partake of the advantages of an established reputation, New England manufacturers are often tempted to put foreign or fictitious stamps on their best fabrics; and thus our country loses its share of credit for the excellence it has achieved, while it must

ment of the rooms appropriated to the different branches of business greatly improved. It is thus in a condition of great efficiency and security, and is believed to be unsurpassed by any similar institution.

"I am, very respectfully,

"Your ob't servant,

"JAMES ROSS SNOWDEN,

"*Director of the Mint.*

"To E. T. FREEDLEY, Esq."

bear the reproach of its defaults. But mechanics in Philadelphia, fortunately, have none of these difficulties to overcome. The same manufacturers, if located here,—and we welcome them,—would find the way clear before them, the prepossessions of people at a distance in the South and the West in their favor, and their products commanding a readier sale in consequence. Every auctioneer will testify that a Philadelphia made carriage will command more spirited bidding, and most probably a higher price, than a Connecticut carriage of equal quality. The stamp, "Philadelphia," is everywhere regarded as *prima facie* evidence of good materials and superior workmanship. A Philadelphia mechanic is everywhere a title of reputable distinction, and a very acceptable passport to employment in every intelligent master-workman's shop. Hence our Manufacturers reverse the practice of their competitors in New England, and put their names and stamp on their *best* products, leaving the *inferior* in some instances to those who choose to adopt them.*

7. Lastly. OPPORTUNITIES FOR ART-CULTURE. Art, in its relations to Manufactures, has not, until quite recently, been appreciated by any considerable portion of the American, or even the English people, to a degree in anywise approximate to its importance. Both have long known, it is true, that certain goods sell better than others—that English and American prints, for instance, would be less saleable at the same price than those of France; yet, even while claiming superiority in the quality of the cloth, neither has been willing to attach any special importance to beauty and originality of design.

* The principal exception to this rule is, that New York dealers sometimes pay such irresistibly tempting prices to have their names affixed, as makers, to articles actually made in Philadelphia, that our Manufacturers forego the honor for the sake of the money.

This is the more remarkable, inasmuch as it must be evident to the least imaginative, how many articles are valued mainly for their style of ornamentation. We might mention carpetings and floor-cloths, carved wood and furniture, curtains, and other hangings; inlaid floors, ornamental glass, stained glass, metal work, grates and stoves, gas fittings, paper and other hangings, porcelain, pottery, works in the precious metals, works in stone, and a great variety of garment fabrics. The French, in the meanwhile, have unceasingly aimed at perfection in the Ornamental Arts. To improve the national taste, they long ago established Schools of Design and National Collections of Art; and to train up a band of skilled workmen, they more recently erected National Manufactories, employing the best painters, sculptors, and designers, as well as men of the most scientific acquirements in Botany, Mineralogy, and Chemistry. In these establishments the cost of repeated failures is totally disregarded, and every effort made to bring to perfection the fabrics wrought in them, both as to the highest excellence in workmanship and materials, and to their embellishment in ornamental design. The result is, that both English and American Manufacturers must admit, as Cobden did before a Manchester audience, "we do not know what we shall have to print, nor what the ladies will wear, till we find out what the French are preparing for the next spring." But with all their schools, Art collections, and national manufactories, we do not believe that the French would have attained any notable success in decoration, if they had adopted the Yankee system of segregation; and instead of carrying on their manufactures in cities like Paris and Lyons, they had sought cheap lots, gentle water-falls, and the mossy banks of meandering streams. Taste is a thing of culture—it is only in isolated instances, if ever, a gift of

Nature. The ability to judge, and especially to execute what is tasteful in works of Art, is the result of long familiarity with good models and constant observation of the master-pieces in Art. The sight of excellence in the products of skilled workmanship stimulates to exertion, and produces excellence in other fabrics perhaps essentially dissimilar. Hence, the great advantage of carrying on the higher class of Manufactures in or near the cities abounding in the best models, and where the eye, if not the hand, may be educated almost imperceptibly to a high degree of artistic perception.

Now, if we were seeking some one of the various cities in which to apply these principles, where, we would ask, is the principal home of the Arts in America? Which contains the finest models in Architecture, Sculpture, and Painting? There could be but one answer—Philadelphia. No other city in the Union contains so many buildings that are models of classic beauty—so many evidences of a cultivated taste—so many eminent artists—and, we may say, so many devotees of Music, for no other city has been able to sustain the Italian Opera with equal success. A procession of those in this city, who make Art their study, would be imposing from its numbers, as well as the talents of its members. At their head, by common consent, we would find the veterans Sully, Neagle, and Peale; and not far behind them, Lambdin, Waugh, Scheussle, Hamilton, Rothermel, Weber, Van Starkenborg, Moran, Schindler, Conarroe, Boutelle, and Bowers; and among the younger men, George C. Lambdin, George F. Bensell, Edwin Lewis, Haseltine, Richards, Furness, and many others, who are entitled to a niche in the temple of artistic fame; while in the ranks there would be many who, when the leaders fall, can fill their places—many Engravers on wood and steel, and Lithographers, who give to our Government's costly

publications their principal value and attractions—many
Designers and Artists in bronze, whose chandeliers and
lamps, at the World's Fair, extorted admiration from
the English and French for "lightness and purity of
design,"—some beautiful women, too, whose cultivated
fancies, stamped on paper or woven fabrics, gladden
the eye in thousands of homes; and sculptors, whose
works in stone and marble grace Galleries and Capitols,
and whose sarcophagi and mausoleums adorn almost
all the Cemeteries in the land. Ornamental Art is without a home in America, if it be not in Philadelphia.
Here then is the proper place for the establishment of a
Normal School of Design, to supply manufacturing towns
throughout the country with competent teachers, who
may aid in elevating the Art-products of America to a
level with those of the most advanced European countries. We trust some one of our men of fortune will in-
herit the blessings of future ages by endowing such an
Institution, and in connection therewith establish a Museum of Art, which shall contain all the best models—
ancient and modern—in every department of Decorative
Art, from a coffee-pot to an original Apollo Belvidere.

There are many other advantages that might be noted
—the law of limited liability in Partnership for instance
—tending to show that Philadelphia ought to attain
eminence in Manufactures. We, however, pass them by,
for they may all be included in one point, viz., *Philadelphia is already a great Manufacturing city*. I hold it to
be eminently safe to infer, that a locality in which manufacturing industry has already taken a deep, permanent
root, particularly if it manifest an indigenous growth,
possesses a soil adapted therefor, whether by analysis
we can perceive the ingredients or not. Moreover, it
seems probable, almost certain, that the spot in this

country now exhibiting the most varied and extensive development of mechanical industry, in conjunction with enduringly favorable circumstances, will remain for a century to come the central and chief seat of the higher and more artistic Manufactures in America, notwithstanding the growth and promise of other places possessing theoretically marked advantages.

To illustrate the present development of manufacturing industry in Philadelphia, I herewith submit the results, not generally of my own observation or knowledge, but that of others, and principally of reports made to me by gentlemen specially employed to report on certain branches—men far more competent and more experienced in mechanical matters than myself—and not one of whom is a native of this city. Months have been occupied in this investigation; but as comparatively few facts,—especially statistical facts,—after due inquiry, could be precisely and accurately ascertained, and none others were desired, the reports give no indication of the labor involved.*

* Numerous attempts have been made at different times to investigate the manufacturing industry of Philadelphia. Several years ago a Statistical Society was organized, we believe, for the express purpose of ascertaining the capital in trade and manufactures, the number of hands employed and wages paid, and the aggregate of production; but its officers, we understand, have not as yet submitted their report. More recently, a committee of highly respectable and trustworthy gentlemen, appointed by the Board of Trade, undertook the commission; but the most important information that they ascertained and reported was, that "inquiries of this kind are exceedingly impertinent and offensive, and they will not be answered; nor can any authority compel a response to them. They will be either treated with silence; or, if replied to, they will elicit no full and reliable intelligence. We do not make this assertion without ample reason." The Board of Trade consequently recommend, and their advice has been ceeded by us, not to extend inquiries beyond what can be precisely and accurately ascertained. If, by this course, a less number of important facts are elicited, many rash or doubtful assertions are avoided. Our conviction with respect to statistics is, that the *mean* of estimates of intelligent

They may also, to a certain extent, be considered the opinions of one or more of the leading men in each branch of industry; for large indebtedness is due to this source, both for original suggestions and confirmation of points otherwise doubtful. The reports submitted are not intended to exhibit the entire manufacturing industry of Philadelphia—to ascertain that would require the purse of Fortunatus, and inquisitorial powers far greater than any possessed by the Pope of Rome, the King of Naples, or the Emperor of all the Russias, or all of them combined—but simply to state the facts that have come within the range of our observation, and submit them in illustration of the position and assertion, that *Philadelphia is already a great Manufacturing city, most probably the greatest in the Union.*

men, familiar with the branch with which they are connected, or with the business of their neighbors, is likely to lead to more reliable aggregate results than any direct personal inquiries of each individual. In the latter case, the small operators who reply at all, are habitually disposed to exaggerate, and the larger ones, who have a mortal aversion to the tax-gatherer and competitors, frequently report a small product and a gloomy state of affairs. It is probable, however, that each succeeding attempt will be attended with more success than the previous ones; and the time will come when it will be possible to exhibit statistically the particulars, as well as the aggregate of the mechanical and manufacturing industry of Philadelphia. At present, the best than can be done is to make a *readable* exhibit.

REPORTS

UPON THE

PROGRESS AND PRESENT DEVELOPMENT

OF THE

LEADING BRANCHES OF PRODUCTIVE INDUSTRY

IN

PHILADELPHIA.

ASSUMING that an Alphabetical arrangement of subjects would be most convenient for reference ; but, deeming it advisable to group together those which have practically some points of affinity, whether through identity of raw material or similarity in uses, we come to—

I.
Agricultural Implements, Seeds, Fertilizers, &c.

The manufacture of Agricultural Implements, we are somewhat astonished to learn, is comparatively a new branch of industry in Philadelphia. It seems almost incredible that her citizens, ever foremost, as we have shown them to have been, in enterprises designed to promote Agricultural Improvement, were, until within a few years, content that the farmers of Pennsylvania and New Jersey *should be dependent upon other States for the improved implements* with which to till the soil. The deficiency, however, is now supplied. Philadelphia now contains some very superior establishments in this branch of industry, as will be seen by the following report, which a gentleman, thoroughly familiar with the subject, has placed at our disposal.

Pennsylvania, so widely celebrated for her Agriculture, did not make within her borders, until within a few years, many of the Implements used in tillage and harvesting. It is true, almost every cross-road had its blacksmith and wheelwright, whose united efforts produced a plow, or a harrow; and many of the former stand, to the present day, unrivaled

11

in the immediate locality of their production; but regular Agricultural Machine-shops are of quite recent establishment, the larger portion of the Implements, formerly sold at the city warehouses, having been imported from New England, whose sterile soil had compelled its energetic sons to seek more profitable occupation than its tillage.

In 1854, we find, was founded the first establishment in Eastern Pennsylvania, for the manufacture of Agricultural Implements generally; prior to that, there were shops located for specific objects, as for instance Grain Drills, of which those made by Steacy, and by Pennock, had acquired marked celebrity; but for the manufacture of Farm Implements generally, we believe none of any moment existed. In the year abovementioned, David Landreth & Son, who, with their predecessors in the house, had for many years kept large supplies in Philadelphia, obtained from various sources, established their Steam-works at Bristol, not only for the supply of their principal warehouse in Philadelphia, and their branch-houses in Charleston, S. C., and St. Louis, but for the trade in general. Shortly subsequent thereto, was likewise established that of Bradfield, the 'Mount Joy Car Manufacturing Company,' Savery's Eagle Plow Factory, that of C. B. Rogers, and Boas, Spangler & Co., of Reading, and more recently Boyer & Brother, each of whom turn out admirable machines, both as regards workmanship and materials; and Philadelphia, once dependent upon other cities for tillage implements, is now not only independent, but capable of ministering to the wants of her sister States; and we trust all from distant points, whom business or pleasure may bring among us, will examine the rural machinery manufactured in and near our city.

GARDEN SEED TRADE. The Seed trade of Philadelphia, though, in comparison with many other branches, one of very limited extent, is nevertheless entitled to consideration, when discussing the industrial pursuits of our citizens. From its nature, it cannot be expected that we should count the amount of sales, in this department, by millions— a few hundreds of thousands, at the most, complete the aggregate; but the reputation which our city sustains in this especial branch, is more worthy of note than the amount of sales, however large they might be. In no city of the Union, is the sale of Garden Seeds conducted as at Philadelphia. In New York, Boston, and Baltimore, the only other points at which the wholesale trade in Seeds approaches a profession, the supplies are mainly obtained from Europe, where the effect of cheap labor upon prices, coupled with freedom from imposts at home, enables the importer to purchase many varieties for sale here at a cost far below the actual expense of production in this country. It is true the humid climate of Great Britain, from which country the major portion are obtained, is not favorable to ripening Seeds, and that many kinds suffer by a sea voyage—so greatly do they swell, that the twine on papered parcels is not unfrequently imbedded, or burst, by the expansion; and in other cases, there is reason to believe Seeds already impaired by age, are shipped to this 'western wilderness.' Still so low-priced are many, in comparison with the American, that the mere dealer, whose study is to buy cheap, imports his stock—not recklessly we hope, but trusting for the best, and anxious to quote low prices to the country-merchant—a fatal policy—to none affording in the end pleasure or profit. The druggist, or merchant, who retails them, enticed by low quotations, is beset by indignant planters; and the market-gardener, who has unfortunately staked his crop upon the issue, finds his land and labor for the season have been cast away—far better for him had he paid the full price for American Seeds, of reliable character. We trust he may have learned a useful lesson, for that must be his compensation.

In the 'Horticulturist,' a periodical of high repute, as associated with Downing, its founder and editor, we find, in the No. for August, 1854, an interesting article on ' *The Seed Trade of Philadelphia*,' attributed to J. J. Smith, Esq., the present efficient editor of that magazine. I presume your limits will not admit of quoting much there said, but refer the reader to the article itself. The fact is there made known, that in the production of American Seeds, Philadelphia stands pre-eminent—if not alone, almost without a rival; and the productions of one establishment, which dates its origin within a few years of the Revolution, are sought for and exported to nearly every country to which American commerce reaches. Tons are annually shipped to the British possessions, to India and South America, the West Indies, and the shores of the Pacific, each of which call for annual supplies. One firm, which is specially alluded to, by reason of its greater prominence, viz., that of David Landreth & Son, has Seed Grounds, (Bloomsdale, near Bristol), embracing nearly four hundred acres, cultivated in drill crops, requiring a large force of hands, twenty head of working stock, and a steam-engine for threshing and cleaning seeds. The estate, in its entirety, exceeds any similar establishment in the world. Robert Buist and H. A. Dreer are also extensive growers; and we proudly claim for Philadelphia a class of seed merchants, worthy the confidence of all who may have occasion to purchase, whether for personal use or purposes of trade.

FRUITS. The market of Philadelphia has long been famous for the quality and abundance of its Fruits—the products of orchards in the vicinity of the city. There might be seen in high perfection the choicest of each class, affording all interested the opportunity of useful comparison, and test of relative value; whilst annually a show of Fruits, held by the Horticultural Society, the accumulated contributions of every quarter, facilitates practical comparison with similar articles, drawn from distant sources. Hence, it may readily be seen that Philadelphia nurserymen have ample and valuable opportunities to determine the kinds and varieties most worthy of propagation.

For some years past the culture of the PEAR has attracted more than ordinary interest; and it is a fact, which should not be passed unnoticed, that Philadelphia and its neighborhood have spontaneously produced some of the most valuable varieties of this fruit : seedling trees not surpassed by any, either of native or foreign origin. Here was the nativity of the *Seckel*, of world-wide notoriety—of the *Kingsessing*, the *Lodge*, the *Tyson*, the *Ott*, the *Philadelphia*, the *Moyamensing*, the *Petre*, and some others of high value; and here is the residence of Dr. W. D. Brinklé, whose indefatigable labors in pomological research have gratified his fellow-citizens and benefited the world at large.

We might extend this sketch of the Agricultural resources of Philadelphia, but perhaps enough has already been said to enable you to illustrate the idea which we desire to express, that the City of Brotherly Love stands unrivaled in this department of industry."

In addition to Garden Seeds, referred to above, Philadelphia is one of the principal distributing points for Clover and other field Seeds, not only supplying the Southern and Western States, but sending largely to New England, Great Britain, and the British Provinces. Within the last two months, 46,180 bushels of Cloverseed were purchased, and recleaned here for shipment, of which 35,000 bushels were shipped to Liverpool and to New York, 3,000 bushels to the South, and the balance to points in the interior, and to the West. A large proportion of the very best Seeds, and noted particularly for cleanness and quality, is grown in the counties adjacent to Philadelphia. The annual sales, we are assured, frequently amount to one million of dollars. One firm, Messrs. P. B. MINGLE & Co., through whose hands an immense quantity of Seeds pass annually, are known probably to all dealers.

FERTILIZERS.

The manufacture of Artificial Manures has become quite an extensive business within a few years. Those made in Philadelphia, are known as Super-phosphate of Lime, Bone-dust,

Plaster of Paris, Poudrette, Philadelphia Urate, or the concentrated and fixed nitrogen of urine, and Bone-black waste.
In addition to these, there are agencies for the sale of the Peruvian
Guano, for a fertilizer known as Blood Manure, and others.   The
popularity of Peruvian Guano was such, that in one year the sales
of the agent in Philadelphia amounted to 22,000 tons, at $40 per
ton, or $880,000 ; but the advance in price checked the demand, and
led to the manufacture of a great variety of Artificial Manures.

The substitute for Guano, that would seem to be in the greatest
demand with the farmers of Pennsylvania, and adjacent States,
judging from the extent of the manufacture, is the SUPER-PHOS-
PHATE OF LIME.   It is said to possess fertilizing properties more
permanent than those of Guano.   Though only introduced fully
to public notice in 1851, its manufacture now forms an item of
some importance in the general aggregate of industry.   It is
a somewhat singular fact, that when first introduced it commanded
a higher price per ton than Peruvian Guano.

There are seven manufacturers of Super-phospate of Lime in
Philadelphia, who are well represented by the two most extensive
—POTTS & KLETT, and MITCHELL & CROASDALE.

The manufactory of POTTS & KLETT is situated near Camden,
but their product is sold exclusively by a house in this city.
This firm are also well-known manufacturers of Chemicals.   The
works of Messrs. MITCHELL & CROASDALE are situated in the
Nineteenth Ward, and cover nearly an acre of ground.   They
produce what they call "Highly Improved Super-phosphate of
Lime," being a compound of ground bones, Peruvian guano, and
other substances.   The bones are first boiled—the fat extracted,
and *pure* bone, free from vegetable ivory, which is merely inert
matter, is alone manufactured into fertilizers.   Additional works
are now being erected for boiling bone.   Their products are
sold by CROASDALE, PIERCE & Co., Delaware Avenue, above
Arch street.

Another preparation of bones, known as *Bone-dust*, is made
to a considerable extent by the manufacturers of Glue, &c., and
by FRENCH, RICHARDS & Co.   The latter firm, and the Phœnix
Mill and others, make *Plaster of Paris* for fertilizing purposes.

With regard to the relative merits of the respective fertilizers,
11*

we know nothing, and can only refer those interested to Agricultural Chemists, or to the pamphlet circulars of the manufacturers, in which the properties of each are duly set forth. We can however assure purchasers, that they can probably procure, in Philadelphia, any fertilizer of value that they may desire, on advantageous terms.

The Statistics of the Manufacture, for 1857, as nearly as can be ascertained, are as follows:

Super-phosphate of Lime, 7,000 tons, or 55,000 bbls., at $45 per ton,	$315,000
Bone-dust, 2,000 tons, at $35 per ton, - - - - - -	70,000
Plaster of Paris, 3,000 tons, at $6 per ton, - - - - -	18,000
Other Fertilizers (see above) approximate, - - - - -	100,000
	$503,000

The annual sales of Fertilizers in the city, including Guano, the refuse of tanneries, morocco manufactories, sugar refineries, &c., will probably amount to a million and a half of dollars. It is, however, much to be regretted, that the sweepings of the streets, and human ordure, are not more carefully economized to aid in restoring to the earth the fertility of which it is robbed by the necessary consumption of a vast city. In Paris, a contractor pays a large sum into the City treasury, for the privilege of removing these fertilizers, and yet derives a handsome profit from the contract.

II.
Alcohol, Burning Fluid, and Camphene.

There are in the city nine establishments engaged in distilling Alcohol and Camphene, or Pine Oil, several of whom make it an exclusive business. *Alcohol*, it is generally known, is distilled from Whisky—nine gallons of the latter making about five of the former. Alcohol, for burning-fluid, is 95 per cent., while Druggists' Alcohol is but 84 per cent., being reduced to that standard after distillation. *Pine Oil*, or *Camphene*, is distilled from Spirits of Turpentine, the well-known produce of the pine forests of North Carolina. This loses in distillation about one gallon in a barrel, or two and a half per cent. *Burning Fluid* is made by the admixture of one gallon of Pine Oil to four gallons of Alcohol.

The Statistics of the business, for the year ending July 1, 1857, are as follows :—

Raw Material.

Whisky, 2,077,000 gallons, average cost 31 c., - - -	$643,870
Spirits of Turpentine, 380,000 gallons, average cost, 47 c.,	178,600
	$822,470

Product.

Alcohol, sold by distillers, - - - - -	395,000 gallons.
Pine Oil, " " - - - - - -	147,250 "
Burning Fluid, - - - - - - -	1,112,000 "
	1,654,250 "

Of the value of $1,022,140, averaging nearly 62 cents per gallon.

There are a number engaged in the sale of Burning Fluid, who purchase the Alcohol and Pine Oil from the distillers, and these are included in the above statement. The product may be stated, in another form, as follows :—

Alcohol, - - - - - - -	1,284,600 gallons.
Pine Oil or Camphene, - - - - -	369,650 "
Total, - - - - - -	1,654,250 "

This does not embrace the Alcohol produced by Powers and Weightman, and other Manufacturing Chemists, the value of which is included in the Statistics of Chemicals; nor that made by Rectifiers, which is known as "High Wines."

Burning Fluid was first known as Spirit Gas, and the discovery patented by Isaiah Jennings, in 1830, who soon after commenced its manufacture in Philadelphia, but subsequently abandoned it. Mr. Locke made it under the Jennings patent, and was the only manufacturer in Philadelphia to any extent, previous to the expiration of the patent in 1844. The merits of Burning Fluid, as a material for light, consist in its brilliancy, cheapness, and far greater cleanliness than Oil; its principal demerit is— liability to explosion. Upon this important point, we have been favored by Messrs. Yarnall & Ogden, one of the principal firms engaged in the manufacture, with the following observations, which deserve attention, both from their intrinsic importance and the experience of those who make them.

" It has been ascertained, that nearly all the accidents attending the use of Burning Fluid originated, either by attempting to fire shavings, or other combustible materials, with a fluid lamp, mostly glass; or, by attempting to fill the lamp while burning. This is by far the most fruitful source of accidents—but thanks to the inventive genius of the American people, several kinds of lamps have been patented, and are now in use, which entirely prevent the possibility of an accident occurring from this cause, for the act of unscrewing the top of the lamp, puts out the flame by the action of a spiral spring which forces up the slides on the tubes, and thereby extinguishes the flame, and entirely prevents the possibility of an accident; and to make it still more complete, the fluid is confined in a gutta-percha sack, so that in case of a glass lamp falling and breaking, the flame cannot possibly ignite the fluid. There are still other improvements in these lamps—one is, that the gas and not the fluid is consumed, thereby making a light equal to gas. Persons using the fluid would do well to introduce these lamps."

The firms engaged extensively in this manufacture, are the following :—

Z. LOCKE & Co., 1010 Market street. This firm are said to be the oldest and the largest distillers of Alcohol in the city. Mr. Locke, the senior partner, commenced the distillation in 1829; and, as we previously stated, was for many years the only one who made it an exclusive business. Besides druggists', and 95 per cent. Alcohol, they make *Atwood's Patent Alcohol*, which, on account of its purity and freedom from any disagreeable smell, is preferred and much used by perfumers. The manufactory of Messrs. Locke & Co. is an important one.

P. BUSHONG & SONS, Broad street, above Race. This firm are very extensively engaged in the production of Alcohol, and the manufacture of Burning Fluid. They combine therewith the distillation of Whisky from grain, having a large establishment therefor at Reading, consuming about 166,000 bushels of Corn annually, and 84,000 bushels of Rye; and thus all the processes are economized, by conducting in one establishment the entire manufacture, from the original raw material to the finished product. The firm employ forty persons, ship goods eastward, and their fluid has attained a high reputation for quality.

YARNALL & OGDEN, 472 North Third street. To this house we previously referred. They are the successors, at their present location, of those who were among the first to introduce Burning Fluid to public notice; and their sales now extend to all parts of the South, as well as to Pennsylvania and States adjacent.

JOHN W. RYAN, Prime street, below Front, has been identified with the business for many years, and enjoys an extensive city trade.

Messrs. PORTEUS & PHILLIPS; ROWLEY, ASHBURNER & CO.; WETHERELL & BROTHER, manufacture these articles largely, in connection with other products; and WM. KING, and J. McINTOSH, also make to some extent.

The quality of Burning Fluid made in Philadelphia is very superior, and in the South readily commands a higher price than that made elsewhere. It is shipped eastward to Providence, Newport, Hartford, New Haven, Boston, Bangor; and southward to all the Southern States, to California and South America.

III.
Books, Magazines, and Newspapers.

The honor of having established the first printing-press in America, must be awarded to Cambridge, Mass. Philadelphia, however, may claim, with laudable pride, that in less than six weeks after the city was founded, a printing press was established, being the second set up in the North American Colonies;* and, moreover, that many of the most important works in American literature bear the imprint of her publishing houses. We shall attempt to trace the progress of Book and Periodical Publishing, chronologically, though the records within our knowledge are so few, and the pressure of engagements so distracting, as to render the task a difficult one.

Prior to the Revolution, and for some years afterward, the most notable issues of the Philadelphia press,—in fact, the American press,—came within Webster's definition of a pamphlet, that is, a small book, consisting of a sheet of paper. The first book published in this country, of which we have any knowledge, was "The Bay Psalm Book," issued from the Cambridge press, in 1640, and this was probably the most successful of any. *Seventy* editions were republished in England and Scotland. The first publication in book or pamphlet form issued from the Philadelphia press was a sheet Almanac, for the year 1687, in twelve compartments: the year beginning with March, and ending with February, as was usual before

* Thomas's History of Printing.

the eighteenth century. A copy of this early specimen of American typography, bearing the imprint of "Wm. Bradford, Printer," is preserved in the Philadelphia Library. His second work was a quarto pamphlet, on the subject of "The New England Churches, by G. Keith," dated in 1689. The name of Bradford continued to be identified with the history of printing in Philadelphia until a very recent period.

In 1699, the press established by Bradford passed into the hands of Reynier Jansen, evidently a Dutchman by name, who managed it until the year 1712. There are now in the Philadelphia Library two very curious pamphlets, bearing his imprint, and so rare that they are probably the only copies extant. The first was published in 1700, and is entitled, " Satan's Harbinger Encountered : his False News of a Trumpet Detected : his Crooked Ways in the Wilderness laid open to the view of the impartial and judicious. Being something by way of Answer to Daniel Leeds, his book, entitled, ' News of a Trumpet Sounding in the Wilderness,' &c., C. P., (Caleb Pusey). Printed at Philadelphia, by Reynier Jansen, 1700." The second bears date 1705, and is entitled, " The Bomb Searched and Found Stuffed with False Ingredients. Being a just confutation of an abusive printed half sheet called 'BOMB,' originally published against the Quakers by Francis Bugg; but espoused and exposed, and offered to be proved by John Talbot. Printed at Philadelphia, by Reynier Jansen, 1705."

The second printing-office in Philadelphia was established by S. Keimer, in 1723. The first publication, bearing his imprint, of which we have any knowledge, is a very curious and rare one, entitled "The Craftsman : a Sermon composed by the late Daniel Burgess, and intended to be preached by him in the High Times, (sic.,) but prevented by the burning of his Meeting-house. Philadelphia : Printed by S. Keimer, (Circa), 1725."

The advertising columns of the journals, for the succeeding quarter of a century, from 1725 to 1750, contain announcements of a number of curious books and pamphlets, of which we append a list below.*

* THOMPSON WESTCOTT, Esq., the author of the "Life of John Fitch," recently made an examination of the Journals, from 1728 to 1750, and noted

In 1735, Christopher Sower published a Quarterly Journal, in German, which was the first work of the kind in a foreign language published in the Colony. The same year he published a Newspaper, the first German Almanac, "Extracts from the Laws of the Province, by William Penn," and several other works. the announcements, in the advertising columns, of the principal publications. The list was presented to the Philadelphia Library, and the courteous Librarian of that Institution placed the same at my disposal. The following are the most important.

American Books and Pamphlets advertised in the Pennsylvania Gazette.

1728. *Dec.* 24.—God's Mercy surmounting Man's Cruelty, exemplified n the Captivity and Redemption of Elizabeth Hanson, wife of John Hanson, f Knoxmarsh, at Keacheachy, in Dover Township, who was taken captive with her children and maid-servants by the Indians in New England, in 1725, etc. To be sold by Samuel Keimer, in Philadelphia, and by Heurtin, Goldsmith, in N. Y.

1729. *Nov.* 30.—A Short Discourse, proving that the Jewish, or Seventh-day Sabbath, is abrogated or repealed. By John Meredith. Printed and old by the printers hereof, B. Franklin and H. Meredith. Price sixpence.

1730. *Feb.* 19.—The Spirit's Teaching Man's Sure Guide: Briefly asserted nd recommended to the sober perusal of all Christian believers. By Charles Woolverton, Senr. The second edition. Franklin and Meredith, Printers.

1730. *Feb.* 3.—An Elegy on the Death of that Ancient, Renowned and Useful Matron and Midwife, Mrs. Mary Broadwell, who rested from her labors, Jan. 2, 1730, aged a hundred years and one day. Sold by David Harry, printer, in Philadelphia.

1730. *Dec.* 29.—Ralph Sandiford, being bound for England, hath printed a second impression of his Negroe Treatise, to be distributed *gratis;* or sold to those who would rather pay, at 12*d.* each.

1731. *March* 4.—Some Considerations Relating to the Present State of the Christian Religion, etc. By Alex. Arscot. Franklin & Meredith, Printers.

1732. *Oct.* 5.—The Minister of Christ and his Flock: a Sermon by David Evans, preached at Abingdon, Pa., Dec. 30, 1731. B. Franklin, Printer.

1734. *May* 23.—The Constitution of the Free-Masons: containing the history, changes, etc. Reprinted by B. Franklin, in the year of Masonry, 734. (Franklin Gazette.) 2*s.* 6*d.* stitched; 4*s.* bound.

1737. *Sept.* 22.—A Treaty of Friendship held with the Six Nations, Philadelphia, Sept. and Oct., 1736. Franklin, Printer. Price 8*d.*

1738. *Aug.* 17.—Benjamin Lay's Book against Slave Keeping. Printed by himself. 2*s.* 6*d.* each.

1739. *May* 10.—The Art of Preaching, an imitation of Horace's Art of Poetry. Franklin, Printer. 6*d.*

At that time all the type used in the Colonies was brought from Europe, and finding this very inconvenient, he commenced a Type Foundry and Manufactory of Printing Ink. This was the first Type Foundry in the country, and the celebrated house of L. Johnson & Co., Philadelphia, claim, through Binney & Ronaldson, to be

July 26.—The History of Joseph, a Poem by a female hand. Franklin, Printer. 1*s.*

1740. *May* 22.—Whitfield's Sermons, 2 vols. : one, Sermons ; one, Journals. Franklin, Printer.

A Letter from Rev. Mr. Whitfield to the Religious Societies lately formed in England and Wales, etc.

A Letter from the Rev. Mr. Whitfield to a Friend in London, showing the fundamental errors of the book entitled "The Whole Duty of Man."

The Danger of an Unconverted Ministry, considered by Gilbert Tennant, etc. Franklin, Printer. 6*d.*

July 3.—The Character, Preaching, etc., of the Rev. George Whitfield, impartially represented and supported in a Sermon preached at Charleston, S. C. By J. Smith, V. D. M. Franklin, Printer. 4*d.*

A New and Complete Guide to the English Tongue, etc., collected by an ingenious hand, for the use of Schools. Franklin, Printer. 2*s.*

1741. *Jan.* 15.—Free Grace, a Sermon by Rev. John Wesley. Franklin, Printer. 6*d.*

22.—Free Grace Indeed! a Letter to Rev. John Wesley. Franklin. 6*d.*

Feb. 19.—Free Grace in Truth, by Rev. John Dylander, minister Swedish church, Wecaco. Franklin. 3*d.*

1742. *Dec.* 21.—A Short Narrative of the Extraordinary Work of God at Camberslang, in Scotland. Wm. Bradford, Printer.

1743. *March* 3.—The Interest of New Jersey with regard to Trade and Navigation, by laying duties. Bradford, Printer.

12.—Every Man's Right to Live: a Sermon by Rev. Lewis of Thurenstein, Monravia. Franklin.

1744. *Jan.*—Oglethorpe's Expedition. Report to Assembly of South Carolina into the causes of its failure. 2*s.* 6*d.*

April.—A Journal of Proceedings in the Conspiracy to Burn New York, by white men and some negroes, etc., in 1742. By the Recorder of the City of New York.

Sept.—A Grand Treaty held at Lancaster, etc. Franklin. 18*d.*

Oct.—Remarks upon Mr. Geo. Whitfield, proving him a man under Delusion. By George Gillespie. Philad., Printed for the Author, and sold at the Harp & Crown, in 3d street, opposite the Workhouse.

Nov.—An Account of the newly invented Pennsylvania Fire-places, etc, with a copperplate, etc. Price 1*s.*

the legitimate successors of Christopher Sower, in the business. In 1743, he printed a quarto edition of the German Bible, Luther's translation, having 1272 pp. This was the largest work which had then been issued from any press in the Colony, and was not equaled for many years after. Copies of this Bible were sold, bound, at fourteen shillings, and are now highly prized by book collectors. About 1744 he resigned his press to his son, and died about 1760. He was a man of large influence among his countrymen, and frequently acted as their representative in their intercourse with Government.

His son, also named Christopher Sower, continued the business of his father on an enlarged scale, printing many valuable Books, and a Weekly Newspaper. In 1762, he printed a second edition of the German quarto Bible of two thousand copies; and in 1776, completed a third edition of three thousand copies. He had by

1745. An Essay on the West Indian Dry Gripes. By Dr. Cadwalader. Franklin.

The Art of Preserving Health. By Dr. Armstrong. Reprint. Franklin. 2s.

Sept.—Mr. Prince's Sermon on the General Thanksgiving occasioned by the taking of Cape Breton; with a Particular Account of the Expedition, etc. Price 1s.

1746. *July.*—The New Manual Exercise, by Gen'l Blakenly, and the Evolutions of the Foot, by Gen'l Bland. Franklin, Printer. 6d.

Reflections on Courtship and Marriage. Franklin. 1s. 6d.

1748. *Oct.*—The Congress between the Beasts, under the mediation of the Goat, for negotiating a Peace between the Fox, the Ass, wearing a lion's skin, the Horse, the Tigress, and other animals at war. A farce in two acts. Now in rehearsal at a new and grand Theatre in Germany. Written originally in High Dutch, by the Baron Huffumbourghausen, and translated by J. J. H. D. G. R., Esq., veluti in speculo. Second edition. To be sold by Gotthard Ambruster, at the German Printing-office, in Arch st. 2s. 6d.

1749. Proposals for publishing a Map of Pennsylvania, New Jersey, and New York, and the lower counties, by Lewis Evans. Price, two Pieces-of-8 each.

Aug.—A Particular Relation of the dreadful Earthquake, etc., at Lima and Callao, in Peru. Translated from the original Spanish. Price 9d.

1750. *June 7.*—A Short Treatise on the Visible Kingdom of Christ. By Thomas James. Franklin & Hall. Price 6d.

Letters from the Dead to the Living, by Plularetes. Franklin & Hall. 9d.

12

far the most extensive Book manufactory, then, and for many years afterward, in the country.   It employed several binderies, a paper-mill, an ink manufactory, and a foundry for German and English types.   He was well educated by his father—was ordained a minister of the German Baptist Society; and as a man of integrity was deservedly esteemed.   He died at an advanced age, in 1784.   He left several children, among whom Christopher, (third), David, and Samuel, were practical printers and publishers.   The name continues to be very popularly represented in the trade, by his descendant, Mr. Charles G. Sower, senior partner of the firm of Sower, Barnes & Co.

In 1782, Robert Aitken published, it is believed, the first American Bible in the English language.   His edition was recommended to public patronage by Congress.   It was projected during the war; but with peace, importation of books began, and Aitken "lost more than three thousand pounds in specie." But his lasting memorial is, that he printed *the first American edition of the English Bible.*   "The very *paper*," said the *Philadelphia Freeman's Journal* of that time, "that has received the impression of these sacred books, was manufactured in Pennsylvania; the whole work is therefore purely American, and has risen, like the fabled Phœnix, from the ashes of that pile in which our enemies supposed they had consumed the liberties of America."

But Book Publishing in the United States, even as late as 1786, was yet in its embryo state.   It is recorded, that in that year *four* Booksellers held a consultation as to the policy of publishing an edition of the New Testament, deeming the matter a work of great risk, requiring much consultation previously to the determination of the measure; but the change in the state of public affairs soon infused life and vigor into the business.   Less than four years afterward one of the prudent gentlemen, above referred to, ventured upon the publication of an *Encyclopedia,* in eighteen quarto volumes.   When the first half volume was published, in 1790, he had but two hundred and forty-six subscribers, and could only procure two or three engravers.   One thousand copies of the first volume were printed; two thousand of the second; and when he had completed the eighth, the subscription extended so far as to render it necessary to reprint the first. He then found difficulty in procuring printers for the work.*

In 1792, Ebenezer Hazard published a quarto volume of "Historical Collections," intended as materials for a History of the United States, and in 1794 another volume. These collections were made under the patronage of Congress.

In the succeeding year, William Cobbett, then a refugee in Philadelphia, commenced his political career by writing, at the solicitation of some friends of Washington's administration, a pamphlet, for the purpose of vindicating Jay's Treaty; and which is said to have had considerable influence in quieting the public mind. In the same year he issued his "Tuteur Anglais," Thomas Bradford, Publisher.

In 1804, Mathew Carey set up the Bible in quarto form, and this is believed to have been the first Bible, kept standing in type, of that size in the world—over 200,000 impressions were published. But the state and condition of the Book Publishing business in Philadelphia, in the early portion of the present century, were so well sketched by Henry C. Carey, Esq., the distinguished political economist, and surviving representative of the house of Mathew Carey & Son, at a Festival given May 24th, 1854, in honor of Mr. Abraham Hart, on his retirement from the Bookselling business, that we cannot do better than submit a lengthy extract from his speech on that occasion.

"For myself, Mr. President, I am a sort of 'remainder' of an edition, a representative of a by-gone race of booksellers, as our friend Hart is of the present race. There are several among our friends here assembled disposed to insist that they carry on their shoulders more years than I do, and they are a property, the possession of which I am not disposed to dispute with them; but there is none of them whose connection with the trade dates back to as early a period as mine. I was, Mr. President, at the first trade dinner ever given in this country, and of all who sat at that table, there is, I believe, no one now living but myself. It was somewhat more than half a century since. Mr. John Conrad, father of our friend Judge Conrad, filled on that occasion a distinguished place; but he has recently passed away, and has left me, as I think, alone. The occasion of that dinner was the holding of the 'Literary Fair,' that was attempted in imitation of the great Leipsic

* "Hopkinson's Oration before the Academy of Fine Arts," 1810.

Fair, and intended to be held alternately in this city and New York. One was held in each city. My father took me with him to the New York one; and although only eight years of age, I was even then a bookseller, perfectly familiar with the contents of our establishment. They called me 'the bookseller in miniature,' and being such, I was the proper representative of the trade of the day, for it was a miniature one, and the gentlemen engaged in it made miniature fortunes, compared with those that, I am happy to learn, are accumulated by the men of our day. We then depended on Great Britain for Latin and Greek, English, French, and Spanish dictionaries, and to a considerable extent, even for grammars. The classics, Cæsar, Horace, Virgil, and Homer, were all imported, as was the case with Rollin, Plutarch, Sully, and a host of other common books. Prices were high, and sales were small. School dictionaries of the size of Walker's abridgment, which now sell, as I am told, for three dollars a dozen, then sold for more than half that price per copy. Schools were few in number, and there was small demand for books.

" Two years later, my father carried into effect his project of getting up the Bible in quarto, with movable type, and it was the first in the world, as I have always understood. It was a gigantic operation; first cost, fifteen thousand dollars—a very large sum in that day—and it was one that he never could have effected without the aid of Mr. James Ronaldson, one of the worthiest men that this city has ever possessed. He was then the sole type-founder for the Union, and supplied the letter required for all the newspapers, magazines, and books, that were printed from Maine to Georgia. All of it came from that small foundry in South street above Ninth, a fact from which, alone, you might judge of the diminutive size of the publishing trade of that period.

"Small as was the trade in foreign books republished, that in domestic ones was still far less. There was then, in fact, no domestic literature. It was half a dozen years later that Irving, Paulding, and Verplanck, made their first appearance on the stage as joint authors of the little periodical newspaper, well known as 'Salmagundi.' Some years still later, Bradford and Inskeep were thought to have displayed remarkable liberality in giving to a young lady of this city a hundred dollars for the copyright of a very clever novel. Fanny Fern would now look upon such a sum with no slight disdain, were it offered in exchange even for a contribution of a dozen pages; and yet the liberality manifested in the case of the Philadelphia novelist was probably greater than that now exhibited by the extensive house in Auburn. American books could not then be sold. It was almost sufficient to insure the condemnation of a book to have it known that it was of domestic origin. My friend, Major Barker, a man of excellent literary ability, dramatized Marmion about the time to which I have referred ; but the manager, Mr. Stephen Price, did not venture to produce it as an American work. It was carefully packed up as coming from England, with imitations of the English post-marks, and was produced as the work of an English author. As such it succeeded ; but the real authorship having soon leaked out, the public thenceforward ceased to find in it the merit that before had been so clearly visible. Under such circumstances, it was scarcely extraordinary that an English writer should find reason for asking the question—' Who reads an American Book ?'

 " We are surrounded by evidences of progress, but among the chapters which record its history, there is none more remarkable than the literary one. That chapter records great changes in the amount of trade, but there are other changes that are perhaps equally remarkable. Even so recently as forty years since, the trade looked chiefly to the South for a market for their books. Messrs. Conrad & Co. had branches in Baltimore, Alexandria, Fredericksburg, Richmond, Petersburg, and Norfolk, but none in the West. Bradford and Inskeep had one in Charleston. Benjamin Warner, one of the most high-minded and excellent men ever connected with the trade in this city, had one in Richmond, and another in Charleston ; and it was in this latter place that our friend, Mr. John Grigg, first exhibited the ability by which he has been since distinguished. My father had a branch in Baltimore, and one in Richmond. The tendency was then to look almost altogether South, while it is now almost altogether West. Chicago, a city that scarcely existed ten years since, now absorbs, I imagine, more books than Norfolk, Charleston, and Savannah united. The consequences of this have been unfavorable to the printers and publishers of Philadelphia, separated, as she has been from the great West by a range of mountains that New York could turn, while she was bound to scale them. The Erie Canal gave to New York and New England a communication with a great and growing market, while that on which this city has chiefly relied became a declining one ; and the consequence has been that the trade has scarcely kept pace with that of other cities, in its growth. It has, nevertheless, increased greatly. You do not publish as many novels as New York, but you present more medical books than all the rest of the Union. You have here, perhaps, the largest distributing house in the world. In conversation a short time since with one of its members, I learned that they employed nearly eighty clerks ; a fact that astonished me, as I knew how large a business we had done with half a dozen. My surprise, however, disappeared when he told me that for many weeks they had sent out an average of

12*

more than ten tons per day, or the equivalent of a thousand reams of printing paper of the size commonly used when I was in business. Were the books they fill all printed ones, there would be no hesitation in asserting that they distributed more literature than any house in the world. It has recently been made a matter of boast that Chambers & Co., of Edinburgh, had sent out ten tons in a fortnight; but we have here as many tons per day in each day of many weeks."

Within the period referred to by Mr. Carey, a large number of very important works were issued. In fact, the first quarter of the present century may be entitled the palmy era of bookselling in Philadelphia. The works published embraced: Dobson's Encyclopædia, 21 vols., quarto; Rees' Cyclopædia, 46 volumes, quarto; Edinburgh Encyclopædia, 18 vols., quarto; Nicholson's Encyclopædia, 12 vols., octavo; Wilson's Ornithology, 9 vols., imperial quarto; Barlow's Columbiad; Pinkerton's Atlas, 1 vol., folio, price $100 per copy; Johnson's Dictionary, 2 vols., quarto, published by Moses Thomas; Gibbon's Rome, 8 vols., octavo; Hume and Smollett's and Bissot's England, 15 vols., octavo; Mavor's Voyages and Travels, 24 vols., 12mo.; British Classics ("Spectator," "Rambler," &c.) 39 vols., 12mo. For the copyright of one book, Marshall's Life of Washington, the sum of $60,000 was paid in Philadelphia. Truly, there were giants in those days.

In 1824, the era of *Trade Sales* was inaugurated in Philadelphia, Moses Thomas, above referred to, being the auctioneer. The catalogue of the first sale, we believe, is still in the possession of that gentleman, who, though past threescore and ten, performs more mental and bodily labor than many men twenty years his junior. The suggestion and arrangement of the plan are credited to Mr. Henry C. Carey, and its adoption was certainly attended by the most happy results to the Book trade. Prior to the establishment of Auction Sales for books, publishers were not permitted honorably to vary from the announced prices of the publication; and their profits were consequently lessened materially by the accumulation of unsaleable stock, for the disposal of which there was no practicable means. By the establishment of Trade Sales, however, this difficulty was overcome; and a publisher who was so unfortunate as not to be able to obtain the nominal price for his book, could sell it to the highest bidder in open market. So popular have Trade Sales become, that now over

$600,000 worth of books are annually disposed of every spring and autumn in Philadelphia and New York, the sales being conducted under the supervision of a committee of leading publishers.

Within the last few years, the demand for minute and exact information in every department of learning has become so pressing, that the subdivisions which may be remarked in mechanical pursuits are also noticeable with respect to the publication of books. Publishers are no longer divided merely into book, newspaper, and magazine publishers, as formerly; but each of the various classes, Medical, Law, Theological, School, Illustrated, German, and miscellaneous books, has its representatives among the publishing houses. We shall instance the more prominent in each department.

1. MEDICAL BOOKS. We are informed that nine-tenths of the Medical Books issued in the United States are printed and published in Philadelphia. There are three firms extensively engaged in this branch, viz., BLANCHARD & LEA, J. B. LIPPINCOTT & Co., and LINDSAY & BLAKISTON; while others publish Medical Books to some extent. The first-named of these houses make this department of the general trade their specialty, and their catalogue contains a more important list of valuable Medical books than probably any in the world. The list of their own publications extends to about one hundred and seventy-five different works, or over two hundred different volumes, besides several Medical journals; one of which, "The American Journal of Medical Science," edited by Dr. Hayes, is among the oldest periodicals of the country. Their cash capital invested in this business is not far short of a quarter of a million of dollars. Messrs. LIPPINCOTT & Co. publish a number of important Medical books, as Wood & Bache's Dispensatory, Wood's Practice of Medicine, Wood's Materia Medica, Smith's Operative Surgery, and many others; and a very valuable periodical, entitled "The North American Medico-Chirurgical Review." Their general operations we shall notice subsequently. Messrs. LINDSAY & BLAKISTON publish a number of text books in Medical Science, and Rankin's Abstract, which has a large circulation. The Homœopathic branch has its representative among the publishers in Mr. RADDE. The contributions which Philadelphia has made to

American Medical Literature, are scarcely less important than her Medical Schools.

2. LAW BOOKS. There are also three houses that make the publication of Law Books their specialty: T. & J. W. JOHNSON, KAY & BROTHER, and H. P. & R. H. SMALL. The first-named publish the Law Library, a reprint of English Elementary works of standard value, now numbering 96 volumes; and the English Common Law and Exchequer Reports, a reprint of the decisions of the Law Courts of England since 1813, now numbering 125 vols., 8vo., and believed to be the largest uniform series issued by any law publishers in the world. Their list, which embraces works on almost every department of law, contains nearly 500 volumes. KAY & BROTHER publish the following valuable works: "Brightly's Analytical Digest of the Laws of the United States," 1 vol., 8vo.; "Wharton's American Criminal Law," 1 vol., 8vo.; "Wharton's Precedents of Indictments": these works comprising the science and the practice of the Criminal Law of the State and Federal Courts of the United States; "Wharton & Stille's Medical Jurisprudence," "Wharton on the American Law of Homicide," "Morris on Replevin," "Troubat on Limited Partnership," 1 vol. 8vo., "Pennsylvania State Reports," 66 vols., and others. H. P. & R. H. SMALL publish Harrison's Digest, Addison on Contracts, Selwyn's Nisi Prius, Williams on Executors, Saunders on Pleading and Evidence, and many other standard works with the legal profession. The business of Law Book publishing, however, is not monopolized by these houses. Messrs. CHILDS & PETERSON, for instance, publish Bouvier's Law Dictionary, and the Institutes of American Law; Messrs. LIPPINCOTT & CO. issue Dunlop's Laws of the United States, and Messrs. KING & BAIRD the Philadelphia Reports, condensed from their Legal Intelligencer, which is the only weekly Law Journal published in this country.

3. RELIGIOUS BOOKS. In this department of the book trade, individual enterprise has been, to a large extent, superceded by incorporated or organized societies, who publish for other objects than the realization of profit. In Philadelphia, we believe, there are seven of these societies, having a publishing department: the Sunday School Union, American Baptist Publication So-

ciety, Presbyterian Boards, (Old and New School), Bishop White Prayer Book Society, Female Prayer Book Society, and the Lutheran Publication Society. As an indication of the magnitude of their operations, we remark the fact, that in 1856, the American Baptist Publication Society printed 16,276,293 pages, equal to 18,478,293 pages in 18mo.

Among the individual houses, in whose catalogues Religious Books predominate, we would instance W. S. & ALFRED MAR-TIEN, H. HOOKER & Co., HIGGINS & PERKENPINE, SMITH, EN-GLISH & Co., JAMES CHALLEN & SONS, and several publishers of Catholic works. Mr. W. S. MARTIEN has been connected with the publication business for nearly a quarter of a century. His list of publications embraces: Scott's Commentary on the Bible, Baker's Revival Sermons, the works of Drs. Alexander, Hodge, Junkin, Burrowes, and numerous other Theological works. Messrs. HOOKER & Co. publish Episcopal Books largely; several edions of the Book of Common Prayer, and many other Religious works. Messrs. HIGGINS & PERKENPINE, we are informed, publish principally Methodist Books.

The Bible is a standard volume, we are happy to say, with nearly all the publishers; but some confine their operations enirely to its issue. One establishment, that of JESPER HARDING & SON, is believed to be the largest Bible publishing house in his country, conducted by individual enterprise.

"On a late visit," says our informant, Dr. James Moore, "we aw over 20,000 copies in different states of forwardness, emracing fifty varieties at different prices, from the copies illus-rated with engravings on wood to those bound in Turkey mo-occo and embellished with fine steel engravings and chromo-lith-graphic illustrations. Five hundred tons of white paper, at from 250 to $300 per ton, worth at least $140,000 ; 500,000 leaves f gold, 40 tons of tar paper ; 20,000 sheep-skins, which the flock f Job, containing only 14,000 sheep, could not have supplied, are nnually consumed at this establishment. It employs 200 persons, book-folding machine, invented by Mr. Chambers, and the only ne ever brought into effectual operation, is at work here daily."

The Rev. THOMAS H. STOCKTON has lately commenced a novel nd important improvement in Bible publication. His edition is distinctive, each book being in separate binding. Its beauty and advantages have been acknowledged by a bold imitation on the part of the well-known Bible house of BAGSTER & SONS; a proof that London sometimes imitates Philadelphia.

Bibles are published in this city at all prices, from 40 cents to $150, and in every style of binding, from the plainest to brown morocco, illuminated, and with painted edges. The styles are generally distinguished by the name of the publisher, as Harding's Bibles, Butler's Bibles, Miller & Burlock's Bibles, Perry's Bibles, Whilt & Yost's Bibles, and Lippincott's Bibles.

3. SCHOOL BOOKS. In this important department of the business, Philadelphia publishers hold a respectable rank, both as to the intrinsic merit of their publications and the amount of their sales. H. COWPERTHWAIT & CO., J. B. LIPPINCOTT & CO., E. C. & J. BIDDLE, E. H. BUTLER & CO., CHARLES DESILVER, U. HUNT & SON, HAYES & ZELL, CRISSY & MARKLEY, and SOWER, BARNES & CO., are all extensively engaged in this branch of the business. The first named house are the publishers of Mitchell's Series of School Geographies, the most popular of all the works on this study published in the United States, and which are said to have met with an annual sale of about 300,000 volumes. They also publish many other popular School books. J. B. LIPPINCOTT & CO. publish—along with numerous works in other departments of literature—a large number of books adapted to various stages of advancement in school studies, and many of which have a large sale, especially throughout the Southern and Western portions of our country. The Complete Pronouncing Gazetteer of the World, edited by Dr. Joseph Thomas and Thomas Baldwin, and published by this house, is a work of such unquestioned merit, that it has found, or will find, a place as a book of reference in almost every school of a respectable grade in the United States, and reflects credit on the publishers. The Messrs. BIDDLE devote their attention almost exclusively to the educational department of the Publishing and Bookselling business, and limit the range of their publications to works adapted to advanced pupils. Crittenden's Series of Treatises on Book-keeping, Cleveland's Series of Compendiums of English and American Literature, and their Series of Class Books of English Etymology

may be mentioned as fairly representing the class of their publications, which have a widely extended and well merited reputation for thoroughness. This firm are known, and deservedly distinguished as Educational publishers. Sower, Barnes & Co., also confine their publications almost exclusively to Educaional works. One of their leading publications is "Pelton's Series of Outline Maps," the demand for which supports two manufactories. They also publish Sanders' Readers, and works upon Arithmetic, Grammar, History, Philosophy, and Chemistry. C. H. Butler & Co. have given special attention to the improvement of School books, in point of mechanical execution. A diamond, they have supposed, deserves a handsome setting. Their exertions in this respect have been very successful, and not limted to the department of School literature. Messrs. Charles Desilver, U. Hunt & Son, Hayes & Zell, Crissy & Markley, and others, publish Educational Books of established reputation, and which are extensively sold. As regards mechanical execution, it may be safely asserted that—owing to the moderate cost of manufacturing in Philadelphia—the publications of her principal School Book publishers combine, in an eminent degree, characteristics of essential importance in the implements with which we are to "teach the young idea how to shoot," viz., *durability, neatness*, and *cheapness*.

4. German Books. The extent of the German population in our city and State, renders the publication of German books a distinct and important branch—there being at least four houses whose attention is principally engrossed by it—Wm. G. Mentz, J. W. Thomas, Ignatius Kohler, and John Weik. The last named is very extensively engaged in the publication of German works; and is the first publisher, within our knowledge, who regularly and successfully exports books from this country to Germany. In the present season, we are informed, he has sent thither three thousand copies of a German Dictionary, to aid in instructing his countrymen in the signification of their language. He is the sole publisher of a complete edition of Heine's works, either this country or in Germany.

Numerous as these subdivisions are, they might with propriety further extended. One class of publishers prepare their works with a view to sale mainly by *subscription*, or *through agents*. We would instance J. W. BRADLEY, who publishes nearly thirty volumes of Arthur's works ; JOHN E. POTTER, GEORGE W. GORTON, and the long established house of LEARY & GETZ. Another class make the publication of Juveniles, or books adapted for the young, a principal feature in their business ; among these, PECK & BLISS, J. P. KELLER, WILLIS P. HAZARD, H. H. HENDERSON, and HENRY F. ANNERS, are prominent. *Music Books* are leading publications with others; and many of the volumes of this description that are standards in the South and West, are published in this city. MILLER & BURLOCK have sold nearly a half million of the " Southern Harmony," and T. K. COLLINS publishes " Aikin's Christian Minstrel ;" and we believe nearly a dozen others. The attention of other publishers is largely engrossed by *illustrated* works. Some of the most magnificent Fine-Art Books in the world are the issues of Philadelphia houses. We would instance RICE & HART's National Portrait Gallery of Distinguished Americans, containing 144 steel engravings, which alone cost $40,000 ; and their McKenney & Hall's History of the Indian Tribes, with 120 colored illustrations, and the North American Sylva, with 277 colored illustrations. E. H. BUTLER & Co.'s publications also embrace a number that are remarkable for the elegance of their illustrations, and for typographical beauty. Murray or Longman might be proud of such books as their editions of Burns, or Goldsmith, or Thomson, or Keble's Christian Year ; or Heber's Poetical works, or Stevens' Parables, or Read's Female Poets, or Macaulay's Lays of Ancient Rome, adorned as they are with all the embellishments of taste and art. Messrs. HAYES & ZELL, and J. W. MOORE, have also issued several most superb works. Lastly, we come to the class who may be designated as publishers of miscellaneous works, or

5. GENERAL PUBLISHERS. Nearly all those whose names we have already mentioned belong to this class. With very few exceptions, we presume none will refuse to aid in the parturition of a book of respectable character and average merit, if they can be assured of its saleability, or guaranteed against loss. Whether it be a novel or a professional treatise, an annual or a commentary, they are indifferent, provided the popular demand is sufficiently significant,

in their opinion, to justify reasonable expectations of profit from its publication.   It would, therefore, be tedious to repeat the names of all in Philadelphia who publish miscellaneous books. We shall only mention, in this connection, those who are prominently distinguished as extensive and wholesale publishers of miscellaneous works.

Foremost in the ranks of general publishers, are J. B. LIPPINCOTT & Co., the house referred to by Mr. Carey, as probably the largest book distributing house in the world.   It was established nearly thirty-five years ago by John Grigg, Esq., long and widely known as the most successful of booksellers, who, with his partners, conducted the business under the style and firm of Grigg & Eliot, and Grigg, Eliot & Co., until the year 1849, when Mr. J. B. Lippincott purchased the respective interests of Messrs. Grigg & Eliot, and in connection with the junior partners of the old firm, established the present.   This purchase was probably the heaviest ever made by one individual in the book trade.

The firm of J. B. LIPPINCOTT & Co. is now composed of six partners, Messrs. Lippincott, Remsen, Claxton, Willis, and two recently admitted, C. C. Haffelfinger and John A. Remsen. Their general business combines that of *Publishers, Printers, Bookbinders,* and *Wholesale Booksellers* and *Stationers.* As publishers, they have frequently set up in a year twenty thousand solid octavo pages of new standard works, besides printing large editions from the stereotype plates of over two hundred different volumes, now in their vaults.   Within the last few years they have issued a number of most costly and valuable books, as for instance, their Gazetteer of the World, at a cost of $50,000 ; Indigenous Races of Mankind, by Nott & Gliddon ; and more recently Blodgett's Climatology, which has been highly eulogized by Humboldt, and other eminent scientific authorities.   The character of their leading publications, as well as the enterprise of the publishers, will be inferred from these ; or perhaps more distinctly, when we state that the original cost of *four* of their works, including their illustrated edition of the Waverly Novels, and the Comprehensive Commentary, was $186,300.   They have recently incurred an important outlay to

13

secure to Philadelphia the publication of Webster's Dictionary, of which they now publish five different editions.

In connection with the Publishing house, Mr. Lippincott has recently erected a six-story building, equipped with new and superior machinery for printing and binding books, and in which about one hundred and fifty persons are constantly employed. The capital invested by this firm in the general business exceeds a half million of dollars; and the copyright money paid by them to authors annually cannot be far short of one hundred thousand dollars.

CHILDS & PETERSON, to whom we previously referred in connection with Law Books, are widely known as the publishers of the Arctic Explorations, 2 vols., 8vo., for which they paid the estate of Dr. Kane the sum of $65,000, as author's proceeds of the *first year's sale*, being, it is believed, a larger amount of copyright money than was ever before paid *for one work* in the world. They have now in press, *Allibone's Dictionary of Authors and Literature*, which will contain a mention of every author who has written in the English language, making in all upward of 30,000 names. It has been in course of stereotyping for the last five years, will be issued in 1859, in one volume, super-royal octavo of 1,800 pages, and will contain twenty per cent. more matter than Webster's quarto Dictionary. The firm of Childs & Peterson was established in 1848, and consists of Robert E. Peterson and George W. Childs.

T. B. PETERSON & BROTHERS have in their possession the stereotype plates of about six hundred different books, small and great, principally novels. They have invested about $50,000 in Dickens' works alone, of which they print twenty-nine different editions : the only complete series in the United States. The sales annually average 50,000 volumes. This firm manifest an abiding confidence in the virtues of printer's ink.

PARRY & MCMILLAN are the successors of A. Hart, late Carey & Hart, and their list of publications, as may be supposed, comprises a number of very valuable books. Many of their more recent issues are of a religious character.

WILLIS P. HAZARD's catalogue is an interesting one, embracing a new edition of the most popular Poets, standard library editions of good authors, fine editions of Shakspeare, and a variety of Juveniles, &c.

The catalogues of H. Cowperthwaite & Co., Charles Desilver, E. H. Butler & Co., Lindsay & Blakiston, James B. Smith & Co., J. L. Gihon, and others, contain the titles of numerous choice miscellaneous works.

The Publishers of Philadelphia occupying, as they do, a more central position than their brethren in New York and Boston, and having peculiar advantages for circulating works throughout the entire Union, are much sought after by the latter as agents for the distribution of their publications. Hence wholesale bookselling is usually combined with publishing. Hence, too, while preferring, as they manifestly do, for their own issues, works possessing some substantial merit aside from mere entertainment, their stocks are generally very miscellaneous, embracing every variety of books, from the most ephemeral to the most substantial. The fact, however, that wholesale bookselling is combined with publishing, renders it difficult even for those who are so disposed to furnish accurately the statistics of their own publications. The best approximation that we can make, with the aid of experienced printers, in estimating the annual business of those who decline to furnish any statement for themselves, toward ascertaining the value of books published annually in Philadelphia, gives a result of $3,690,000; and of capital employed, $2,500,000. This is exclusive of books issued by printers, bookbinders, and authors who occasionally assume the risk of becoming their own publishers.

NEWSPAPERS—MAGAZINES.

The first newspaper published in Pennsylvania was printed in 1719, by Andrew Bradford, in association with John Copson, and entitled "The American Weekly Mercury."* The second newspaper was started by Samuel Keimer, in 1728, (immortalized by Franklin, in his autobiography, as a "great knave at heart"), and bore the title of "The Universal Instructor in all Arts and Sciences, and Pennsylvania Gazette." It was a folio sheet. After the return of Franklin from England, he, in 1729,

---

* Mease's "Picture of Philadelphia, in 1811," and "Newspaper Record."

united for a short time with Hugh Meredith, and continued Keimer's paper, on a whole or half sheet, as occasion required. In 1739, as we previously stated, Christopher Sower established at Germantown a German quarterly, but changed it to a weekly in 1744, under the title of "The Germantown Gazette." What however may be called the third paper in Pennsylvania was "The Pennsylvania Journal," issued December 2d, 1742. It was continued till 1800. "The Chronicle and Universal Advertiser" was the fourth in Philadelphia, and the first with four columns on a page. It was an influential sheet, and lived till 1773. Seven Journals in the English language, and six in the German, were started in Philadelphia before the Revolution.

The first daily paper published in the United States was "The Pennsylvania Packet, or General Advertiser," established in Philadelphia as a weekly, by John Dunlap, in 1771, but converted to a daily in 1784. At this period, Mr. David C. Claypoole became associated in its management. To him Washington presented the original manuscript of his farewell address, which was recently sold by his executors, through Messrs. Thomas & Sons, and purchased by Mr. Lennox of New York, for a sum exceeding $2,000. The first daily evening newspaper established in Philadelphia was "The Philadelphia Gazette," by Samuel Relf, in 1788. In 1790 Mr. Bache published "The Aurora," afterward purchased by William Duane. 1791, Mr. E. Bronson originated "The United States Gazette," which is now continued under the title of "The North American and United States Gazette." The other daily newspapers that flourished in Philadelphia, in the early part of the present century, were "The True American," established in 1797, by Mr. Bradford; "The Freeman's Journal;" "The Register," commenced in 1804, by Mr. Jackson; "The Democratic Press," established in 1807, by John Binns, and "Poulson's American Daily Advertiser," which was the successor of "The Pennsylvania Packet."

The first paper that habitually treated Letters and Arts in connection with commercial and political matters was "The Daily National Gazette," originated at Philadelphia in 1820. According to Dr. Griswold, in his history of American Literature, the establishment of this paper was an era in our national mind.

Philadelphia was the second city in the Union to encourage penny papers; and the "Ledger" has now a larger uniform circulation, it is generally believed, than any *daily* newspaper printed *in the world*, the London Standard perhaps excepted. It is printed on two of Hoe's Last Fast eight-cylinder Printing Machines, capable of printing 20,000 impressions per hour.

At the present time twelve Newspapers are published daily in Philadelphia, as follows :—

Name.	Published by	When Established.	Remarks.
MORNING.			
N. American & U. S. Gazette.	Morton McMichael..........	1791	Independent.
Pennsylvania Inquirer.......	Jesper Harding & Son....	1829	"
Public Ledger..................	Swain & Abell..............	1836	"
Pennsylvanian.................	William Rice..............	1832	Democratic.
Press	John W. Forney............	1857	"
Daily News....................	J. R. Flanigen	1848	American.
Philadelphia Democrat......	Hoffman & Morwitz........	1838	German.
Free Press.....................	F. W. Thomas..............	1848	German Independent.
AFTERNOON.			
Evening Bulletin..............	A. Cummings & G. Peacock........................	1847	Independent, successor of the American Sentinel, established in 1812.
Evening Journal	Grayson, Irwin & Montgomery	1856	Independent.
Evening Argus	Joseph Severns & Co......	1851	Democratic.
Evening Reporter.............		1857	Successor of Mor'g Times.

The following Newspapers are published weekly :—

Saturday Evening Post,
The Dollar Newspaper,
The Weekly North American,
Philadelphia Saturday Bulletin,
Weekly Pennsylvanian,
Pennsylvania Inquirer, tri-weekly
Forney's Weekly Press,
Fitzgerald's City Item,
Dollar Weekly News,
National Argus,
Commercial List,
United States Business Journal,
U. States Rail-road and Mining Register,
Southern Monitor,
Sunday Dispatch,
Sunday Transcript,
Sunday Mercury,
The Public Mirror,
The Tattler,
Banner of the Cross,

Episcopal Recorder,
The Presbyterian,
The American Presbyterian,
Christian Chronicle,
Christian Observer,
The Catholic Herald,
The Friend,
Friends' Weekly Intelligencer,
Friends' Review,
The Moravian,
The Woman's Advocate,
Germantown Telegraph,
Frankford Herald,
The Legal Intelligencer,
Real Estate News Letter,
Masonic Mirror,
The New World,
Vereingte Staaten Zeitung,
Republican Flag,
Philadelphia Wochenblatt.

Every merchant and every farmer throughout the Union should subscribe for at least one newspaper from a city so important as Philadelphia; and no one who reads aright can fail to obtain an amount of practical, useful information, that will amply repay the trifling expenditure.

In addition to the Journals above enumerated or referred to, there are about fifty Periodicals published in Philadelphia, including Medical, Legal, Scientific, and denominational organs. Of the strictly Literary Magazines there are four—Peterson's Magazine, Arthur's Home Magazine, Graham's Magazine, and Godey's Lady's Book, established in 1830. The last is specially designed for the ladies; and from its devotion to their tastes, and the valuable supply of fashion plates, it has attained an enormous circulation. It is a creditable fact, worthy of being noticed, that an oath has never been registered in its pages. The Philadelphia Magazines have long been celebrated for their high moral tone and their devotion to the Fine Arts—making exquisite illustrations a leading feature.

We regret that we are unable to furnish any reliable Statistics of Newspapers and Periodicals. We therefore proceed to consider the most important of the industrial branches to which the creation of Books and Periodicals has given rise.

IV.

The Book Manufacture and its Kindred Branches.

Livingstone, the celebrated traveler, in the Preface to his Journal of Explorations, states that those who have never carried a book through the press, can form no idea of the amount of toil it involves; and adds, that the process has increased his respect for authors a thousandfold. Again, in the Introduction, he says, " I think I had rather cross the African continent than write another book." But books are the embodiment, not merely of the labor of the author, but of a vast number of persons occupied in seemingly diverse branches of industry. A volume, however insignificant in itself, represents a portion of the labor of rag-gatherers, carters, bleachers, paper-makers, paper-machine makers, miners, furnace-men, type-metal makers, type-founders, compositors, pressmen; bookbinders, publishers; manufacturers of printing presses, of printing ink, of gold leaf, of bookbinders' tools, and bookbinders' muslin and leather; sometimes engravers on wood or steel, lithographers, stereotypers or electrotypers; marble-paper makers, and others. All those who are directly concerned in the production of books are largely congregated in Philadelphia, with many in each of the kindred branches, as blank books, maps, envelops, &c. To notice the present development of the more important and prominent of these is our present purpose.

1. TYPE METAL.

Type-metal is composed of block-tin, lead, and antimony. There is one large establishment in this city for manufacturing Type and Stereotype metal, in connection with Babbitt metal and packing, and anti-friction metals of every description. In this foundry eight furnaces are used; one of which is constructed on a new principle, and is used for smelting lead and antimony from the ore, as it comes from the mine. This furnace is of the largest size. The proprietor, Mr. H. W. Hook, has for some years directed his attention to the production of a durable metal for Type-founders, and now manufactures an article, of which the ingredients are not disclosed, but which is claimed to be the best in the United States. It is known as Hook's Adamantine Metal; and when cast into the smallest type may be driven, it is said, without injuring its face in the least, into the hardest type heretofore known. It produces type in sharpness, clearness, and delicacy, comparing favorably with electrotype. Mr. Hook also produces a metal for *hardening*, which may be used as a substitute for tin.

2. TYPE-FOUNDING.

Type-Founding, in or near Philadelphia, dates from 1735, when Christopher Sower established a printing-office at Germantown, and cast his own types. Soon after the close of the war, Mr. John Baine, of Edinburgh, established a Type-foundry here; and was, it is believed, the first who regularly carried on the business of Type-founding in the United States. In 1790 Baine died, and Messrs. Archibald Binny and James Ronaldson established another foundry, unconnected with any other business, and were eminently successful. It is to them the world is indebted for the first real improvement in Type-founding since the days of Peter Schœffer; this was in the type mould : enabling a caster to cast 6,000 types in a day, as easily as he could have done 4,000 by the old process. It is known in Europe as the American Mould.

In 1808, Mr. Wm. L. Johnson patented a machine for casting type, by which he was enabled to give a sharper outline and better face to the letter, by using a pump to force the liquid metal into the mould. This idea subsequently passed through many modifications and improvements, and the casting capacity of Binny and Ronaldson's Mould is now multiplied threefold.

The quality of Philadelphia type bears a favorable comparison with that of Europe, and is cheaper. The metal used is a mixture composed chiefly of lead, antimony, and tin, in proportion to the kind of type required. There are now four establishments in the business—L. Johnson & Co., Collins & McLeester, Pelouze and Son, and A. Robb ; and another firm, Starr & Co., produce type for marking linens. The amount of capital invested is stated approximately at $500,000, and the aggregate product at $420,000. Messrs. Johnson & Co. employ in this department of their general business, which includes also Stereotyping and Electrotyping, about two hundred and twenty-five persons, and produce annually about 600,000 lbs. of type, averaging fifty cents per pound, or $300,000. This firm publish very elegant specimen books ; which may be referred to as a sample not merely of their own product, but of all the Philadelphia Type-founders.

3. STEREOTYPING.

Stereotyping is the mode of casting perfect fac-similes, in metal, of the face of movable types. The plan is simple. After arranging the type in pages, and getting it perfectly smooth and clean, it is placed in a frame, the surface being thoroughly oiled to prevent the mould from adhering, when liquid gypsum, or Plaster of Paris, is poured over the page. The mould thus taken, if perfect, is dressed with an instrument, and a hole made to admit the metal. It is then dried, after which it is put into an iron casting-box, and the whole immersed in liquid type-metal. Twenty to thirty minutes usually suffice for casting. The box is then swung out of the molten mass into a cooling-trough, in which the underside is exposed to the water. When hard, the caster breaks off the superfluous metal, and separates the plaster mould from the plate. It is then picked, the edges trimmed, the back shaved to a proper thickness, and made ready for the press.

Stereotyping is of comparatively recent introduction into the United States. Mr. L. Johnson, to whom we have already referred, was among the first to practice the art in Philadelphia, about thirty-five years ago. For many months he was able to execute all orders with his own hands. There are now seven Stereotype foundries in this city, employing about one hundred and eighty hands, and having a capital invested of $150,000.

4. PRINTING.

The processes of Printing are familiar to the majority of persons, and are generally the same in all the offices: nevertheless, the results produced, singular to say, differ very widely. The characteristics of the typography of different printers are almost as marked as their nasal protuberances, and as readily distinguishable by close observers. Some of the literati of Washington, recently admiring the typography of a work, credited it to their own printers; when a publisher present, though he had never seen nor heard of the work, confidently attributed it to Philadelphia; and subsequent information confirmed his sagacity. In fact, a large portion of fine work, bearing the imprint of Southern printers, and particularly of the Washington establishments, is executed in this city. Some of the most beautiful specimens of which the Art in America can boast, have issued from the Philadelphia press. The typographical beauty of such works as the Indian Biography, the National Portrait Gallery, Schoolcraft's Indian Tribes, and Gliddon's Types of Mankind, is acknowledged everywhere, and is the result, not of accidental circumstances, but of long experience, unsurpassed facilities, and unremitted vigilance and care.

There are now about fifty Printing-offices in Philadelphia, exusive of the newspaper press, employing from three to one hundred persons each; and executing work of every description, from a card to a quarto. The compositors in the city, who are members of the Printers' Association, are about four hundred; while the force engaged in ordinary times is nearly one thousand persons. Many of the employees who tend presses are females, whose earnings average $4 per week. Power-presses are in use in all the leading establishments, five having seventy-four in constant operation, the cost of which was not far short of $150,000

As an evidence of the capacity of our Book offices for executing work with expedition, we may state, that the average of each week's work done in one single establishment is about fifteen hundred tokens for fifty-nine working hours, which is equivalent to nine million duodecimo pages; or, in other words, the presses of one firm turn out about *twenty-five thousand* duodecimo volumes of ordinary size each week. By working at night, the quantity could be doubled. As an illustration of the facilities possessed for printing in *different languages*, we subjoin the following extract from a note received from one of the leading firms :—

"We set up for the Stereotypers, during the last year, Dr. Jayne's Medical Almanac, in the following eight languages :—*English, German, Hollandische, Swedish, Norwegian, French, Spanish,* and *Portuguese.*

"We have set up and printed from type six volumes Latin Theology, octavo, with Greek and Hebrew Notes; a Hymn Book in *Cherokee*, from movable type; several German Dictionaries, from one of which we have sent three separate editions to *Germany*—in all six thousand copies. Also, a German Grammar, two editions of which went to Germany. Last month we completed a small work in French on Prosody —one half the edition was ordered, and shipped by us direct to Paris, on account of the author, Victor Value, Esq.   K. & B."

Ornamental Printing, or *Printing in Colors*, is carried on by the principal firms as a branch of their general business, and it is also made an exclusive business by several. This department of the Printer's Art has been developed wonderfully within a few years. In England its importance dates from the commencement of the present century, when the rivalry existing between the proprietors of various State lotteries, induced them to invoke its virtues for making their advertising schemes attractive. It was subsequently applied principally to Playing-cards. On the day of the Coronation of Queen Victoria, an edition of the Sun newspaper was printed entirely in *gold*. This was considered a great feat. At the present time, printers in Philadelphia frequently execute orders for bank checks in gold; and fancy show bills, and especially Druggists' and Perfumers' fancy labels, demand all the tints of the rainbow. A great deal of work is done in this branch upon orders from New York.

Printing for the Blind has been executed in Philadelphia with a fair share of success, and its importance is evident, inasmuch as the statistics show, that in America one out of every two thousand is blind. The process adopted was what is known as printing in relief—the alphabet being Roman capitals, in small, compact, sharp type, and the relief produced by heavy pressure on thick paper between two sheets of copper, having the letters deeply cut. The embossing is thus on both sides. Several works, including nearly all the Gospels and Epistles, have been printed at the Institution for the Blind, and the typography was complimented by the London juries as exceedingly well executed, and comparing "most favorably with the best of the Glasgow books."

PRINTING INKS. In printing, especially plate printing, the tone and color of the Ink used are all important. Without good Inks it would be impossible to attain any marked success in letter-press. There are now five establishments in this city engaged in making Printing Inks—Chas. E. Johnson, Lay & Brother, Caleb Pierce, Woodruff & Co., and L. Martin & Co.* The first named is probably the oldest established concern in the United States. Some of these establishments confine their manufacture to Black Inks solely—others make all the usual varieties of Colored Inks, which have been compared with the English and pronounced superior. Within the last few years a great impetus has been given to

---

* The youngest of these enterprising firms, Messrs. L. MARTIN & Co., have very politely proposed to furnish a fair sample of Philadelphia Ink for use in these pages. This firm claim to have freed their Inks entirely from those empyreumatic oils, the presence of which so often endangers the best efforts of the printer, by causing the work to change to a dingy brown. Mr. Adams, the author of the well-known "Adams' Typographia," is associated in this firm, and has succeeded in producing a very superior article, of which the one-dollar quality has been used side by side with English four-dollar Ink, and the difference is imperceptible to an unpracticed eye. They state, "true economy in business is one of the chief elements of success; but in relation to Printing Ink particularly, the lowest priced is not always the cheapest; but that which, containing the most color, will require the least ink, and as a consequence, produce the finest work."

In connection with Printing Inks, Messrs. L. Martin & Co. manufacture Lamp Black extensively—many tons being exported to Europe every month. Their manufactory is said to be the largest in the Union. A few years since, America depended for Lamp Black entirely upon foreign supplies.

this business, and a great improvement effected in the manufacture ; and, at the present time, the quality of Ink made in Philadelphia is said to be superior to any in the country.   Product about $160,000.

5. PAPER.

The extent of the publishing interest in Philadelphia necessarily makes this city one of the great Paper marts of the Union. There are now at least thirty-five houses in Philadelphia engaged more or less extensively in dealing in Paper and Paper-maker's materials ; and on Sixth, Commerce, and other streets, there are immense warehouses filled, ream upon ream, with paper of all kinds, sizes, and colors, the product of European, New England, and Pennsylvania mills, consigned for sale, and sold in some instances at lower prices than the same paper could be purchased for cash from the manufacturer.   In the vicinity of the city, along the Brandywine, in Delaware and Chester counties, the Paper-mills are very numerous, some of them the oldest in the country, as for instance, the Ivy Mill, which produced, it is said, the first Printing Paper made in America ; and also, the first Bank Note Paper, viz., that for the old Continental money.   But within the limits of the consolidated city, there are only nine Paper mills— four in Manayunk, three on the Wissahicon, one in West Philadelphia, and one on the Schuylkill, near the Soapstone quarry. These mills, however, are generally of the first class, some of them new Steam-mills, (see APPENDIX) ; and produce, including those of two other parties, whose head-quarters are in this city, and whose products are sold *exclusively* here, an average annual value of $1,250,000.   The City Mills are occupied principally in executing orders for paper of a particular quality, or unusual sizes.

The products of the Philadelphia mills include, besides the usual varieties of News and Book Paper, some of the rarer descriptions. The Writing and Letter Paper at least of one manufacturer is very celebrated ; the Bank-note paper of another received the premium of superiority from the Boston Bank Association, in 1855 ; Plate and Lithographic Paper is made by Messrs. Magarge ; and paper from Straw by a firm in Manayunk—the Ledger and Dollar Newspaper alone consuming about 50,000 reams, or

24,000,000 sheets per annum.* Pasteboards and card-boards for printers, paper-box manufacturers, and others, are manufactured extensively—one firm, Messrs. A. M. Collins & Co., making annually 5,000 gross, or upward of 700,000 sheets, and believed to be superior in style of finish to any other in the Union.

The advantages that the manufacturers of Paper in Philadelphia have, including cheap coal, superior water, a first-rate Paper-making Machine establishment, and cheap supplies of raw materials, are so manifest, that no better locality could be selected for the manufacture, particularly of the fine kinds of Paper.

6. BOOKBINDING.

Bookbinding may be considered both as an Art and as a manufacture. As an Art, it was practiced two thousand years ago, and we are told that no expense was deemed by some too great for the decoration and preservation of rare manuscripts. Only within the last ten years, however, has ornamental binding received much attention in the United States; and except for the superior quality of Philadelphia workmanship, the national reputation, in this particular, would be far from enviable. The credit of American Book-binding was saved, both at the London World's Fair, and at the Crystal Palace in New York, mainly by the favorable specimens sent from this city. There are now establishments in Philadelphia occupied almost exclusively upon Fine Art binding for private individuals, and largely upon orders from New York, Albany, and Boston. In this city, if nowhere else in the country, men of wealth and taste can procure their rare volumes bound in accordance with the true principles of the art— that is, to adapt the style of the covering to the contents of the volume.

---

* A resident of this city, Mr. G. A. Shryock, claims to have been the first in this, or any other country, to manufacture by machinery Paper and boards from various kinds of straw and grass. The discovery of straw as material for Paper was patented by Col. William B. Magraw, of Meadville, Crawford County, in 1828. In the succeeding year, Mr. Shryock purchased cylinder machine, and adapted it to the manufacture of Paper from this material, and his success was noticed in the American and European journals of that period.

As a manufacture, modern Book-binding is distinguished for the extent to which machinery has been employed, and the consequent rapidity of production. In our large establishments, occupied principally with orders from Publishers and Publication Societies, nearly all the leading processes are executed by machines. Mechanism is applied to block-gilding, blind-tooling, and embossing; hydraulic presses are used instead of the old wooden screw-presses; Riehl's cutting-machine supersedes the plow; cutting-tables, with shears, are now used for squaring and cutting mill-boards for book-covers; and machines have recently been invented for backing and finishing. As a result, 1,000 volumes can now be put into boards in *ten hours*, and 2,500 volumes are with some an average day's work, in busy seasons. The Binders of this city have peculiar advantages in being able to procure the requisite materials direct from manufactories in their midst. The very best Morocco that can be obtained is made in this city; Marble paper, of unsurpassed quality, is made here; and Tar boards are supplied from the immediate vicinity. Messrs. GASKILL, COPPER & FRY, supply the best Bookbinders' tools and ornamental brass work used in New York and Boston, as well as in this city; and the whole trade looks to Philadelphia for the most efficient Bookbinders' machinery.*

Of *Marble Paper*, Philadelphia has the principal manufactory in this country, that of Mr. CHARLES WILLIAMS. He claims to have been the first in the Union who made the Antique Dutch and Drawn patterns, and also the only one that has succeeded in matching the celebrated "Papier de Annonay," of France; and still is the only manufacturer in the United States of all of the English, Dutch and French patterns. He supplies Boston, New

---

* Book-binders' Muslin is not made in this city, and we believe not in this country, excepting by one house, Messrs. N. M. ABBOTT & Co., of New York city. This house is entitled to very great credit for their successful and persevering efforts to produce an article of American manufacture, fully equal in all respects to the imported, and cheaper. Their sales now exceed one half the entire present importation. We are assured by practical binders, who have thoroughly tested Messrs. Abbott & Co.'s muslins, that the trade would find an advantage in using the American in preference to the foreign article; and we trust patriotism and good sense will induce them to heed the suggestion. This volume we shall order to be bound in American muslin.

York, and the chief cities of the whole Union with nine-tenths of the fine papers that are used on extra work, and go by the name of English and French paper. " There is no manufacturer in this country," he says, " that has equaled my styles of work, and none that has even matched my common papers for durability and finish."

The binding of BLANK BOOKS is a distinct but important branch of the general trade. In the manufacture of Account Books, foreigners acknowledge that Americans excel all others; and those who are conversant with the best workmanship executed in this city, will acknowledge that our makers are not surpassed by any. Every variety and description of Blank Books for merchants, banks, and public offices, are made here, from the cheapest to those distinguished as *indestructible*, which are bound in Russia, paneled, and edged with brass, varying, in full sets, from $75 to $300. The leading establishments employ the most improved machinery, and no expense is spared to expedite and cheapen the processes of manufacture. One enterprising individual recently paid $17,000 for a limited monopoly of one machine for *Ruling*. By means of this invention, known as the McAdams' Ruling Machine, horizontal and vertical lines, in red and blue inks, are ruled on both sides of the paper by a single passage through. Seventy-five to one hundred reams may thus be ruled per day; while eight or ten reams are considered rapid work by the ordinary hand-ruling machine. The pens used in the McAdams' invention are of gold, tipped with rhodium, which prevents wearing of the points, and insures uniform and unbroken lines. It is estimated that these pens will endure six years. Any variety of width of heading is made possible by an improvement in lifting the pens. Another invention, in general use, is a machine for paging Blank Books, introduced by the enterprise of Mr. Willard. The method consists in the simultaneous application of numbered types to each side of the sheet; and so great is the working capacity of the machine, that it is capable of producing thirty or forty thousand impressions per day, requiring only one operator to move the treadle. Purchasers now generally stipulate that their Blank Books shall be delivered *paged*.

The business of Account Bookbinding and Ruling, in this city, amounts to about $260,000; while the entire Bookbinding, including Blank Books, exceeds a million of dollars—as nearly as we can ascertain it, $1,210,000 ; and employs, in favorable times, about seven hundred males and one thousand females.

6. ENGRAVING.

The arts of Engraving may be divided into two principal classes—*Engraving in Relief*, of which Wood Engraving is the principal representative, and *Engraving in Basso*, as Line engraving, Mezzotinto engraving, engraving in Stipple and Aquatinta.

Engraving on Wood has become exceedingly popular within the last twenty years, as a means of illustrating natural or familiar scenes. So great has been the demand for Wood Engraving, that the art in many places has degenerated into a mere mechanical trade. But in Philadelphia, fortunately, there is not an overbearing demand for indifferent styles to illustrate cheap publications; and consequently, the artists have been employed in branches which called for their best powers, and developed the highest capabilities of the Art. Government reports, Scientific works, and orders from individuals other than Periodical Publishers, have been the principal sources of employment to the Wood Engravers in Philadelphia.*

* Philadelphia contains so many names of celebrity, as Wood Engravers, that it would be difficult to instance any as special representatives of the general excellence. We shall therefore merely refer to those whose Engravings are in this volume, and of whose merits the reader can judge.

The principal firm selected is that of BAXTER & HARLEY, whose engravings may be known by the mark. The establishment occupied by them was founded by Mr. R. S. Gilbert, for some time the only Engraver in Philadelphia, and probably the best in the country. The firm subsequently included Mr. William Gihon, since ·deceased, of whom Mr. Baxter was the pupil, and is the successor. Mr. Harley was also a pupil of Mr. Gihon. Mr. Baxter is said to be one of the best designers in the city, and has had a large experience, and been peculiarly successful in engraving Anatomical subjects.—Wilson's Anatomy, Carpenter's Physiology, Gliddon's Types of Mankind, &c., were illustrated by this firm, and are fair samples of the accuracy with which they execute engravings for Medical and Scientific works.

Engraving on *Pine* has been brought to great perfection by Messrs. Baxter & Harley : some of their large show-bills, &c., are really works of Art.

Other of the Wood Engravings in this work were executed by Mr. ED-WARD ROGERS, an enterprising young artist, rapidly rising in his profession.

Line Engraving is the general term for the process of engraving on the two metals commonly employed—Copper and Steel. The manufacture of Bank Notes affords occupation to a large number of the very best Engravers on Steel; but aside from this, the demand for Steel Plates in this city has called into existence several important establishments. Philadelphia now contains what is believed to be the largest Plate Engraving establishment in the United States. It employs about seventy persons, including eighteen Engravers, runs thirty-five presses, and turns out an average annual product of $50,000.*

Mezzotinto Engraving, so far as the history of the art in America is concerned, is inseparably associated with the name of JOHN SARTAIN, of this city, who first introduced it here about a quarter of a century ago, during more than half of which period he was alone in the practice. In his hands it underwent a change in its application, and consequently in its methods, by adapting it to the production of small book embellishments, for which it had not been used before. From the broad effects of large framing prints, it was forced down to the expression of the most minute details, on the diminutive scale of Pictorial Books; and we count by hundreds the Steel Plates engraved in this style during the period referred to, all the product of one prolific hand. The facility of its execution, its inexpensiveness, the richness and softness of its effects, all tended to extend its popularity; and its use, doubtless, hastened the diffusion of that rapidly growing taste for prints in this country, everywhere observable.

The other processes of Engraving enumerated, are—*Engraving*

---

* The establishment alluded to is that of J. M. BUTLER, in Jayne's Granite Building. We extract from our Reporter's Notes:—"Mr. Butler has now in course of publication Washington at Valley Forge, and Franklin before the Privy Council, in Whitehall Chapel, London, 1774. The Painting is by C. Schuessele, Artist, at a cost of $1,800. The Engraving is being executed by Whitechurch, one of the best English Artists, at a cost of $5,000. It will take four years to complete it. Hamilton, the Marine Painter, (for some years associated with Butler in designing), is now making illustrations for Col. Fremont's Explorations. Butler's establishment shows to what a high pitch of perfection the noble art of the Engraver is carried by Philadelphia Artists."

14*

in Stipple, and *Aquatinta Engraving.* Engraving in Stipple is **a** costly method of Engraving, seldom used except in portraiture. The largest, and one of the finest Stipple Portraits ever executed in this country, is a copy of Gilbert Stuart's Washington, executed by Mr. T. B. Welch, of this city. Aquatinta Engraving is a secondary method, unsatisfactory where minute and accurate details are required.

7. LITHOGRAPHY.

This ornamental art, of so much service to the useful arts, is so nearly allied to Engraving, that it might be treated as a branch thereof—being, in fact, Engraving on Stone, or Surface Engraving. The stone used possesses, in a high degree, calcareous qualities similar to limestone, and absorbs to a certain extent the oily substances that are used to give the drawings sufficient adhesiveness to resist the friction of printing. These are Lithographic chalk and Lithographic ink. They are composed of tallow, virgin wax, soap, shellac, and colored with lamp-black. The principal styles in Lithography, are *Linear and Crayon Drawings, transfers on stone from steel or copper-plate engravings, wood-cuts,* or *from Lithographic drawings themselves.*

The second Lithographic establishment in the United States, was opened in Philadelphia in 1828 ; and here many of the most important processes and improvements in the art have had their origin. Works have been executed here that would do credit to the artists of any city or place in the World.*

* In illustration of the correctness of this assertion, we may refer to plate XXI. of the seventh edition of Lieut. Maury's "Sailing Directions," published by Messrs. E. C. & J. BIDDLE, of this city, for the U. S. Hydrographical Department. The plate referred to, by a series of differently colored lines—some continuous and others broken—extending across two quarto pages, exhibits the percentages resulting from the average of observations for 46,000 days, of fair gales, head gales, total gales, calms, fogs, and thunder and lightning, met with in each month of the year, in every five degrees from 10° to 75° West latitude, along Lieutenant M.'s proposed track for steamers bound from America to England; and also, along his proposed homeward track. The precision required both in the engraving and in the printing of this plate was such as almost defied the powers of lithography ; but it was executed to the satisfaction of Lieutenant Maury,

P. S. Duval & Son
22 & 24 South 5th St

Lithographers
ab Chesnut Philad.ª

There are now two hundred and thirty-five Lithographic presses in operation in this city.

8. MAPS. Map manufacturing, it is believed, is conducted on a more extensive scale in Philadelphia than in any other city in the Union. One establishment alone turns out twelve hundred Maps weekly. Connected with it are two Lithographic Printing-offices, having twenty presses ; coloring rooms, in which thirty-five females are employed ; twenty engravers are occupied in the engraving office, and sixteen men in the mounting rooms. Maps of

and of all good judges. Again, some years ago, Messrs. Lippincott & Co., published an Annual, called "The Iris," having chromo-lithographic illustrations, representing various subjects of Indian Life and Wild Scenery, printed in eleven colors. This volume was so much admired in England, that the Queen ordered a dozen copies for her own household, expressing by letter to the agent in London, her admiration of the manner in which the book was illustrated, and saying, "that it was the prettiest book she had seen from America, and reflected great credit on the city of Philadelphia."

The firm who executed the above works now conduct business under the style of P. S. DUVAL & SON. The senior is now, we believe, the oldest established Lithographic Printer in the city. Mr. S. C. Duval, the son, is also an accomplished Lithographer, having just completed a three years' practice in some of the principal establishments of Paris. Among the numerous works recently executed by this firm, we might mention a large and beautiful view of Baltimore, in 1752, printed in colors ; a book of specimens for Messrs. Cornelius & Baker, containing Drawings printed in colors of the principal styles of Chandeliers ; and one hundred and fifty plates for a Government report of the Pacific Rail-road and Mexican Boundary.

The art of Photographic Drawing on Lithographic Stone has been attempted by this firm with probabilities of success. The process is simple, and much more economical than drawing on stone.

The art of transferring Plate Engravings to stone, of so much importance in the publication of Maps, Outline Drawings, &c., was introduced, it is said, into this city by Mr. F. BOURQUIN, who, for a long period, occupied the position of foreman in Mr. Duval's establishment. To this day he has few if any equals in the art. An Isothermal Map, lithographed by him for Blodgett's Climatology, is equal in delicacy and accuracy to copper-plate. The firm with which he is connected, Messrs. F. BOURQUIN, & Co., have executed for this work the view of the Bridesburg Machine Works.

Another Lithographic establishment, in this city, includes and represents a combination of master-artists. We refer to that of T. LEONHARDT & Co., and F. MARAS, 609 Chestnut street. The duties are subdivided thus :

Mr. THEODORE LEONHARDT, whose skill as a Lithographer is well known

all the counties in Maine, and others of the New England States, have been executed in this city ; and one firm in Philadelphia has the contract for making a Map of the State of New York, in which $50,000 has already been expended. The product amounts to about $400,000 annually.

9. ENVELOPES. The British Post-office Statistics state, that in 1850, three hundred and forty-seven millions of letters were posted, and of this large number three hundred millions were inclosed in Envelopes. In this city, many millions of letters are mailed annually ; hence it will be seen, that the manufacture of Envelopes must necessarily have some importance. There are now four establishments engaged in the manufacture, the largest having five machines for folding and gumming, each of which can turn out eighteen thousand Envelopes a day : though, of course, the full power is seldom exerted. Not only plain, but embossed and enameled Envelopes are made ; and one firm makes twenty sizes and styles.

in this and other cities, has special charge of the Engraving Department. He gives particular attention to the execution of Bonds, Diplomas, Certificates of Deposit, Checks, Notes, Drafts, and all other mercantile work, having no superior in this line. His productions with the Ruling-machine are elegant in design, and have been in many cases pronounced equal to copper.

F. MARAS, who has had twenty-two years experience in the leading establishments of Europe, and who is well known in London and Paris as a first-class artist, personally superintends all the artistical branches and the color printing.

Mr. J. H. CAMP, celebrated as a practical printer, and transferrer from Stone, Steel, and Copper Plates, leads the Printing Department. His productions are esteemed perfect by the principal artists ; and in combination with the above-named gentlemen, contributes largely to the reputation and excellence of these establishments. ' In union there is strength."

Mr. THOMAS SINCLAIR has executed some very fine Lithographic Engravings for the Harpers and other publishers ; and Messrs. WAGNER and McGUIGAN, and Mr. ROSENTHAL, have had a good share of Government and other work, which they have executed in a masterly style.

Mrs. BOWEN continues the establishment of her late husband, a celebrated Lithographic Engraver in the department of Natural History ; and M. H. TRAUBEL, and others, execute Lithographic Engravings of great excellence.

The product of the Book Manufacture, and its kindred branches, is stated approximately as follows :—

Type-Metal, Type Founding, and Stereotyping, - - - -	$650,000
Printing—Book and Job, (including Fancy Printing,) - - -	1,183,000
Printing Inks, - - - - - - - - -	160,000
Paper, - - - - - - - - - -	1,250,000
Book Binding, including Blank Books and Marble Paper, - -	1,230,000
Engraving—Wood and Steel, and Lithography, - - - -	570,000
Maps, - - - - - - - - - - -	400,000
Fancy Stationery and Envelopes, - - - - - -	150,000
Total, - - - - - - - - -	$5,593,000

V.
Boots and Shoes.

The manufacture of Boots and Shoes would be treated by Playfair as a subdivision of the general subject of Wearing Apparel. Those who have never heard of the machine, invented by a Pennsylvanian, which will peg a Boot or Shoe, two rows on each side, in three minutes, and cut its own pegs, might be disposed to classify the trade with Stationery. But within the last few years, so great an advance has been made in the science of pedology, and so many improvements have been made in the mechanism expediting the manufacture, that it is entitled to distinct and separate consideration ; and, as respects first-class Boot-makers, particularly those of Philadelphia, there is now neither wit nor justice in comparing them to their predecessors in the time of Simon of Joppa, nor in ranking them, as some one has done, among the great scourges of humanity.

The chief seats, in the United States, of the Wholesale Manufacture of Boots and Shoes, are Lynn, Mass., and Philadelphia. The former makes cheap, common work its specialty—the latter, fine Boots and Shoes, particularly Ladies' Shoes. The former has had the advantage of Boston enterprise in scattering broadcast the particulars of its industry ; the latter has produced annually an equal product of really, though not nominally, cheaper work ; but has taken such precautions to guard the secret that but few persons have even accidentally heard of it. Mr. Edward Young has made the manufacture, in Philadelphia, one of the objects of thorough investigation for this volume, and the following extract from his report exhibits its present condition.

With respect to the reputation of Philadelphia Shoes, I first heard of it while traveling in the South. I noticed that a Petersburg retailer offered his customers ' Great inducements ! Miles' celebrated Boots only eight dollars a pair.' As I had heard of Boots in New England for something less than that per pair, I was led to inquire who Mr. Miles was, and received an account, which was amply confirmed when I recently visited his establishment on Fourth street. An inspection of his work profoundly impressed me with the sagacity of the Petersburg merchant.

The results of an extended personal examination and inquiry into the manufacture of Boots and Shoes, in Philadelphia, are these. In the first place, *the quality is most superior.* This superiority may be ascribed in part to the advantage which the manufacturers have in this market for purchasing leather and skins—sole-leather, calf, goat, and sheep-skins and especially for obtaining morocco, of which they have the first choice from the large stock made in this city;—and can also obtain them in such quantities as they desire. Secondly, *to the skill* of the workmen. A large number of the journeymen are foreigners, chiefly Germans, many of whom are first-class workmen.

Some of the work in both branches excels any I have seen, of either European or American make. The character of the work may be judged by the following scale of prices, paid by the best houses for making—

Men's Dress Boots—Fitting,	75	cents.
Crimping,	10	"
Bottoming,	$2 25	"
Heeling,	12	"
Total,	$3 22	"
Ladies fine heel Gaiters—Cutting,	3	cents.
Binding, &c.,	33	"
Making,	$1 00	"
Total,	$1 36	"

There are about 7,000 men employed, equal to the constant labor of 5,000—average wages $6 per week for 50 weeks, or $300 each,	$1,500,000
2,000 females, not fully employed, averaging $100 per annum,	200,000
Total Wages,	$1,700,000

In addition, there are 165 Sewing Machines in constant use.

Making Men's wear, and making Women's wear, are distinct branches; although several are engaged in both, having, however, separate establishments. The *men's men*, and *women's men*, as the workmen are distinguished, have separate organizations, and neither know nor mingle with each other. Which is the higher caste we do not know—gallantry would say the *women's men*. Besides, they are the most numerous, there being forty-two hundred of the latter to twenty-eight hundred of the former. The average wages is about the same in each branch—some earning but $5 per week, while superior and fast workmen obtain $8 to $10, and occasionally $12 per week. It is generally known that the work is cut in the establishment, and given out to the men who work at their homes.

The Statistics of the Capital and Product are as follows :—

Capital invested by the regular manufacturers,   - - $1,650,000

Product, viz. :

18	Manufacturers, being all those whose annual product exceeds $50,000 per annum, make - - - -				$1,689,000
8	make over	$40,000	per annum,	$335,000	
11	"	30,000	"	330,000	
4	"	25,000	"	100,000	
9	"	20,000	"	180,000	
7	"	15,000	"	105,000	
2	"	14,000	"	28,000	
4	"	12,000	"	48,000	
20	"	10,000	"	200,000	
12	"	8,000	"	96,000	
				———	1,422,000
80	who produce annually from $3,000 to 7,000, average $5,000 each, - - - -			400,000	
100	" " " " $2,500			250,000	
180	who sell to dealers over and above customer work, on an average, 1,000			180,000	
				———	830,000
					$3,941,000

Made in Prison and other Public Institutions, about - -			100,000
" " Burlington, N. J., for Phil'a dealers, at least, -			100,000
Total, - - - - - - -			$4,141,000

The amount then of Boots and Shoes, made in Philadelphia, exceeds, it will be seen, four millions of dollars annually, being more in *value*, though not in *number of pairs*, than the whole production of Lynn, where shoes are supposed to grow spontaneously. In addition to these, there are a large number whose operations, though in the aggregate important, cannot easily be ascertained. They are known by a term, more expressive than euphonious, "*garret bosses*," who employ from one to twelve men each ; and having but little capital, make Boots and Shoes in their own rooms, and sell them to jobbers and retailers in small quantities at low rates, for cash. One retailer, who sells $20,000 worth per annum, buys three fourths of his stock from these makers.

The manufacture of this article, in Philadelphia, owing to superior facilities, could be greatly extended if the jobbers had more of the *amor patriæ*, and would purchase Philadelphia-made work instead of the Eastern made, the sales of which in this city annually amount to nearly *ten millions of dollars ;* but, unfortunately, the latter affords more profit. As consumers, too, we are not blameless ; for were we willing to pay a remunerative price for Philadelphia work, instead of a dollar or two for Yankee-made, which appear to be leather-soled, but which two weeks wear may discover to be paper, this branch of industry would be doubled. Recently, some efforts have been made to compete with the Eastern in price, at the same time excelling them in quality. One maker in particular, E. P. Molineaux, has turned his attention to this branch, and has been eminently successful in selling a better article at the same price as the best Eastern Boots and Shoes. As a specimen, his leather Boots for women, at $1, is superior to the Eastern made at the same price—with the lower qualities, say at 90 cents, he does not compete.

At Martin Bellows' establishment, I noticed a very superior article of grained hunting Boots, made of Pennsylvania leather, blacked on the grain side. Being tanned with oak-bark, the leather is more pliable, and the Boots are almost impervious to water. They are well suited to the West and South, where the hunter has to wade through water.

Since the introduction of Sewing Machines, the manufacture of *Gaiter uppers* has become a distinct branch, and gives employment to hundreds of females.

The entire trade of Philadelphia in Boots and Shoes, including

City manufacture and Eastern work, is stated approximately at *fifteen millions of dollars.*

VI.

Brass and Copper Manufactures.

Of Brass there are two principal varieties, distinguished as Yellow and Red Brass. Yellow Brass is composed of seventy parts of copper, and thirty of zinc ; and Red Brass is produced by using not more than twenty per cent. of zinc. Though these are the proportions generally observed, manufacturers in many instances adopt special precautions to render the alloy homogeneous. In Philadelphia, it is usual for the workers in brass to procure the materials and compound their own metal—the manufacture of Pig Brass, as a general rule, being limited to inferior qualities, made from scraps and filings from the shops. There are at least four concerns in Philadelphia engaged in this branch of the business, and known as brass smelters and refiners.

The uses and applications of Brass are so numerous, that while its manufactures are extremely important, it is very difficult to trace them in their details. In the production of ornamental brass-work, and especially in that department distinguished as Lamps, Chandeliers, and Gas Fixtures, the manufacturers of Philadelphia are declared, by the best foreign judges, to have no superiors in the world. (See LAMPS, &c.) The same may be said of those who convert it into various *Military, Odd-Fellows, Firemen's,* and *Theatrical Ornaments,* and other light and artistic forms. At one establishment, that of Samuel Croft, *Sheet Brass* is made quite extensively, though less so than we should think the demand would justify. *Bells* of every description, from the smallest to a full chime, are made by one firm extensively, and by three others to some extent. Several foundries are chiefly devoted to making *castings* in brass, of every kind of article that may be ordered, from the largest to the smallest, either for brass-workers and finishers who finish up the foundry products, or for use in connection with other manufactures. Brass is used largely by the manufacturers of Marine Engines and Locomotives, and in Ship-work. Castings of this description are supplied by the founders to a large amount ; while many of the engine and propeller

15

builders have Brass foundries as a part of their works. Another form in which Brass is largely used, in connection with steam apparatus, is the manufacture of *Gauges*, &c. One brass-founder, Mr. M. A. Dodge, gives special attention to moulding Steam and Water Gauges for boilers, and *oil-cups* of peculiar construction, for Locomotive and Steam-Engines. Messrs. Hook & Pritchard are engaged extensively in the manufacture of *Brass Boxes*, or *Composition Bearings*, for cars, &c. One firm in Kensington make, yearly, many tons of *Composition Nails* and *Spikes*, for copper sheathing and other copper work; besides *Rudder Braces, Pintles, Dove Tails, Side Lights, Ventilators, Port Hinges*, and other ship-joiners' castings, and *Castings for machinery* generally—the two branches constituting their principal business. They make besides Bells of all sizes, and Carriage and Harness Mountings, &c. *Brass Gun Mountings* constitute a principal item in the business of one shop; and another is occupied principally in producing *Brass Tubing*, or tubes for Philosophical, Optical, Mathematical, and other instruments, as Telescopes, Spy-glasses, Cameras, Air-pumps, &c., &c., requiring the same nice polish interiorly as externally. *Hose Screws* and *Branch Pipes*, (for house and garden hose), employ, either wholly or in part, another manufacturer. *Brass Book mountings* and *ornaments*, as *Locks, Clasps, Bands*, &c., of superior quality, are made by several different persons, and form the exclusive business of at least one manufacturer. *Castors*, for furniture, are made to a limited amount by at least *three* persons, and of excellent quality; but they complain that our dealers do not sustain our own manufacturers.

Besides four Lamp and Chandelier establishments, there are several smaller ones, that make *Lamps* of every variety for domestic use, chiefly of brass; and for the consumption of the various patent oils, and other illuminating substances, besides common oils, &c. *Moulds* of great variety, as for dentists, bottle-makers, and for pressing Sperm or Adamantine candles, for pattern-makers, confectioners, &c., are made by several. Locks, Keys, Door Plates and Knobs, Hinges, Fenders, Andirons, Fire Irons, Candlesticks; and the nameless varieties of house-keeping articles in brass, as Pans, Kettles, Coal-hods, &c., are nearly or quite all made here to some extent at least.   But the distinctive feature of this department of industry, as respects this city, is the manufacture of *Brass Cocks*.   This branch is said to have originated in Philadelphia; and those who are now engaged in it—and there are several quite extensively—occupy a prominent and leading position.

The "Philadelphia Brass Works," WILER & MOSS, proprietors, are occupied largely in the manufacture of *Stair Rods*.   This firm, it is believed, are more extensively engaged in this branch than any other in the United States.   In addition to Stair Rods, and their appendages, they manufacture Brass Mouldings, Brass Nails, Trunk Bands of all widths, Step Plates, Curtain Tubing, Curtain Wires, &c., of every description.   They are also the patentees of a cheap and useful article for lighting gas, known as the "Patent Taper Holder," of which they have already sold to an amount exceeding $10,000.   The Patent Holder is made of brass, rolled like a tube, with a turned handle ; and so constructed, that the wax-taper is slipped in, or thrown out, at pleasure.   In these works 56,000 lbs. of sheet brass, and 70,000 lbs. of hoop iron, were consumed during last year.

The miscellaneous articles in brass, made in Philadelphia, it will thus be perceived, are quite numerous; nevertheless, there certainly is room for very considerable extension of the manufacture. Brass Wire is not made, to our knowledge ; besides many other articles that form prominent items in the industry of Waterbury, Conn.   The locality we think especially worthy the attention of the enterprising.   Intermingled as Brass work is with Iron Founding, Gas Fitting, Plumbing, &c., scarcely a satisfactory approximation can be made to the annual product ; but the mean of estimates, by experienced men, gives a result of $830,000 per annum.

COPPER, like Brass, is applied to a great variety of purposes—its principal use being the manufacture of Brewing Coppers, Sugar Cans, Teaches, Clarifiers, Evaporators, and every article used for making or refining Sugar ; Stills for turpentine, alcohol, &c. ; Pumps, Dye Kettles, Mineral Water Apparatus, Bath Heaters, Drying Machines for manufacturers.   It is also used largely by Locomotive and Stationary Engine builders, and by Plumbers and Gas Fitters ; and for lining Bath Tubs, and as a base for

Tinning, &c. The department of Coppersmithing, in which the manufacturers of Philadelphia excel, is the production of heavy Copper work for Sugar Refiners and Sugar Planters. It is believed, that at least one of the establishments in this branch is unequaled in extent and quality of workmanship by any similar one in the Union. The annual product in Copper, in Philadelphia, is about $400,000 per annum.

VII.
Brewing—Ale, Porter, and Lager Beer.
ALE AND PORTER.

"Beer," says the author of the "Picture of Philadelphia, in 1811," whom we have before quoted, "was brewed in Philadelphia for several years before the Revolutionary war; and soon after peace, the more substantial Porter was made by the late Mr. Robert Hare. Until within three or four years the consumption of that article has greatly increased, and is now the table-drink of every family in easy circumstances. The quality of it is truly excellent: to say that it is equal to any of London, the usual standard of excellence, would undervalue it, because, as it regards wholesome qualities and palatableness, it is much superior; no other ingredients entering into the composition than malt, hops, and pure water. A fair experiment has shown, that even so far back as 1790, Philadelphia Porter bore the warm climate of Calcutta, and came back uninjured. In 1807, orders were given by the merchants of Calcutta, after tasting some of it taken out as stores, for sixty hogsheads. Within a few years Pale Ale of the first quality was brewed, and justly esteemed—being light, sprightly, and free from that bitterness which distinguishes Porter."

The reputation of Philadelphia Ale has but strengthened with the lapse of years; and at the present time the Malt liquors made in Philadelphia take precedence in every market in the Union. The qualities for which they are distinguished are purity, brilliancy of color, richness of flavor, and non-liability to deterioration in warm countries—qualities, the result in part of the peculiar characteristics of the Schuylkill water—in part of the intelligence, care and experience of our brewers, conjoined to the use of apparatus possessing all the best modern improvements made in England and in this country.

The following is an outline of the processes adopted in the manufacture :—

Preparatory to the process of Brewing, the barley is converted into malt. This method consists of four processes, viz.: steeping, couching, flooring, and kiln-drying. Great care is taken, and no expense is spared, to secure the best grain from this and the adjoining States. The grain is first steeped in water contained in wooden or stone cisterns; the water being frequently drawn off and a fresh quantity supplied, to cleanse the grain. When sufficiently saturated to admit of its being crushed between the thumb and finger, it is then drained of the water, and spread over a cement floor to the depth of six or eight inches, and left, with occasional turning, until it sprouts.

In the process of germination, a peculiar azotized substance is evolved, called diastase, which acts as a powerful agent in converting starch into dextrine, and ultimately into saccharine. The maltster continues to turn the barley, at intervals, so as to produce a uniform growth, upon the floors. When the barley has sufficiently sprouted, a stage determined by the sweet taste and the chalky appearance of the inside of the grain, it is dried rapidly, in order to retain the starchy matter, which, in a long growth of the sprouts and rootlets, would be wasted. This drying is done in kilns; here the heat destroys the germ of the grain, expels the moisture, and converts it into a sweet and friable grain called malt. It is then passed through a cylindrical sieve, separating it from all stones, beans, straws, &c. ; and subsequently crushed by rollers. When the brewing is commenced, the ground malt is conducted into a large vat, infused in heated water, and thoroughly mixed by a machine adapted for the purpose : there it remains at rest until the starch is converted into sugar, and then drained into boiling coppers, additional water being sprinkled upon the grain until the saccharine is extracted, which is ascertained by an instrument called the Saccharometer. In these boiling coppers the clear extract, or wort, is boiled with hops, for the purpose of imparting to it an aromatic, bitter flavor, and the property of keeping without injury. This accomplished, it is drained into shallow vessels,

15*

and cooled (by an apparatus called a Refrigerator), to the temperature at which the brewer desires the fermentation to commence. Thence it is conducted into a vat, and mixed with yeast of a previous brewing, where the fermentation is carried on. This process continues from three to five days, during which the temperature of the fermenting body rises, and a rapid disengagement of carbonic acid takes place. To prevent the creation of too high a temperature, which would cause acidity of the worts, it is racked off from the fermenting vats into puncheons of one hundred and twenty to one hundred and fifty gallons capacity, where it purges itself of its yeast. The fermentation being now completed, and the Ale or Porter perfectly clear, the sediment or yeast remaining settles at the bottom ; it is racked off from the puncheon into casks of convenient size for use, or stored in large cedar vats for future consumption.

There are now nine extensive Brewers of Ale and Porter in Philadelphia, viz. : MASSEY, COLLINS & CO., FREDERICK GAUL, ROBERT SMITH, W. C. RUDMAN, ROBERT NEWLIN, GRAY & STALEY, DITHMAR & BUTZ, W. B. TAYLOR, and JAMES MOORE. The oldest Brewery in this city is probably that upon the corner of Sixth and Carpenter streets, which was built about one hundred years ago, by William Gray, a native of Philadelphia. The most noteworthy Brewery is probably that belonging to MASSEY, COLLINS & CO., situated at the northwest corner of Tenth and Filbert sts. ; it was originally erected by the farmers of Chester and Delaware counties, Pa., and purchased from them by the Brewers' Association of Philadelphia ; they subsequently sold the establishment to M. L. Dawson, a member of the Association, and whose ancestors had been prominent Brewers for a period of eighty years. Poultney & Massey, the predecessors of the present firm, in the year 1855, greatly enlarged the buildings, which have recently been increased by the present owners. The buildings, as now erected, form a hollow square of one hundred and fifty feet each way, making an extent of buildings of six hundred feet, seven stories in height, with extensive cellars and vaults underneath the whole, eighteen feet in depth, which are furnished with large vats containing from two hundred to four hundred barrels each, and sufficient for the storage of ten thousand barrels of

Ale and Porter. Their Brewing Apparatus has been put up within the past three years, of the latest and most approved description ; comprising large Mash Tubs, capable of brewing nine hundred bushels of malt daily ; boiling Coppers heated by means of steam-pipes ; large Coolers, and Refrigerators, and Fermenting Tuns, the capacity of the latter being forty-five thousand gallons. Attached to the Brewery are malt-houses, which are designed for the malting of one hundred thousand bushels of barley. From seventy-five to one hundred men are employed about the establishment. The firm is extensively engaged in the manufacture of Pale and Amber Ales, and Porter, for draught and bottling ; Brown Stout, and XX Ale, for all the markets upon the coast, from Maine to Louisiana, also for the numerous markets of the West Indies and South America.

The greatest cleanliness is required in this establishment ; every cask returned to the Brewery being unheaded, scalded, and scrubbed with hickory brooms by hand ; and lime is used frequently to purify the utensils.

The capital invested in the Brewing of Ale and Porter is $1,500,000, and the annual product exceeds one million of dollars.

2. LAGER BEER.

The manufacture of Lager Beer was introduced into this country about eighteen years ago, from Bavaria, where the process of brewing it was kept secret for a long period. Its reception was not a very cordial or welcome one ; and about twelve years elapsed before its use became at all general. Within the last few years, however, the consumption has increased so enormously, not merely among the German population, but among the natives, that its manufacture forms an important item of productive industry. The superior quality of that made in Philadelphia has, no doubt, increased the demand, and by diminishing to some extent the use of fiery liquor, has effected partial good.* Lager

---

* The following report by " Our Reporter," contains some important facts.

"SIR : You entrusted the investigation of the Lager Beer manufacture to one who wants every essential qualification for the task. I can neither speak German, eat Sauerkraut, nor drink Lager. Before undertaking the commission, I wished to ascertain for my own satisfaction, without practical

signifies "kept," or "on hand;" and Lager Beer is equivalent to "beer in store." It can be made from the same cereals from which other malt liquors are made; but barley is the grain generally used in this country. The processes resemble those of brewing Ale and Porter, with some points of difference, and the brewing generally forms a separate and distinct business.

experiment, whether Lager Beer will intoxicate. I procured the evidence before the King's County Circuit Court (Brooklyn), and the following synopsis of the testimony on the part of the defense satisfied me, at least, if not the Jury. One German testified, 'that he had on one occasion drank fifteen pint glasses before breakfast in order to give him an appetite.' Another, Mr. Philip Kock, testified that ' once, upon a bet, he drank a keg of Lager Beer, containing seven and a half gallons, or thirty quarts, within two hours, and felt no intoxicating effects afterward. He frequently drank sixty, seventy, eighty, and ninety pint glasses in a day—did it as a usual thing when he was "flush." ' Others testified to drinking from twenty to fifty glasses in a day. One witness testified to seeing a man drink one hundred and sixty pint glasses in a sitting of three or four hours, and walked straight. Dr. James R. Chilton, chemist, testified to analyzing Lager Beer, and found it to contain three and three quarters to four per cent. of alcohol, and did not think it would intoxicate unless drank in extraordinary quantities. 'He had analyzed cider and found it to contain nine per cent. alcohol; claret, thirteen per cent.; brandy, fifty per cent.; Madeira wine, twenty per cent.; and Sherry wine, eighteen per cent.'

" Lager Beer was first introduced into Philadelphia in 1840, by a Mr. Wagner, who afterward left the city. It was a lighter article than that now used. The first who made the real Lager was Geo. Manger, better known as ' Big George,' who, in October, 1844, had a small kettle in one corner of the premises still occupied by him in New street, above Second. The beer used in the winter is lighter, and may be drawn five or six weeks after brewing; but the real Lager is made in cold weather, has a greater body—that is, more malt and hops are used—and is first drawn about the first of May. It is much improved by age and by keeping in a cool place. When first drawn it is five months old; and as it is usually made in December, it is ten months old when the last is drawn. The vaults are probably the most interesting 'sights' connected with the business. The firm that constructed the first vault is that of ENGEL & WOLF—a firm that ranks among the most extensive, accommodating, and enterprising of our brewers. The vaults are built in the vicinity of Lemon Hill, near the Schuylkill, and consist of solid stone exterior walls. These are subdivided by brick partitions into cellars or vaults of about twenty by forty feet, and communicate with each other by a door large enough to admit a puncheon; in this

There are now about thirty brewers of Lager Beer in Philadelphia, having a capital employed of $1,200,000.

The Statistics of the entire Brewing business in Philadelphia, for 1857, are as follows :—

Product.

Ale, Porter, and Brown Stout, 170,000 barrels, averaging $6,	-	$1,020,000
Lager Beer, 180,000, " " $6,	-	1,080,000
Other Beer, say - - - - - - - - - -	200,000
Total, - - - - - - - - -			$2,300,000

Raw Material consumed, viz. :

Barley or Malt, 750,000 bushels, at $1.40,	-	-	-	1,050,000
Hops, 800,000 lbs, at 15 cents,	-	-	-	-	120,000
Total, - - - - - - - -		$1,170,000

The capital invested in Ale, Porter, and Lager Beer brewing, including *Malting,* is $3,050,000; being, it will be perceived, a

is a smaller door or aperture, about two feet square, barely sufficient to allow the passage of a keg.

"After the brewing has commenced, say in December, unless cold weather occur earlier, the most remote cellar or vault is filled—the ground tier, consisting of large casks, usually three rows, is placed on skids or sleepers perhaps a foot from the ground, the rows far enough apart to permit a man to walk between. On these two rows of casks are placed; and above these, if the vault is high enough, one row of smaller casks or kegs are stowed. The other vaults are filled in like manner. After each is filled, the door is closed, and straw, tan, and other non-conductors are placed to keep out the external heated air of summer. The vaults are ventilated, and the temperature kept as low as possible. Should it exceed 8° Reamur, or 50 Fahrenheit, the beer spoils. One only is opened at a time.

"Messrs. Engel & Wolf, before referred to, have seven vaults, in five of which 50,350 cubic feet were cut out of solid rock. The bottom of the vault is about forty-five feet below ground. This firm have an agency in New Orleans, and sell to nearly all the South, including Texas.

"One of the peculiarities of Lager Beer is the flavor imparted to it by the casks. The casks, previous to use, have their interior completely coated with resin ; this is done by pouring a quantity of melted resin into the cask while the head is out, and igniting it. After it has been in a blaze for a few minutes, the head is put in again, which extinguishes the blaze, but the resin still remains hot and liquid; the cask is then rolled about, so as to coat every part of the interior with it; any resin remaining fluid is poured out through the bung-hole. This resin imparts some of its pitchy flavor to the beer."

larger amount in proportion to the product than probably in any other business.   This arises from the necessity of occupying large plots of very valuable ground, from the extent of the buildings, and from the great number of vats and casks required.   The casks alone, exclusive of vats, in use by Philadelphia brewers, cost $320,000.

VIII.

Bricks, Fire-Bricks, Pottery, &c.

The objects of which Clay is the principal raw material are exceedingly varied in their uses, as well as in appearance; and range from the least ornamental to nearly the highest in the department of Art—from Bricks to Porcelain, from a Clay Furnace to a Terra-Cotta Vase.   Commencing with the least artistic, though the most important, judging from the extent of the manufacture, we are led first to the consideration of Bricks.

1. BRICKS.   The manufacture of Bricks, in Philadelphia, is carried on, like many other important branches of industry, mainly by individual enterprise, the business expanding or contracting according to the current demand, without much concert of action between the producers, and without any very large establishments, at least compared with those in Vienna, or even in Massachusetts.*   The statistics of the trade are given, and its present

---

* Vienna has the honor of containing, undoubtedly, the largest and most remarkable establishment for Brick-making in the world.   The description states that the main factory, occupying a space of ground of two hundred and sixty-four and three-fourths English acres, has twenty-four thousand nine hundred and thirty feet in length of drying sheds, for the manufacture of ordinary Bricks, and eight thousand three hundred and four feet of moulding sheds, for the manufacture of Tiles and facing and ornamental Bricks; besides forty-three kilns, calculated to burn forty-five thousand to one hundred and ten thousand Bricks per kiln, or to burn at one time *three millions five hundred thousand.*   There are in connection with this establishment, infant schools for one hundred and twenty children, a hospital with fifty-two beds, a tool workshop, a wheelwright and carpenter shop, and great watering and kneading pits for red and white ornamental Bricks. Besides this, the proprietor, Mr. MIESBACK, has six other factories in the immediate vicinity, and provided in the same proportion.   In 1851, he supplied *twenty millions of Bricks* for the great tunnel through the Som-

condition is sketched in the following report from a practical Brickmaker; and his conclusions, having been submitted to others familiar with the subject, are approved.

"As nearly as I can ascertain, there are about fifty Brick Yards within the limits of the consolidated city—say twenty-five in the south end, and the balance in the north end, including Germantown, and across the Schuylkill. Those in the southern part of the city will average two and a half millions of Bricks a year each; but the fair average for the whole would be, I think, two millions per year. About thirty hands are employed in each yard—the men's wages ranging from $26 to $60 per month; and the boys' wages, of whom there are six to eight in each yard, are from $15 to $20 per month. The prices of common Bricks range from $6 to $10 per thousand, and pressed Brick from $13 to $18. It takes one third of a cord of wood to burn one thousand Bricks, and wood is worth from $5 to $6 per cord. The capital invested in each yard is from $8,000 to $10,000.

"There are few, if any, now made here by machinery. Our Clay is not adapted for machinery. In Washington, where great quantities of Bricks are made by Brick machines, they do better; but Bricks thus made are never equal in quality to hand-made Brick, which bring in the market $1 per thousand more, and this is about equal to the difference in cost. As to *quality*, Philadelphia Bricks rank as the best made in the country, and those of Baltimore next. Philadelphia has better Sand and Clay, which gives to the Bricks a better color than those produced elsewhere. Yet Baltimore Bricks bring in New York a little higher price than ours, because the Baltimore Clay being purer, and therefore stronger, stands more burning, which renders the Brick harder, and able to bear transportation with less breakage and damage. But they do not look near so well as those of Philadelphia."

There are at least four yards in Philadelphia that produce *five* millions of Bricks per annum each, and I am therefore disposed to

ering, on the Austrian railway; and filled another contract for *forty millions* for public works in Vienna; and these were merely additions to the ordinary make. Number of persons employed in the establishment, *two thousand eight hundred and ninety.*

The largest Brick-making establishment in the United States is supposed to be that located in North Cambridge, Mass. When in full operation it produces, on an average, one hundred and eighty-seven thousand Bricks per day or about *twenty-four millions* during the season. The clay is taken from a pit which is about forty feet deep, and elevated in a car on an inclined plane by steam power. The shafting reaches a quarter of a mile.

regard the average stated above as a low one; but assuming it to be correct, the result is that about one hundred millions of common Bricks, worth about $700,000, are produced annually. In addition, there are about eight millions of fine-pressed Bricks made, worth say $14 per thousand, or $112,000; and the total product is $812,000.

The pressed Bricks of Philadelphia have a deservedly high and extended reputation. One firm, during the last year, exported to Cuba 200,000, and has now 150,000 on hand ready for shipment; and another maker sent, in 1856, *one million three hundred thousand* to New York city.

2. FIRE-BRICKS, &c. The use of Fire-Clay is comparatively of recent date, but has greatly increased within the last few years. It is now employed, not merely for Fire-Brick, but for Chemical Ware, Drain Pipes, Gas-house Tiles, &c.

Philadelphia has probably the first established Fire-Brick manufactory in the United States. The father of Mr. Abraham Miller, whose establishment is on Callowhill street, commenced the business, we are informed, nearly one hundred years ago.

The present Mr. Miller was the first manufacturer, we understand, of the Clay Furnace now so largely used in Summer. There are at least four establishments that use steam for grinding the Clay—GEORGE SWEENEY & CO., J. & T. HAIG, THE HAYWOOD FIRE BRICK and TILE COMPANY, and Messrs. NEWKUMET, and MELICK. Messrs. Sweeney & Co., 1310 Ridge Road, are extensively engaged in making Fire-Brick, Stove Linings, Cylinders, and Bakers' Tile. Messrs. Haig make, besides common ware, Stone-ware, Chemical-ware, Crucibles, &c. The Haywood Company have a very extensive establishment about 2½ miles north of Richmond, owning an extensive bed of superior Clay, and have machines that will produce five hundred pieces of drain pipe, two feet long and four inches in diameter, in an hour. The concern has produced some tubular or hollow Brick, which is now extensively manufactured, we are informed, in England, and by which it is said a saving in brick-work may be effected of twenty-five to thirty per cent. on the cost, with a reduction of twenty-five per cent. in the quantity of mortar, and a similar saving in labor; besides promoting ventilation and freedom from dampness.

Messrs. NEWKUMET & MELICK, the other firm referred to, are extensive manufacturers of Fire-Brick, Gas-house Tiles, &c. This firm has peculiar advantages—one of the partners, Mr. Melick, being part owner of the celebrated Fire-Brick Clay deposit, at Woodbridge, New Jersey, whence the best material is obtained. They have two kilns, employ thirty men, and their works have a capacity for turning out a product of $50,000 per annum. Gas-house Tiles are made by them to suit all the different plans in use, and of a quality superior, as they claim, to any in the United States. Extra nine-inch Fire-Brick are also produced, equal to the best English Bricks. Though but recently established, they have supplied large orders from Cuba and different parts of the United States, and have every requisite facility for filling expeditiously any special demand.

The Pottery art is carried on by several in the city—sometimes in conjunction with the manufacture of Fire-Brick, and by others as a distinct business. Earthenware of all the ordinary description, including Chemical-ware, is made by MORO PHILLIPS, at his factory, in West Philadelphia; and Stoneware Jars, Jugs, Beer Bottles, Ink Bottles, and Stone Pipe for heated air, &c., are made extensively by N. SPENCER THOMAS, at a factory adjoining his Chemical works.

The *General Manufactures in Clay* include, besides those above-mentioned, China-ware, Artificial Stone, Architectural Decorations, Cements, Plasters, Terra-Cotta, Scagliola, Mosaics, Paving Tiles, Roofing Tiles, Draining Tiles, and Drain Pipes, Smoking Pipes, &c. All of these, with the exception perhaps of Mosaics, are made at least to some extent in Philadelphia; Tiles, Pipes, &c., are generally made in connection with the manufacture of Fire-Bricks, and have been already referred to. *Terra-Cotta ware*, as Chimney Tops, Garden and Hanging Vases, Caps and Brackets for churches and private dwellings, glazed Heating Pipe, is made at several establishments, and to an amount exceeding in the aggregate $100,000 per annum. The manufacturers claim to make Terra-Cotta equal to the imported, and at much less cost.*

---

* The manufacture of Terra-Cotta ware requires a Clay of great purity,

16

The "Gloucester China Company," having an authorized capital of $200,000, has made ware possessing the qualities of being not only semi-transparent, but very strong. The articles are such as are required in every household, and the product compares favorably with the European. Decorating Porcelain and China-ware, which had been imported plain, is done in one establishment, to an amount exceeding $75,000 per annum. Of *Calcined Plaster* about sixteen thousand barrels are made yearly, and consumed principally in Stucco work, Architectural Decorations, and in the manufacture of figures, in which a large business is done by Italians. *White Clay Smoking Pipes*, of all lengths, are made here at least by one person, and of very good quality. He has recently sent to England to procure additional assistance, and this branch of fictile manufactures, now very small, will probably soon be extended.

The miscellaneous manufactures in Clay, of which we have any account, and including Fire Bricks, which amount to nearly one half the sum, furnish an aggregate product of $647,000.

IX.

Carriages.

"Comparing the state of the art of Carriage building," say the London Jurors, in their report on Carriages exhibited at the World's Fair, "of former and not very distant times, with that of the present, we consider the principles of building in many respects greatly improved, and particularly with reference to lightness, and a due regard to strength, which is evident in Carriages of British make; and especially displayed in those contributed by the United States, where there is commonly employed in the construction of wheels, and other parts requiring strength and lightness combined, a native wood (upland hickory), which is admirably adapted to the purpose. The Carriages from the

---

resembling that used for Pipe-making and Potter's-ware, containing but little iron, and made up with a quantity of crushed pottery and calcined flint; the whole being well mixed, and burnt to a very high heat. It thus approaches in its nature to what is called Stone-ware; but the fusion of the material is not effected.

Continental states do not exhibit this useful feature in an equal degree."

Comparing the state of Carriage Building in various cities, states, and countries, it will be found by those who make the comparison, that in the art of constructing light Carriages, particularly with reference to combining lightness with strength, and attaining durability in conjunction with beauty of appearance and high finish, no builders, either in this country or in Europe, have been so uniformly successful as some in Philadelphia. The quality of Philadelphia Carriages is indisputably superior. It is true that here, as elsewhere, there are carriages made, like Peter Pindar's razors, to sell; but we fearlessly claim that the general quality is above the ordinary average, and that those who desire a perfect vehicle, will be likely to attain a nearer approximation to perfection in this city than they can anywhere else. The first-class builders have studied, and know to exactness, the proper proportions of every part of a vehicle, and never use more nor less material than is required. They risk nothing to make it light; nor add any unnecessary weight to give it strength. All materials that are in the slightest defective, or that interfere in the least with the purposes of a good Carriage, are promptly rejected. The leading and important parts of a Carriage, as the wheels, axles, &c., are generally made on the same premises, and rigid supervision exercised over every part in the construction. The prominent builders have attained a high and wide-spread reputation, entirely too valuable to themselves to be risked lightly through carelessness, neglect, or indifference. In addition to unremitting vigilance, combined with long experience, their efforts are greatly aided and facilitated by having at hand the very best materials, and in being able at all times to command the very best workmen. The growth of hickory, and oak, and ash, in the vicinity of Philadelphia, is so superior for Carriage purposes, that we might say without exaggeration, that no first-class vehicle can be built without coming to this vicinity for the materials.

There are now thirty establishments, great and small, within the limits of Philadelphia, that make pleasure Carriages. They have a capital invested of about $500,000—employ on an average eight hundred hands, all males, and turn out an average annual

product of $900,000. About two thirds of the Carriages made
are for use in the city, and its vicinity—the remainder being exported to the South and West, the West India Islands, some to
New England, and a few light vehicles are sent every year to
Europe. In many portions of Europe, Philadelphia Carriages
have excited great attention, as wonderful specimens of combined
strength and lightness; and orders from London, Paris, and
other European capitals, are now far more frequent and important than formerly, manifesting a growing appreciation of American superiority in this branch of manufactures. It will be remembered, that to a Philadelphia Carriage a prize-medal was
awarded at the World's Fair.

In the construction of Carriages, the builders of Philadelphia
either adhere to what has received the sanction of experience, or
originate improvements for themselves; very few, if any, of the
many hundreds of patent rights for improvements in Carriages,
that are now on file in the Patent-office, being adopted by any,
and none in general use. The designs of light Carriages have
been made by Philadelphians, perhaps to a greater extent than
by any others. The "Germantown Wagon," which satisfies the
demand for a light, strong, convenient, yet cheap vehicle, had its
origin here, and its advantages are every year becoming more
generally appreciated. These vehicles are constructed to hold
four, six, or eight persons, and lose nothing in elegance and style
by increase in size. Trotting Wagons, weighing not more than
eighty-five pounds, are probably an invention of our builders, for
no one, we should think, would venture upon the experiment except those who could command the white hickory of Montgomery.

The prices of Carriages made in this city vary of course with
the style, quality, and degree of ornamentation. Those for Pres-
idents, Postmaster-Generals, and "such like folk," are costly in
proportion to their elegance; while those that are made for auction sales are so cheap, that New England dealers frequently
purchase large lots to resell, being able to obtain a better article
at a less price than at Bridgeport or New Haven. One maker,
whose product in 1856 was eighteen Coaches and one thousand
one hundred and eighty-two light Carriages, sold about $10,000
worth to New England. Economy in manufacturing is largely

promoted in Philadelphia, by the fact, that the principal constit-
uent parts of a Carriage, from the raw materials of which they are
composed to the finished product—the bolts, screws, springs, hubs,
axles, as well as iron, steel, and wood—are all made in this city,
with every facility for making them economically, and on a large
scale.*

The extent of some of the establishments, and the method of
construction, will be best illustrated by a detailed description of
a first-class establishment. (See APPENDIX.) The varieties of
Carriages made, include every description that purchasers may imagine or can desire, though light Carriages form the bulk of the manufacture. One firm makes the construction of *Private* and *Hackney
Coaches* a principal feature in their business, and have been very
successful in satisfying the requirements of good taste. *Omni-
busses* are made by one or two of the Omnibus proprietors for
their own use; and by one maker for sale, who, however, in consequence of the introduction of Passenger Railways, apprehends that
his "occupation's gone." It is manifest, however, that the manufacture of public Carriages has not attained its proper development; and upon inquiry as to the causes, we are informed, that
our citizens have not given this branch that encouragement, preference, and patronage to which its intrinsic importance entitles
it. It should be remembered that the Carriage manufacture is
an important interest, wide-reaching in its ramifications, affecting the prosperity of an immense number of trades : the various manufactures of iron—rolled, wrought, and cast; of steel
springs, nails, screws, bolts; oil and other cloths, patent leathers, paints, varnishes, and glue; of wheel-makers, painters,
blacksmiths, trimmers, carvers, silver-platers, wood-turners, the
makers of tools for all these, and others. Every purchase then
of a Carriage abroad is a discouragement of industry at home.

For other vehicles, see WAGONS, CARTS, &c.

---

* We are informed that Patent or Japanned Leather is not made to any
extent in Philadelphia; and that a manufactory of the kind would be well
supported, and could not fail to do a successful business, if properly
managed.

16*

X.

Chemicals, Paints, Glue, &c.

The manufacture of Chemicals, in the United States, may be said to date from the war of 1812. The commercial restrictions which preceded that war caused such a scarcity and dearness of Chemicals, that the preparation of the more prominent articles offered an attractive field for enterprise. Previous to that period, however, a Philadelphian had established successfully a manufactory of Sulphuric acid. This was Mr. John Harrison, the first successful manufacturer of oil of vitriol in the United States, and the founder of the well-known house of Harrison Brothers. He had spent two years in Europe in acquainting himself, as far as he could gain access to them, with the processes used by the chemists ; and after his return to America devoted himself to the manufacturing of Chemicals. How much earlier he succeeded, we have no means of ascertaining, but in 1806, he was fully established as a manufacturer of oil of vitriol and other Chemicals, in Green street, above Third. His leaden chamber was a small one, and capable of making about forty-five thousand pounds, or three hundred carboys of oil of vitriol per annum. So successful were these operations, that in 1807 he had built a leaden chamber eighteen feet high and wide, and fifty feet long, capable of making three thousand five hundred carboys per annum. The price which the acid then brought was fifteen cents per pound.

The application of Platinum to the concentration of sulphuric acid, was also first attempted in Philadelphia by Dr. Erick Bollman, who had distinguished himself by a gallant and all but successful attempt, in company with Francis K. Huger, of South Carolina, to rescue General Lafayette from his guards, during his imprisonment at Olmutz. Dr. Bollman was a Dane, a man of powerful and versatile mind, a physician, a chemist, a political economist, and a general scholar. Among other pursuits, he had turned his attention to the working of crude platinum, of which there was a considerable quantity in this country, and for which there was no demand. He had brought from France the method then lately discovered by Dr. Wollaston, for converting the crude

Samuel Sartain, Engraver.

Powers & Weightman. Manufacturing Chemists. Philadelphia.

grains into bars and sheets; and, in 1813, he had wrought it into masses, weighing upward of two pounds, and into sheets more than thirteen inches square. One of the first uses to which he applied these sheets was the making of a platinum still for John Harrison, for the concentration of his oil of vitriol. This still weighed seven hundred ounces, contained twenty-five gallons, and continued in use fifteen years.

This early application of Platinum to the concentration of sulphuric acid is highly creditable to the American manufacturer, for its use for this purpose was then a novelty in Europe.

Charles Lennig was the first Philadelphian who largely manufactured oil of vitriol by putting up extensive leaden chambers, and concentrating the acid in platinum vessels so arranged as to be kept constantly at work, while discharging a steady stream of concentrated acid.

At the present time Philadelphia contains the most extensive Chemical manufactories in the United States. Messrs. POWERS & WEIGHTMAN, for instance, are among the largest manufacturing Chemists in the world. They have two establishments—one at the Falls of Schuylkill, where they make oil of vitriol, aquafortis, nitric and muriatic acids, Epsom salts, copperas, blue vitriol, and alum, all on a large scale. At their establishment at the corner of Ninth and Parrish streets, Philadelphia, they manufacture sulphate of quinine, which is their staple article; mercurials, morphias, and Medicinal Chemicals generally. (See APPENDIX.)

Their Chemicals have an enviable reputation for purity, exactness, and beauty; and the firm is well-known for its liberality, fairness, and reliability. *The house was founded about forty years ago* by two intelligent foreigners, Abraham Kunzi and John Farr; and the reputation acquired by them has been maintained, and, if possible, increased by the present proprietors.

NICHOLAS LENNIG & Co. are believed to be the most extensive manufacturers of oil of vitriol in Philadelphia. Their works are at Bridesburg, and occupy over twelve acres of ground. Their list of manufactures includes soda-ash, alum, copperas, aquafortis, nitric and muriatic acids; all the various preparations of tin for the use of dyers, such as tin crystals, oxymuriate of tin, pink salt, &c.

HARRISON BROTHERS & Co. make white and red lead, litharge, and orange mineral, oxide of zinc, white and brown sugar of lead, alum, copperas, oil of vitriol, aquafortis, muriatic acid, iron liquor, red liquor, &c., &c. To the founder of this house we have already referred, the works being commenced and erected in 1807. The productions of this house enjoy a high character for purity and genuineness.

ROSENGARTEN & SONS, formerly Rosengarten & Denis, are largely engaged in the manufacture of sulphate of quinine, and other pharmaceutical preparations. This house was established in 1823, and was among the first to manufacture the valuable vegetable alkaloids in this country. Their laboratory is well-known, and is one of the most important in the United States.

BUCK, SIMONIN & Co. are the successors of Wm. Coffin & Co., in the manufacture of copperas, metallic nickel, and the oxide of cobalt, so highly prized in painting porcelain and queensware. These are of very great importance in the arts, and the last has not heretofore been manufactured to any extent in this country. We are informed, that all Messrs. Buck, Simonin & Co. make of this article is exported to England, for use in the porcelain man-ufactories. This firm are also extensive manufacturers of bichro-mate of potash by a superior process, patented both in this country and in England.

SAMUEL GRANT, JR., & Co. have extensive Chemical works at Manayunk, where they make muriatic, nitric, and numerous other acids; aquafortis, bleaching salts in large quantities, sugar of lead, soda-ash, and various articles used by dyers and printers, to which we will subsequently refer.

POTTS & KLETT manufacture oil of vitriol, muriatic and nitric acid, Paris, Prussian and soluble blues, pulp lakes and sienna, paper-makers' and paper-stainers' colors generally.

BURGIN & SONS are extensive manufacturers of bicarbonate of soda, sal soda, soda saleratus, Rochelle salts, and Seidlitz mix-ture, &c.

MORO PHILLIPS, at the "Aramingo Chemical Works," makes oil of vitriol, aquafortis, nitric and muriatic acids, copperas, &c. Mr. Phillips has the contract for supplying the United States Mint, and its branches, excepting those at New York and San Francisco, with nitric and sulphuric acids.   His office in Philadelphia is at 27 North Front street.

SAVAGE & MARTIN, at their "Frankford Chemical Works," manufacture oil of vitriol, aquafortis, nitric and muriatic acids, aqua ammonia, nitrate of iron, muriate of tin, tin crystals, blue vitriol, &c.   Their office is at 18 North Front street.

There are several establishments in the city, engaged principally in making various preparations for coloring purposes, and have been successful in attaining excellence in a manufacture where excellence is rare.   The oldest Color establishment is that of CHARLES J. CREASE, who makes Prussian blues, chrome greens, chrome yellows and reds; and besides these, he makes nitric acid, aquafortis, muriatic acid, &c.

JOHN LUCAS & Co. also make Prussian and ultramarine blues, chrome yellows and reds, zinc greens, &c., both dry and in oil.

BREINIG, GATTMAN & BREINIG, at their works at Fairmount, make several of the chromes.

GEORGE W. OSBORNE & Co., 104 North Sixth street, manufacture Osborne's American Water Colors.

Daguerreotype and Photographic Chemicals are made extensively, and of a very superior quality, by GARRIGUES & MAGEE, 108 North Fifth street.   This firm give especial attention to the manufacture of these Chemicals—pure nitrate of silver, Becker's chloride of gold, collodion, gun cotton, also Becker's rotten-stone for polishing, &c., being leading productions.

BENJAMIN J. CREW & Co. make cyanide of potash, collodions, chemically pure acids, hypophosphites of soda, lime, potash, and ammonia.   The house was established by the senior partner as a manufactory of Photographic and Ambrotype Chemicals; but recently they have largely increased their laboratory and facilities for the manufacture of Artistical and Medicinal Chemicals generally.   No firm has been more successful in the production of the finer kinds.

HENNELL STEVENS & Co., a house recently established, give their attention particularly to the manufacture of fine and rare Chemicals, of undoubted purity.   Their list comprises over two hundred different Chemicals, many of them, they state, can be had of no other parties, and some of them they believe to be of great prospective importance. Within the last few years this firm has diligently experimented on crude *Glycerin* from soap-waste, with a view of rendering this available, as well as of bringing Glycerin, by lowering the price, into more general use in the arts; making it, for instance, a substitute for molasses in the formation of printing rollers, and facilitating its incorporation in printing paper, thereby rendering the latter always soft and pliable, and requiring no wetting before use. They have succeeded, as we learn from the Journal of Pharmacy, January, 1858, in producing from the concentrated fetid liquids of the soap-makers, by apparatus involving its distillation, Glycerin, almost tasteless and odorless, and equal to that of " Price's Candle Company," which, it is well-known, is made from pure Palm oil. A young firm, aiming, as Messrs. Stevens & Co. do, to check the importation of rare Chemical products, by manufacturing them of superior quality, is deserving of every possible encouragement.

Yellow Prussiate of Potash, so largely used for dyeing purposes and making Prussian blue, is made by CARTER & SCATTERGOOD, who are now the sole manufacturers. The annual production in Philadelphia was 400,000 lbs. per annum, worth, say 30 cents per pound.

HENRY BOWER, on Gray's Ferry Road, makes *sulphate of ammonia*, and a variety of Chemical products.

In these establishments, which represent a capital of two and half millions of dollars, much the larger proportion of the best Chemicals used in the United States are made. The factories, which are in many instances immense structures, are generally located out of the city proper—at Tacony, Bridesburg, Frankford, the Falls of Schuylkill, and some in or near Camden ; but the capital belongs to the city, and their products centre here as a point for redistribution. Some idea of their extent and importance may be derived from the fact, that they consume 2,400 tons of sulphur, 800,000 lbs. of saltpetre, 1,500 tons of salt ; and produce daily of sulphuric acid 45,000 lbs., or over 16,000,000 lbs. yearly ; of alum, 20,000 lbs. daily ; of muriatic acid, 15,000 lbs. ; of nitric acid, 8,000 lbs. ; of copperas, 15,000 lbs. daily ; of nitrate of silver, 150,000 ounces annually; besides the numerous preparations before enumerated, and used in the manufacturing arts and in medicine. The consumption of *Quinine* fluctuates of course with the state of health in the West; but it is said, that in one year 250,000 ounces were made in Philadelphia.

In addition to these, and probably other manufacturers of Chemicals, there are several manufacturing Chemists engaged in the preparation of Medicinal and Pharmaceutical preparations. N. SPENCER THOMAS, for instance, is extensively engaged in the manufacture of Medicinal extracts, conducting the evaporation in *vacuo*, by a very superior and perfect apparatus. His extracts, medical and fluid, are certainly remarkable for beauty, strength, and reliability; and, as he claims, not equaled by any in this country or in Europe. His vacuum apparatus is capable of making 100,000 lbs. of extracts per annum. In addition to these, he prepares also, in vacuo, the concentrated Eclectic medicines; and manufactures *blue mass*, mercurial ointment, glycerin, &c.; and prepares powdered drugs of very fine quality, by what is denominated the dusting process. His supply is always full and ample. E. H. HANCE is also engaged in the same business, and manufactures Extracts and Syrups to a considerable extent.

Many of the Apothecaries carry on, in addition to their regular business, the manufacture of a few select Chemicals.

THOMAS J. HUSBAND has for some years prepared what is known as " Husband's Calcined Magnesia," which has obtained a very considerable reputation, and is extensively used. In the Twentieth Report of the Franklin Institute, the judges of Chemicals assert, that this magnesia " is believed to be the best in the United States;" and some of the most distinguished professors and practitioners of medicine have pronounced it quite equal to the genuine Henry's magnesia.

CHARLES ELLIS & CO. are large manufacturers of Extract of Magnesia, which is extensively used in the United States. They are also large manufacturers of spread adhesive plaster, and roll plasters; and have a Laboratory in the southern part of the city, in which they make many of the officinal Chemicals, extracts, &c.

Some of the Wholesale Druggists prepare, with or without the sanction of the Medical Faculty, one or more domestic remedies, the popularity of which, in some instances, establishes a considerable manufacturing business. B. A. FAHNESTOCK & Co. might be referred to as illustrating this assertion; or GEORGE W. CARPENTER & Co., who compound a list of domestic remedies which are widely known, and have an extensive sale. This firm, however, are particularly distinguished for having provided a great depot of supplies for druggists and physicians; probably the greatest and most wonderful, for variety and comprehensiveness of stock, that this country affords. Every article pertaining to the business of a druggist or a physician, from the rarest Surgical instrument, or the most complete collection of Anatomical preparations, Chemical and Philosophical Implements and Apparatus for colleges, through the entire range of simple or prepared Drugs, Medicines, and Chemicals, to the minutest article required by either at the outset of their profession, not excepting shop-furniture, medicine chests, saddle-bags, medical text books, &c., may be found in this Chemical warehouse, which is a Drug emporium in itself.

The "Essence of Jamaica Ginger," prepared by FREDERICK BROWN, has almost entirely superseded the use of ginger-tea, and powder, so long regarded as popular remedies in domestic practice for various complaints of the stomach and digestive organs. This preparation is recognized and prescribed by the Medical Faculty, and has become a standard family medicine of the United States. The Chemists of late years have, in a great measure, overcome their professional aversion to prepared remedies, adapted to the various ills that flesh is heir to; and as those of this city have every advantage for procuring the recipes of the most celebrated physicians and medical professors, it is safe to infer that every preparation of the kind announced by a reputable established pharmaceutist of Philadelphia, possesses some considerable merit. About one third of the Apothecaries of the city of Philadelphia are members or graduates of the Philadelphia College of Pharmacy; and we do not fear to say that, as a body of men, intelligent and skillful in their profession, they are unsurpassed in any community; while we could name individuals among them who have few if any superiors in their profession, as regards scientific knowledge and practical skill, in any metropolis in Europe.

The business that, in connection with prepared prescriptions, approaches more closely to a manufacturing pursuit, and therefore, though denounced by the schools as irregular, is for our purposes the most regular—is the manufacture of what has been denominated PATENT MEDICINES. The individuals and firms engaged in this business are both enterprising themselves, and the promoters of enterprise in others. How many paper-mills, glass factories, printing and engraving offices, lithographic establishments, paper-box manufactories, &c., would be tenantless—how many journals that are now brilliant lights in the firmament of journalistic literature would have gone out, leaving the world in partial darkness, except for the material aid afforded through the popularity of Patent Medicines! When to these benefits we add another, viz., that the preparations in many instances are beneficial, and as respects almost all, entirely harmless, the manufacture would seem to be entitled to a larger share of respectful consideration than it has hitherto received.

Philadelphia, though it has not entirely escaped, has been preserved in a great measure from the visitation of those whose sole aim is to speculate on human distress. The remedies of the established firms have much weighty testimony in favor of their excellence; and the popularity, and consequent saleability of a few, are truly remarkable. The enterprise of at least one Philadelphia firm has made their preparations known, not only throughout this country, but in the islands of the Atlantic and Pacific oceans; Burmah, Siam, India; and almost every nationality in Europe. They expend annually over *one hundred thousand dollars* in advertising alone. They keep eight double-medium, and two single-medium, and eight steel-plate presses in operation throughout the year. Their consumption of printing paper, during the last year, was 14,000 reams, costing $39,782 96; and during the present year, they will print 2,600,000 Almanacs for gratuitous distribution. The rooms in the upper stories of an immense structure are occupied—one as a laboratory, another as a printing-office, a third as a binding and packing-room, and a fourth as a pill manufactory.*

* In the last-mentioned room we saw pills arranged in pyramidical form, dry, sufficient, one would think, to physic "all creation," with some to

17

About eighty persons are furnished constant employment in that establishment. For eight months of the year the expenditure of the firm referred to, for *postage*, is $25 per day. Wherever a few backwoodsmen have reared their lonely cabins, an agency for these preparations is established; and so remote and isolated are some of the frontier posts, that a box shipped hence cannot reach its destination in a year.

The total annual sales of all the Patent Medicines—bitters, syrups, cattle powders, &c., made in Philadelphia, cannot be ascertained; but it is the opinion of half a dozen of the principal manufacturers, that they might safely be stated at one million of dollars, net prices. At "long prices," the basis on which statistical statements are made in neighboring cities—the sum would be doubled.

The preparation of *Dye Stuffs* is made a specialty, or at least a prominent branch of their general business, by several manufacturers, viz.: BROWNING & BROTHERS, SAMUEL GRANT, JR. & CO., and J. M. SHARPLESS.

BROWNING & BROTHERS are the proprietors of the well-known "Aroma Mills"—a stamp which, on Extracts of Dye-Woods, is everywhere recognized as an assurance of excellence. This firm are also manufacturers of Paints, in the preparation of which they state they use only the pure linseed oil, and are careful to have them faithfully and finely ground.

SAMUEL GRANT, JR., & CO. have very extensive Chemical Works at Manayunk, occupying seven acres, where they manufacture the Chemicals beforementioned; and in addition, prepare Dye-Woods largely, ground, chipped, and extracts, and every article used by dyers. They manufacture several products that are not made elsewhere, it is believed, in the country, and are con-

---

spare, for the inhabitants of the planetary systems. Pill Machines, we are told, have not as yet been found to perform satisfactorily; and Pills are made by passing the prepared material, which is in long strips, through grooved rollers, with much the same hand-motion as women roll dough into cakes. The motion, we presume, is precisely the same when Bread Pills are made.

tinually adding new ones to their list; as for instance, Gelp salts
made from Indigo.

They recently engaged in the manufacture of liquid chloride
of lime, used by paper-makers and bleachers; surrogate of alkali, used in the place of soda-ash for cleansing wool; silicate of
soda, used by calico-printers; and muriate of manganese, a mor-
dant, which is used in dyeing cotton and wool *together*, instead
of separately, as previously done. For these preparations, as well
as for the machinery for making them, a patent is applied for by
Mr. Prentiss, one of the firm, who is known already as the patentee of a lubricating oil. The store of Messrs. Grant & Co. is
at 139 South Water street.

J. M. SHARPLESS makes the usual Extracts of logwood, fustic,
and quercitron; and also grinds and chips the same, and other
Dye-Woods.

J. ANDREYKOVICZ, a Polish Chemist, located at 28 Franklin
Place, makes Extracts of Indigo, distinguished as Indigo Paste
and Carmine; and is prepared to make a new dye-stuff known as
Archill.

There are other mills that, in addition to grinding Dye-Woods,
or disconnected therefrom, are engaged in grinding, powdering,
and refining Drugs. The oldest is that of Charles V. Hagner.
The mills of CHARLES VANHORN & Co., one of the principal firms in
this branch, were twice destroyed by fire—in 1852, and again in
1856; but since their last destruction they have been greatly extended and improved. They have now a capacity for producing,
and frequently do produce weekly, 6,000 lbs. of Drugs, 36,000 lbs.
of Spices, 14,000 lbs. of Founder's Facings, and 35 tons of Dye-
Woods.

WHITE-LEAD—PAINTS.

The production of Paints, particularly of the Salts of Lead,
which enter so largely into their manufacture, has added greatly to
the Chemical and Manufacturing reputation of Philadelphia.

Of White Lead there are four manufactories, viz., those of WETH-
ERILL & BROTHER, JOHN T. LEWIS & BROTHERS, HARRISON BRO-
THERS & Co., and E. DAVIS & RIGGS. The works of Messrs. Wetherill & Brother were established during or before the Revolution,
by the grand-father of the present proprietors, who, it is said, intro-

duced the manufacture into the United States. They are situated on the west side of the Schuylkill, employ a steam-engine of eighty-horse power, and consume daily 18,000 lbs. of Pig Lead. The article manufactured by this firm has always maintained a high reputation; and is sent to every part of the United States, and exported to the West Indies. JOHN T. LEWIS & BROTHERS are the successors of Mordecai Lewis & Co., who founded the works in 1819. At the period of their establishment Pig Lead cost $7\frac{1}{2}$ cents per pound, and White Lead sold for 15 cents. During the last year the raw material cost within one cent of the price above-named, and the manufactured article sold for $8\frac{1}{2}$ cents. The present firm have nearly a half million of dollars invested in the manufacture, and produce annually about 4,500,000 lbs. of White Lead, besides Oils, &c. Messrs. HARRISON, BROTHERS & CO.'s establishment dates from 1812; to them we previously referred. Messrs. E. DAVIS & RIGGS are also a well-known firm.

The capital invested in this business is nearly $1,000,000, and the annual product $960,000. It is to be regretted, that nearly all the raw material used is imported—English and Spanish Lead being principally employed—but it is gratifying to know that the American manufacturers, particularly those of Philadelphia, have effectually succeeded in stopping the importation of the finished product. No painter will use the foreign if he can obtain the Philadelphia White Lead.*

* The process of manufacturing White Lead is described as follows :—
"The Pig Lead is melted and converted into sheets by a very simple process. Each workman is supplied with a flat piece of board, of about three feet in length and five inches in width, which has raised edges, to prevent the metal, in a melted state, from passing off at the sides. Standing by the side of the furnace with this board, held by the handle in one hand, and with a ladle in the other, the metal is poured over it. Being held at a considerable inclination it passes rapidly off into the kettle, except what adheres to the bottom, which forms the sheet. This is not thicker than the fiftieth part of an inch. Being instantly cooled, it is turned over the edge of a board raised to a level with the hand, when the mould is returned at once to the edge of the kettle; and the ladle, which the workman still holds, is again filled. Thus the operation goes on from morning till night. This is the first process in the manufacture of White Lead. The sheets are next rolled loosely together, in a sufficient number to fill a pot six inches in diam-

Another branch of the Paint manufacture consists in grinding White Lead and Colored Paints, and the Chromes and other colors in oil, in connection with the manufacture of Putty. The principal firms engaged in this business are, GEORGE D. WETHERILL & CO., JOHN LUCAS & CO., ROBERT SHOEMAKER & CO., BROWNING & BROTHERS, FRENCH, RICHARDS & CO., C. SCHRACK & CO., and JOHN D. SPEAR & SON. Some of these Paint Mills are most complete establishments, and have every appliance for carrying on the processes successfully and advantageously. The annual product is at least $770,000.

One of the firms mentioned, Messrs. JOHN LUCAS & CO., in addition to grinding Paints, &c., at the Eagle Mills, in the city, are also the proprietors of the New Jersey Zinc and Color Works, Gibsboro', N. J., established for the manufacture of an Oxide of

eter. Before being placed in the pot a pint of strong vinegar is put into it, which the metal is not permitted to touch. The pots are then stacked in the following manner: first a layer of manure is laid, then a row of pots. On the top of the pots, boards are laid, on which there is a covering of manure; then the pots again; and so on to the roof, about twenty feet in height. A stack usually comprises from twenty to thirty tons of the sheet lead, besides the weight of the pots, vinegar, boards, &c.

"After being closed up, a stack is left undisturbed for about two weeks, during which period a somewhat complicated chemical process goes on. The manure throws off heat, which raises the general temperature to 180° Fahrenheit, and the vinegar slowly evaporates. The acid vapor, acting upon the lead, it first becomes an oxide; then an acetate, by combining with the acetic acid vapor; and this is transformed to a carbonate by carbonic acid arising from the manure. Very little of the lead remains when the stacks are taken down. The contents of the pots are now found to be a dry bluish mass, which crumbles at the touch. This is White Lead in its rough state. Before it leaves the hands of the manufacturer it goes through a variety of processes. First, it is passed through one or more sieves, to separate from it what little of the lead remains. It next undergoes a number of washings in troughs, to free it from impurities; after which it is put into a kiln and dried. From this it is conveyed to a mill and ground, into which, at the same time, linseed oil is led by means of pipes. Out of the mill the White Lead comes forth in its pure state, not white at first, however, though it soon becomes so after exposure to the atmosphere. It is then put into kegs and barrels, and is ready for home consumption or transportation. The smallest kegs hold twelve and a half pounds; the largest barrels fifteen hundred pounds."

17*

Zinc, the introduction of which, as a White Paint, was pronounced by the London World's Fair Jurors, one of the most remarkable events in the recent history of the Chemical Arts. The works were considerably extended during the last year, and have now facilities for turning out annually upward of 2,000 tons of White Zinc and Colored Paints, Chrome Greens, Chrome Yellows, Chinese and Prussian Blues. The senior member of the firm now resides at the Works, and gives his whole attention to the Manufacturing and Grinding department. They have recently brought out a Zinc Green, fully equal to the article manufactured in France, and at a much less cost; it has a body equal to the best Chrome Green, is less poisonous, more brilliant and durable. The members of this firm came to Philadelphia from England, in 1849; since which time they have been unremitting in their endeavors to establish, successfully, the manufacture of all the European Painters' Colors.

GLUE, CURLED HAIR, ETC.

These manufactures are essentially, though not nominally Chemical. They subserve a peculiarly useful purpose, by converting substances that would otherwise be almost worthless, into products of commercial value. The refuse and offal from tanneries, morocco factories, and slaughter-houses, used by Glue and Curledhair manufacturers, are not generally available for other purposes; and without consumption in this way, would be troublesome to remove or prove nuisances to the community.

In Philadelphia there are three firms extensively engaged in the business, viz. : BAEDER, DELANEY & ADAMSON ; H. GERKER, SON & Co. ; and KESLER & SMITH. The works formerly owned by Charles Cumming & Co. have recently been purchased by the first-named firm, who are now, without doubt, the most extensive manufacturers in this branch in the United States. (See APPENDIX.) The product, as made up by us, is as follows :—

12,500 barrels Glue, at $22, - - - - - -	$275,000
Curled Hair, - - - - - - - -	300,000
Raw-hide Whips, - - - - - - - -	50,000
Miscellaneous, viz., Gelatine, Sand Paper, Isinglass, Plastering Hair, Bristles, &c. - - - - - -	150,000
Total, - - - - - - -	$775,000

The capital invested approximates $600,000, as extensive buildings and expensive fixtures are required ; and nearly four hundred persons are furnished constant employment, receiving about $85,000 annually, in wages. The consumption of coal by the three establishments is about two thousand tons yearly, and of lime thirty thousand bushels.

Philadelphia has peculiar advantages for these manufactures. The climate is favorable, and the Tanneries of Pennsylvania, of which there are an immense number, furnish an abundant supply of raw material ; while from South America, the importation is direct—several hundred bales being imported annually. The articles produced are distributed throughout the country, from the East to the West ; and are exported to the West Indies, South America, and the Canadas.

VARNISHES.

Gum Copal, which is the chief article in the preparation of Copal Varnishes, is a singular kind of resin, that exudes naturally from different large trees in the East Indies and other places, and is imported in a crude state, principally to Salem, Mass., where it is cleaned and prepared for use. Of Varnishes there are four principal kinds, designated as Coach, Cabinet, Japan, and Spirit Varnish, though of each there are many qualities. There are also four principal manufacturers of Varnishes in Philadelphia— C. Schrack & Co., B. C. Hornor & Co., G. S. Mayer & Co., and H. R. Wood & Co. The first-named firm are the oldest established in the manufacture in the United States, and now produce a Coach-body Varnish, which is pronounced by competent judges to be in all respects equal to the best English Varnish, for the same purpose. Since the decease of Mr. Schrack, the business is conducted by Mr. Joseph Stulb, who has had twenty-two years experience in the profession.

The prices of Varnish range from 90 cents for Japan or Iron Varnish, to $4, for Coach-body ; averaging $2 per gallon. The production in 1857 was 115,000 gallons, worth $230,000.

Our statistical summary of Chemicals, and the products of Pharmaceutical processes, is as follows :—

Chemicals, including Dye Stuffs, Chrome Colors, and Extracts, -	$3,335,000
Medicines—prepared remedies of Druggists and Chemists, (estimated,)	300,000
" Patent or Proprietary, "	1,000,000
White Lead, - - - - - - - - -	960,000
Zinc Paints, and products of Paint Mills, - - - - -	770,000
Glue, Curled Hair, &c. - - - - - - -	775,000
Varnishes, - - - - - - - - -	230,000
Total, - - - - - - - - - -	$7,370,000

XI.

Clothing—Ready-made.

Within the last quarter of a century a most important and complete revolution has been effected in the Tailoring business, by
the introduction of Ready-made Clothing. Some twenty-five
years ago the only Clothing kept for sale was that which is known
as " Slop Clothing," for seamen. But the inconvenience attending delays and misfits on the part of tailors—the advantages of
procuring a wardrobe at a moment's notice—the ability of merchants to manufacture and supply Clothing equally as good, but
much cheaper, at wholesale than to order, led to the establishment
of this as a distinct branch of business. In 1835, the wholesale
manufacture of Clothing in the United States was first entered
into, to any considerable extent, principally in the city of New
York ; but many of those who then engaged in it were prostrated
by the commercial disasters of '37. In 1840 the trade was re-es-
tablished and increased ; and since then has continued to enlarge
and increase, until its present extent exceeds ordinary belief.
We need, however, only point to the number of stores devoted to
the business, to illustrate the popularity of the system.

One great benefit to the community, resulting from the success
of the Clothing manufacture, is the immense field of employment it
opens for the poor, especially for females. The poor of our large
cities are thus supplied with a never-failing source of occupation.
Some of the other cities have a large portion of their stock manufactured in the rural districts ; but Philadelphia Clothiers deem
it better policy to employ the population of their own city, and so
far as possible to have the work done in their own establishments,
being certain of having it better and more neatly done than in the
country. The prices paid to employees, it is true, are not a very

munificent remuneration for labor; but by respectable Clothiers no advantage is taken of the necessities of the helpless. Exceptional cases there undoubtedly are, in which the poor are oppressed; but we are convinced the business principles of our respectable Clothiers, accord with the principles of humanity, and that the females they employ are paid reasonably fair prices. To this conclusion we are not led by mere assertion of the manufacturers: we are convinced by an examination of their books. At the leading establishments we found that women earn from $3 to $6 per week. Those who make but three dollars make the coarser articles, or are unexperienced in needle-work. Women of neatness, industry, and taste, can make $5 to $6, on fine vests. The average earnings are about $4. For making a silk vest, 62½ cents to $1 is paid. For the commonest pants, which are thrown together, 25 to 37 cents are paid; and two pairs a day is the average product.

Coats, and finer kinds of work, except vests, are made during the dull seasons of the year by tailors, who at other times are employed in fashionable shops at higher rates. This ensures good work at cheap rates. The wages earned by these vary from $6 to $10 a week; but as most of them have families, the earnings of their wives and children always amount to something in addition. The cutting is a trade in itself, and requires talents of a peculiar kind. In the good Clothing warehouses, the men employed in this department are all of long experience and undoubted ability.

One feature of the Ready-made Clothing manufacture, peculiarly deserving of commendation, is the thorough system with which the operations are conducted. In the large establishments every thing is carried on with the regularity of clock-work. As soon as a piece of cloth has been received into the store, it is carefully examined, and the blemished portions, if any, withdrawn. After this examination, each piece is taken to the superintendent, with a memorandum of the quantity it contains, its cost, of whom purchased, &c., all of which is entered in a book; also, the number and description of garments to be made; how trimmed; name of cutter, price of making, &c. It is then passed to the cutter, who receives directions as to the kind, style, and size of the article to be made ; and after being cut, the pieces are handed over to the trimmer, who supplies buttons, thread, lining, &c. The goods are then received by one of the foremen, who gives them out to be sewed and finished ; and on their return they are examined by him, and forwarded to the sales department.

The extent of the Dry-Goods manufacture, in the vicinity of Philadelphia, particularly of that class of goods which forms the raw material of the cheaper kinds of Clothing, gives the Clothiers great advantages in procuring materials on the most favorable terms, direct from the manufactory, without charges for transportation. In several descriptions of Ready-made Clothing, therefore, the prices in this city are considerably below those in any other market. Between the respective dealers, however, no difference is said to exist.*

The advertisements of the trade constitute a novelty in themselves, and a new department of literature. History, Metaphysics, Poetry, and Science, are made to contribute to the sale of coats and trowsers. Milton becomes a salesman, Shakspeare canvasses for Stokes, and Thomas Carlyle has buttoned up his profoundest philosophy in Clothing. The "Bard of Tower Hall"

---

* If we were a tourist and disposed to jump at conclusions, we would say there was less rivalry and jealousy existing among the manufacturers of Clothing than in any other branch of industry with which we are acquainted. In reply to our inquiry, at two of the leading establishments, Messrs. ARNOLD, NUSBAUM & NIRDLINGER, and A. T. LANE & Co., as to the distinctive features of their business, neither, to our great surprise, were willing to admit that they differed in any very important particulars from their neighbors. Messrs. LANE & Co. stated that they did not deal in piece goods, and that their facilities for procuring materials on favorable terms, perhaps through a connection with a leading Dry-Goods house, were undoubtedly unsurpassed; but their business, in its general features, was like that of their neighbors. Messrs. ARNOLD, NUSBAUM & NIRDLINGER, stated that probably a larger proportion of their stock was made on their own premises, under more immediate personal, careful, rigid supervision, than is customary; and their connections with various portions of the country were quite as extensive as any other; but they would not desire to be mentioned at all, except as representatives of the general trade.

This language being so different from that which we are accustomed to hear, was as refreshing as a cup of cold water to a weary traveler, or a gleam of sunshine to the storm-tossed mariner.

is one of the most popular of modern poets, and weekly exhibits the Muse in plaid pants, and a swallow-tailed coat. One method of advertising, which originated in this city, is the publication of a Bulletin of Fashion, which our principal establishments furnish gratuitously to their distant patrons. This is a large and beautiful lithograph, containing the latest styles of about twenty-four garments; and also the styles of the previous season for those who do not wish the very latest; each garment being numbered to facilitate orders. We have been assured that country merchants have found this sheet of so much service, both in making their purchases and sales, that it may justly be considered an advertisement of as much benefit to the buyer as to the seller.

Still another means adopted by Clothiers to attract public attention, is the production of novelties in Clothing. One has made up a coat of double pilot-cloth, adapted for wearing either side outward, of a different color. Another has manufactured a suit of clothes from black-dyed and prepared sheep-skins; a third has produced an Alpaca coat, sufficiently light and portable to be carried in the pocket; a fourth has attached a shirt-collar to a waistcoat; and a fifth has imitated the richly-embroidered and fur-lined leather coats, in use in the northern parts of Europe. Notwithstanding all this exercise, the genius of invention, we have reason to believe, is not exhausted.

The goods which form the bulk of the manufacture in Philadelphia, are those styles, sizes, and qualities, peculiarly adapted to the wants of distant sections—the West, and Southwest. To conduct such a business successfully necessarily requires a large capital, for the manufacturing must be commenced some four months before the selling season; and as the term of credit usually given is six or eight months, the Clothier cannot realize from his investments in a less average time than a year. It is remarkable, therefore, and evidence of the general solvency of the trade, that so few succumbed to the late severe monetary pressure.

The minor subdivisions of the Ready-made Clothing manufacture, deserve some consideration. The principal are—Boys' Clothing, Shirts, Collars and Bosoms; and certain kinds of Ladies Clothing, as Mantillas, Corsets, &c.

In several of the establishments, Youths' and Boys' Clothing is a department of the general business; but it is also a manufacture in itself, with its own fashions, styles, stores, and customers. The fashions and styles are generally original with the makers; and so highly are many of them appreciated abroad, that it is no uncommon thing for French *Modistes* to transfer them to their own fashion-plates, claiming them as of their own invention, and purely Parisian. This class of Clothing is well worthy the attention of Country merchants, who will not only find a ready sale for it, but have the satisfaction of introducing improved patterns to a whole neighborhood.

The manufacture of *Shirts* and *Shirt Collars,* is now a distinct organized and extensive branch of industry. In Philadelphia it furnishes at least three thousand persons with constant employment—counting solely the wholesale establishments, and those retailers who do partly a wholesale business. The Shirts made include every variety, from the cheapest—and it is claimed by disinterested persons, that the low-priced article is cheaper than that made in New England—to the "Shoulder Seam" Shirts of WINCHESTER & Co., Chestnut street, of which the price ranges from $60 down to $12 per dozen. This firm also makes *Collars* of the better qualities.

The manufacture of Shirt Collars and Bosoms is often a business disconnected from that of Shirts, and has attained a rapid development since the introduction of Sewing and Stitching Machines. Hand needle-work would be totally incapable of meeting the demand. Besides, the machines perform with more uniformity and durability than is possible by hand, and relieve females of the most laborious, unhealthy, and least lucrative portion of the work. In enameling Collars and Bosoms, at least one house in this city adopts the method peculiar to Troy, N. Y., imparting a rare and distinctive gloss.*

* The house alluded to is that of EDWIN A. KELLEY, 16 Bank street. In his establishment, which is one of the most complete in the country, about six hundred hands are employed throughout the year, and forty Sewing machines kept constantly running, manufacturing Shirts from $5 to $40 per dozen. His attention, it will be perceived, is given mainly to the finer grades of goods, but prepared exclusively for the Wholesale Jobbing trade. He has also a large establishment in Troy, N. Y., where he manufactures Collars, enameling them according to the Troy method, which is described

Of Ladies Clothing, the two articles which can properly be said to form a department of the Ready-made Clothing trade, are *Mantillas* and *Corsets*. The manufacture of the latter has, within a few years, become a considerable branch of industry. Large quantities are woven by machinery, and in some instances without seams. They are also combined with Anatomical Bandages or Supports; and generally it may be said, that the shape and make have been very much improved, while the price has been much reduced. The manufacture of *Cloaks* and *Mantillas*, as a wholesale business, dates its introduction into this country within the last ten years. So popular, however, has the system become that many Country merchants, instead of purchasing velvets as formerly, now purchase Cloaks, Talmas, and Mantillas, made in the latest styles in the centres of fashion.

The statistics of the Ready-made Clothing manufacture, in Philadelphia, are stated approximately as follows:—

Capital invested,	$3,300,000
Wages paid annually,	2,800,000
Product, as follows:	
Sixty-seven firms, or all whose annual manufacture of Clothing exceeds $40,000 per annum, make to the amount of	6,040,000
All others, (estimated by a leading manufacturer,)	3,600,000
	$9,640,000
Shirts, Collars, and Bosoms,	937,500
Gentlemen's Furnishing Goods,	250,000
Mantillas and Corsets,	330,000
Total,	$11,157,500

as follows. The apparatus used for ironing consists first of a grooved roller, suited to the shape of the collar, and covered with flannel. The iron is beveled to fit the groove, and is warmed by a red-hot heater placed in a cavity; this iron is secured to the short arm of a lever, which is attached to another lever or treddle, one end of which is fastened to the floor. The attendant, by pressing with her foot upon one end of the lower lever, is enabled to use great power, while she turns the wooden roller on which the collar is placed. This great pressure aids in giving the gloss; though great care and skill, and materials of the best quality, are requisite to ensure the highest polish.

XII.

Confectionery.

The word Confectioner, and the term Confectioneries, occur in the Scriptures in a form denoting, that the making of sweet preparations was an established art in the time of Samuel. (1 Samuel, viii. 13.) The business of preparing them, however, it seems, was then, and until within two centuries ago, confined to physicians and apothecaries, who used honey or sugar, principally for disguising disagreeable medicines, and pharmaceutically in making syrups, electuaries, &c. We presume that the separation which has taken place between the arts of preparing conserves and the compounding of drugs, was originally instigated by the ladies or the juveniles, both of whom, like saucy boarders, prefer their flies on a separate plate.

The manufacture of Confectionery, in its modern development, as practiced in England and the United States, bears the distinctive artistic characteristics of French ingenuity and invention. In no other country does the preparation of sugar, as a luxury, absorb so much mental attention, and afford a livelihood to so many persons. It is a long established custom for French gentlemen to present the ladies of their acquaintance, on New Year's Day, with a box of sweetmeats; and so faithfully and generally does the custom continue to be observed, that in Paris two thousand persons find regular employment in making Confectioner's fancy boxes, the most of which are distributed on that single day. The ingenuity and invention of the French manufacturer, says some one, are inexhaustible; "Every season he produces some novelty, and for years this competition has continued between himself and his rivals, and yet there is no abatement of his ardor or his success; now his production consists of a new box; now of some intricate interlacing of fruits; now of some wonderful crystallizations, and now of some new mode of concealing the motto; but in most cases, his art is exerted tastefully to introduce a looking-glass." But the competition that has existed between himself and his rivals, though it may not have abated his ardor, has induced him to resort to some very reprehensible practices. To give a more exquisite flavor to his essences, or to secure vividness and durability of color to his confections, he has not hesitated to use the most noxious and poisonous substances—as verdigris and other poisons. An eminent English physician testifies that he detected, by post-mortem examination, the essential oil of bitter almonds in the stomach of one who had suddenly died after partaking of some French sweetmeats. To such an extent had the use of deleterious mineral substances been carried in the manufacture of Confectionery, particularly for exportation, that the French Government interfered, prescribing what colors the Confectioners might use. This list of permissible substances, however, contains so many of suspicious origin, that henceforth we much prefer, and declare for, the more pure and safe, if less brilliant Confectionery made in Philadelphia.

REPORT.

"Permit first a word of explanation. When you did me the honor to compliment my detective powers, by stating they were in demand to unravel the mysteries of the Confectionery business, I must confess that I had supposed there would be no difficulty in ascertaining who are manufacturing Confectioners. I was even verdant enough to suppose, that the advertisements in the newspapers would, at least, furnish some indication, whether there were many or few; and I took up my evening paper, the *Bulletin*, with confident expectation of acquiring considerable information on the subject. I was delighted to observe at the first glance, well displayed, the announcement—" New Confections; Oriental Nongat, (one dollar per pound); Sherbet Drops; Banana Drops, (fifty cents per pound). Stephen F. Whitman, *Manufacturing Confectioner*, 1210 *Market st., West of Twelfth.*" I looked further, column after column, and would you believe it, found not another Confectioner's advertisement. Imagine my perplexity. Could it be possible that there was only one manufacturing Confectioner in Philadelphia. If so, what a nabob he must be. Thirty millions of people who consume each at least fifty cents worth of candy in a year— that is, fifteen millions of dollars a year : and a fair proportion of the quantity purchased is known to be obtained in Philadelphia. Can it be possible this Mr. Whitman supplies them all ? I called on Mr. Whitman, and he frankly told me that though he did a fair share of business, as he deserves to do, and believed he was one of the largest manufacturers of fine Confectionery in Philadelphia, there were many others, mentioning Rennels, Richardson, Miller, Henrion, and others. I called on them, and they informed me of others ; and these again of still more ; until sick and surfeited, that night I saw in my dreams a delegation of Confectioners, with Whitman at their head, coming to souse me in a caldron of boiling candy. The results of my observations, continued, however, for a long period subsequently, are as follows:

"There are about two hundred Confectioners in Philadelphia, the most of whom manufacture to some extent—the business being done not by a few very large concerns, but diffused among a number of small ones. The makers, in most instances, know who will probably be the purchasers and consumers of their candy, and therefore take pains to have it pure and first-rate in quality. Sixteen of the wholesale manufacturers used, in 1857, 1,400,000 lbs. of sugar, costing, say $147,000, and which made 1,400,000 lbs. of Candy, worth on an average 18 cents per pound, or $252,000. A fair average product for the others—some making much more, and some less, is $2,000 per year ; or for all $368,000. About one half of the Confectioners in Philadelphia operate in the finer branches of Ices, Jellies, *Pieces Montées*, &c., to the extent, on an average, of $4,000 each, or $400,000 for all ; and the number of persons employed in said one hundred establishments will average five each, or five hundred in all. In addition to the regular trade, there is an immense business done, during seven or eight months in the year, by the country people, who bring in Ice Creams by thousands of gallons, which they vend in the markets, or serve to the hundreds of cake-shops, and other occasional depots ; and thus diverting large quantities of material from the customary objects of milk, cream, butter, &c., causing a great increase in the prices of those articles. This however cannot be enumerated, and the product is stated as follows:

Sugar Confectionery, including Molasses Candy,	$620,000
Pieces Montées, &c., - - - - -	400,000
Total, - - - - - -	$1,020,000

"In point of Wholesale Candy Manufacture, New York, of course, is far in advance of Philadelphia ; but in the ornamental branches, *Pieces Montées* in particular, the quality of the Ices, Jellies, &c., and *Patisserie* in general, Philadelphia is unquestionably superior to the former city, or any other in the Union. The French Confectionery, made in this city, is also of surpassing excellence and beauty.

"Now, as a partial compensation for the trouble I have given the Confectioners, I desire to offer them a hint, borrowed from my Turkish experience. In Turkey, there is a preparation known as *Rahatlocoum*, in great favor with the Turkish ladies, from its alleged property of developing those proportions of figure which, in that enlightened country, are deemed a most essential attribute of female beauty. The preparation is of the most agreeable flavor, and composed of the following innocent materials : one part of wheat starch, six parts of sugar, and twelve parts of water. These are boiled together for some time; and when the mixture has lost so much of the water by evaporation that it will congeal to an elastic jujube-like mass, it is run into a flat tray and allowed to cool ; sometimes blanched almonds are mixed with it. About six hundred tons of *Rahatlocoum* are made annually in Turkey. There is a fortune in the suggestion for some of our Confectioners.

" *Flavoring Extracts,* for flavoring Pies, Puddings, Cakes, &c., are made to a considerable extent, and are said to possess all the freshness and delicacy of the fruits from which they are prepared. *Artificial Essences* for flavoring Syrups, &c., are also made, but from less agreeable and desirable materials, as the makers can testify."

The branch of the conserve art, for which the United States received the most credit at the World's Fair, in London, was the preservation of soft fruits in brandy. The Peach is the favorite conserve ; and in this city, which has unsurpassed facilities for procuring the best fruits, the business is carried on largely and successfully, considerable quantities being exported every year to England and other countries.

The London Fancy Cake Bakers occasionally make some very successful attempts to produce gigantic Bride cakes ; and exhibited at the World's Fair at least three, varying in value from $150 to no less a sum than $750. One it is said possessed the advantage of movable ornaments; so that after the cake has disappeared, the sugar may be transmitted, like the silk dresses of our ancestors, as an heir-loom from the grandmother to her grand-daughter. But the greatest achievement, in the way of large cakes, we think, was that made some years ago by Mr. Parkinson, of Philadelphia, for a Franklin Institute Exhibition. It was about the size of an ordinary cart-wheel, and weighed about 1,200 lbs. The ingredients were as follows, viz. : 120 dozen eggs, 150 lbs. butter, 150 lbs. flour, 150 lbs. sugar, and 500 lbs. of fruits ; besides the icing and ornamentation.

Within the last few years, the demand for costly banquets has tested the inventive genius of our Confectioners, and called forth

18*

some wonderful displays.   Among the remarkable and expensive festivals, we recall to recollection the following :—

The dinner to Capt. Matthews, of the pioneer Steamship City of Glasgow, at the Chinese Museum, cost - - - -	$4,900
Kossuth banquet, at United States Hotel, - - - -	2,000
Henry Clay ball supper, - - - - - - -	1,500
Consolidation ball supper for 4,500 persons—cost, exclusive of wines, &c., - - - - - - - - -	3,500

But the *model* festival of all, perhaps, ever got up in modern times, was one furnished by Mr. Parkinson, at his present establishment, on Eighth st., in the spring of 1852, it being a return complimentary entertainment, given by fifteen gentlemen, " merchant princes" of Philadelphia, to a like number of "eminences" of New York, making thirty persons in all.   No price was named, but a *carte blanche* given to the accomplished caterer, who set his wits to work—procuring green peas and strawberries from the South, salmon and other rarities from the East, and every luxury and epicurean delicacy from the earth, air, and flood ; while a fourth element was scientifically employed to adapt the whole to the gratification of the human palate.   The saloon was decorated in the most elegant manner ; while gold, silver, china, and glass of the most costly and beautiful styles, flashed and glittered on the board.   The feast was composed of twenty separate courses, each with its appropriate liquors, wines, and liqueurs, designated in a bill of fare, or rather *programme*, which of itself was a perfect curiosity of beauty and taste, comprising a highly ornamented and illuminated page for each course.   The cost of this memorable entertainment was exactly $1,000 !

XIII.

Distilling and Rectifying.

The consumption of Spirituous Liquors, both as a luxury and in the arts, is so vast that their manufacture necessarily involves considerations of great commercial importance.   According to the census of 1850, the manufacture of Malt and Spirituous Liquors employs a capital of $8,334,254 ; consumes 3,787,195 bushels of barley, 11,067,761 bushels of corn, 2,143,927 bushels rye, 56,517 bushels oats, 526,840 bushels of apples, 61,675 hhds of molasses, 1,294 tons of hops; furnishes employment to 5,487 persons, and produces 1,177,924 barrels of Ale, &c., 42,133,955 gallons of Whisky and High Wines, and 6,500,500 gallons of Rum. The centre of the Whisky manufacture is probably Cincinnati, Ohio, for we notice that, in 1856, there were distilled in that city and vicinity, 19,260,245 gallons of proof Whisky; consuming, if we allow one bushel of corn to every three gallons of Spirits, 6,420,082 bushels of corn. In Philadelphia there are but five concerns engaged in distilling Whisky from rye, corn, &c. They have a capital employed of nearly $500,000, and in 1857 produced 2,100,000 gallons, worth on an average 30 cents per gallon, or $630,000. The largest distillery produces 750,000 gallons per annum; a product which, though large, is less than that of some in the interior of the State. The distilling of Spirits from molasses, which forms a large item in the manufacturing industry of Boston, is not carried on to any extent in this city.

The leading business connected with the manufacture of Spirituous Liquors in Philadelphia, is *Rectifying* Whisky. There are at least eight firms very extensively engaged in this pursuit; and many others, who rectify from five to forty barrels per week. The capital invested is $1,250,000; and the product, in 1857, was 7,650,000 gallons, which, at 33 cents per gallon, amounted to $2,524,500. The principal firms are, JOHN GIBSON, SONS & CO., KIRKPATRICK, DE HAVEN & CO., A. C. CRAIG & CO., B. F. & H. HUDDY, WHITE & VANSYCKEL, WM. WALLACE, COLLINS, ROCKAFELLOW & CO., H. & H. W. CATHERWOOD, and A. J. CATHERWOOD. The first-named, Messrs. Gibson, Sons & Co., are the most extensive Rectifiers, having a capital exceeding $350,000 employed in this business, and in the manufacture of their well-known superior Monongahela Whisky, at their extensive works recently erected on the Monongahela River. Their trade lies chiefly in the principal cities of the Southern States; and perhaps no firm has been more active and liberal to extend trade in that section of the country than they. But all the firms named control a large capital, and can keep their liquors in store until time imparts that flavor which it is said age alone can give. Besides Whisky and Spirits, Cordials and Baywater are made to the amount of at least $200,000. The conversion of Whisky into Alcohol and Burning Fluid we previously considered.

XIV.

The Dry Goods Manufacture.

The trade in Dry Goods, considered as a branch of commerce, is the most important of any now existing in this country. It controls a greater amount of capital, employs a larger number of persons, and distributes a greater value of commodities, than any other branch of mercantile pursuit. The list of Dry Goods merchants in our large towns is far longer than will be found engaged in the sale of merchandise under any other heading; while throughout the interior the very name of "merchant" is associated with one who, whatever else he may sell, is a Dry Goods dealer. There are certainly "merchant princes" among those engaged in mercantile pursuits; but in capacity, energy, and aggregate wealth, the dealers in Dry Goods, as a class, are emphatically THE MERCHANTS of our day and country.

The variety of articles embraced in the term Dry Goods, is seemingly exhaustless; but the materials of which they are composed, are principally Cotton, Wool, Flax, and Silk. All of these, with the exception of the last, are natural, or at least leading products of this country; all of them, with perhaps the same exception, are bulky in their raw and unmanufactured state. Hence one would naturally suppose, that the mills for manufacturing them would be situated in the same country as the place of their production, if not in the same district. No one would certainly suppose, even hypothetically, that a free, civilized, and ingenious people, would rely upon foreign countries for the supply of their necessities, or be persistently guilty of the gigantic folly of going four thousand miles to mill. It is indeed difficult to reconcile such a course of conduct with the traditionary notions of American independence and American sagacity; but happily, the day is gradually passing away when any exposition of the anomaly will be necessary.

The first regular Cotton Factory established in the United States, was located in Beverly, Mass., and went into operation in 1787. In 1789, it received a visit from President Washington, then on a tour through the Eastern States. At that time the British government, defeated in a war just closed, took its revenge in the only manner possible, viz., by prohibiting, with severe penalties, any exportation of machinery, or even drawings of machinery, from that country. A handsome set of brass models of Arkwright's machine was secretly prepared for shipment, but was seized at the Custom House. Mr. Samuel Slater, who had served a regular apprenticeship to the business in England, came out in 1789; and although he was without models or drawings of the machinery needed, he succeeded in starting at Pawtucket, R. I., three cards and seventy-two spindles, on the 20th of December, 1790. These were the first Arkwright machines operated in this country. The first Cotton Factory started in Massachusetts, with the improved machinery, was located near Pawtucket, on the other side of the river, and commenced operations about 1795.

The first Cotton Mill established, as we are informed, in the county, now the city of Philadelphia, was situated at La Grange Place, near Holmesburg. The machinery was supplied by Alfred Jenks, who had been a pupil and colaborer for many years with Samuel Slater, and who established his manufactory of cotton machinery in Holmesburg, in 1810. The oldest established Cotton Mill, now in operation, is the Keating Mill in Manayunk, owned by J. C. Kempton.

The first Woolen Mill started in the State was at Conshohocken, by Bethel Moore, a name that continues to be identified with the manufacture. It would be desirable to trace, chronologically, the successive steps marking the progressive development of the manufacture of textile fabrics, in this city; but, unfortunately, there are no records within our knowledge containing sufficient and reliable data for the purpose. In 1824, we find a list showing there were thirty-three Cotton and Woolen factories in the city and vicinity, worked by water or steam-power; and twenty of them had no less than 28,750 spindles in operation, and the number increasing. A few years subsequently, an English writer announced that Philadelphia *was the great seat of hand-loom manufacturing and weaving.* But beyond such isolated statements as these, the growth of this important interest seems to have attracted but little historical recognition; and we can only conjecture that it was overwhelmed by the flourish of trum pets which attends the erection of a factory in New England, though it may produce less in a month than the hand-looms of Philadelphia produce in a week.

Looking then at present circumstances only, without attempting to account for their existence, we are astonished by the undeniable revelation, that Philadelphia *is the centre of a greater number of factories for textile fabrics than any other city in the world.* We do not desire to be understood as saying, greater number of looms, or greater value of production ; but simply what we state, a greater number of distinct, separate establishments fairly entitled to be called factories. No other city in the world, within our knowledge, is the centre of two hundred and sixty Cotton and Woolen factories, and containing, besides, hand-looms in force and production equal to seventy additional factories of average size. Moreover, we claim that Philadelphia is the centre of a larger production of indispensable domestic goods, than any other city or place in the United States. In making this claim, we do not desire to be understood as saying all descriptions of goods, but of domestic goods, indispensable particularly in the South and West. If this be true, the inference is unavoidable, that Philadelphia is the cheapest market in which the merchants of the South and West can purchase such goods. These statements lead us to the consideration of two points ; first, *the description of fabrics made here,* and secondly, *the extent of the production.*

The textile fabrics made in Philadelphia might be considered as of two classes—one, designated "Philadelphia goods," and the other "imported"—the former comprising a variety of heavy articles essential in domestic use, and the other, delicate, ornamental fabrics, sold in New York, and frequently in this city, as Parisian or German goods. We however shall adopt for convenience the usual subdivisions, viz. : Cotton goods, woolen and mixed, Hosiery, Carpetings, Silks, &c.

1. COTTON GOODS.

The application of the wonderful natural product, which has been called by some vegetable wool, to the manufacture of articles of utility and of ornament, is one of the most interesting records of industrial achievement. In Philadelphia, this application has principally been directed to the production of articles calculated to promote the comfort of the masses—the artisan, the farmer, and the mechanic—and very great credit is due to the fabricants for having brought many unpretending articles of this description to a high degree of perfection. *Tickings* are made in large quantities, and of a far better quality than those made in New England. They are distinguished for having more stock, and less starch in them. Mr. Wallis, one of the English Commissioners to the American World's Fair, thus speaks of certain goods of this class that came under his notice. "They are 36 inches wide, 1100 reed, No. 30 warp, and No. 35 filling or weft, with 140 picks to the inch. It is scarcely possible to conceive a firmer or better made article; and the traditionary notion that really good Tickings can only be manufactured from flax receives a severe shock, when such Cotton goods as these are presented for examination." The varieties of Tickings made in Philadelphia, and its vicinity, are far more numerous than elsewhere; and the prices range from 7 to 24 cents—those at the latter price being a most superior article.

Of *Apron* and *Furniture Checks*, Philadelphia may be said to have the monopoly in the manufacture; none being made elsewhere, as we are informed, to any extent. They are of various grades, ranging in price from $7\frac{1}{2}$ to 17 cents. These goods are well-known, and it is therefore needless to add that they are of the first class. A superior Check for miners' shirting is made, worth from 12 to 20 cents.

Ginghams are made of all qualities, ranging from $8\frac{1}{2}$ to 16 cents. These goods, for strength and durability of fabric and colors, and neatness and beauty of styles, are, at the low prices at which they are produced and sold, the cheapest article, probably, for women's and children's wear in the whole range of the Dry Goods manufacture. They are much preferable to the Scotch at the same prices, and are free from the dressing which adds so much to the apparent weight of the latter.

Of Cotton goods classed as *Pantaloonery*, *Cottonades*, &c. a great variety of kinds, qualities, and styles are made. The manufacture of these is conducted on a large scale, the production of one manufacturer alone having reached three and a half million of yards in a year. They are now made almost entirely of fast colors, as the demand for the very low priced (of fugitive colors) is yearly diminishing. They are from 25 to 29 inches wide, and range in price from $8\frac{1}{2}$ to 25 cents. Philadelphia *Cottonades* are favorites with Jobbers and Clothiers throughout the country.

Heavy wide *Brown Sheetings* are made in the vicinity of the city, probably heavier than any other in this country; some two yards wide, made of yarn, No. 14, count 50 by 56, has been specially recommended as adapted for the purpose for which they are designed. They are goods which, in consequence of the cheapness of cotton, can be produced cheaper in this country than English goods of the same quality. Heavy blue Mariners' *Shirtings*, formerly designated in the West as "Hickory Shirtings," are made largely; the prices ranging from 8 to $10\frac{1}{2}$ cents. *Denims* are made to a large extent, specially adapted for plantation use, being heavier than any made elsewhere. Other goods, particularly adapted to the Southern trade, and known as *Negro Plaids*, Chambrays, or Crankies, are a prominent article of production with many. *Nankeens*, 28 inches wide, are made from the Nankeen cotton grown in Georgia and South Carolina; price about 10 cents for plain, and 13 for heavy twilled. Several mills also produce *Ducks*, *Osnaburgs*, and *Bagging*, some of which is of excellent quality. *Prints* are made of all grades, from the highest to the lowest, in Madder and Steam colors; and some descriptions, as black and white, and half-mourning prints, are made here exclusively. The prices range from $4\frac{1}{2}$ to 10 cents—those at the latter price bear favorable comparison with the well-known Merrimacks. It is no exaggeration to say that our Calico Printers are unexcelled by any. *Printing Cloths* are made at two or three factories, and the production, although limited, is quite successful —it is believed that this branch of manufacture will increase.

Cotton Hosiery will be referred to subsequently; and of the minor narrow textiles, as for instance, *Stay Binding* or *Twilled Tape*—white, black, and in colors, the production could be expressed only by millions of yards.

For the production of Cotton Yarns there are several mills; but a large quantity used by the manufacturers of Cotton goods is brought from Paterson, N. J., and also from Augusta, Georgia, and from other parts of the South. The production of this article, in Philadelphia, should be at least equal to the wants of the manufacturers.

2. WOOLEN AND MIXED GOODS.

Wool is described by an eminent scientific authority, in the following lucid manner. It is a peculiar modification of hair, presenting, when viewed under the microscope, fine transverse or oblique lines, from 2,000 to 4,000 in the extent of an inch, indicative of an imbricated or scaly surface, on which, and upon its curved or twisted form, depends its remarkable felting quality and its consequent value in manufactures. The Woolen manufacture, in its narrow or restricted meaning, applies only to Cloths made of short wool, and such as possess the quality of felting together, and elasticity; the other branch is called the *Worsted manufacture*, in which long wool, and such as possess no particular tenacity of fabric, is used. The former term, however, is rarely used in the strict sense; and in considering the leading manufactures of Philadelphia in this department, we shall apply it according to its popular signification.

The principal varieties of Woolen goods made in Philadelphia, are *Cassimeres, Satinets, Kentucky Jeans, Shawls, Flannels*, and *Linseys*, or *Woolen Plaids*.

Cassimeres are made to a considerable extent, both all Wool and Cotton and Wool, of various grades. The finest, in imitation of the French, are nearly equal in quality of wool and excellence of finish to any foreign goods, while they are much lower in price. The *Satinets* range from 30 to 75 cents, and are largely produced. *Kentucky Jeans*, of unsurpassed quality, and of great variety of colors, are a leading article of production. They are 27-inch goods, of various grades, from 13 to 40 cents. The better qualities have all wool filling. *Twills* and *Tweeds*, of various patterns and colors, and having a diversity of names, are also made in large quantities: prices from 20 to 33 cents. Most of these have all wool filling. Philadelphia-made Jeans, Twills and Tweeds, are staple goods; and like the Checks, Ginghams, and Cottonades, have a high and deserved reputation, especially at the West, where they are in great demand.

19

Shawls, chiefly all wool—both long and square, plain and fancy colors, greatly diversified in patterns, are made to considerable extent.   The Medium-long Shawls bring from $2 up to $8; while the Square, are from 75 cents to $3½.

Flannels, of various colors and qualities, both all wool and domet, are also largely produced.   An article, all wool, termed *Welsh* Flannel, and used largely by miners, glass-blowers, and foundry men, for shirts, is made by several, and highly esteemed.

Linseys, or *Woolen Plaids*, are made of various qualities; some one half, others one third wool : prices, from 10 up to 33 cents.   Very large quantities are sold in the West, as far as the new Territories and the Rocky Mountains; the heaviest being used there for the clothing of laborers and backwoodsmen.   They are also very extensively sold in the South for clothing for domestics; while some are used for linings.   The higher grades are very superior, and all are desirable goods and in constant demand.   Many are woven in hand-looms.   They are largely shipped to New York, Boston, and Baltimore.   A superior article of 6-4, all wool plaids, price about $1, is also made.

Of Mixed Goods there is considerable variety, principally however the product of hand-looms.   *Coverlets* of cotton and wool, red and white, and other patterns, belong to this class, and are a favorite and serviceable article.

Damask, Birdseye, and *Huckabuck Diapers*, from 5-4 to 11-4, both brown and bleached, are largely made.   They are heavy and very serviceable goods : prices from 10½ to 26 cents.   Some linen Table Cloths and Toweling, of superior quality, are made on Jacquard machines.   It is claimed that the *Damask Table Cloths* are equal to the very best patterns of the imported; while they are superior in durability.   One firm is making Marseilles of excellent quality.   *Bed Spreads*, both bleached and brown, Stair Crash, and a variety of similar goods, are also made in hand-looms.

Union Checks, half linen and half cotton, are made of very superior quality : price from 14 to 20 cents.

Worsted Braid, or " *Ferreting*," occupies many looms; and *Carpet Bindings*, of cotton and wool, are with many leading arti cles of production. Of men's, women's, and children's mixed blue-and-white *Hose,* and *Half Hose,* ten thousands of dozens are annually made.

3. CARPETINGS.

The production of Ingrain and Venitian Carpetings, in Philadelphia, is so important a branch of the general manufacture, that it deserves at our hands special and separate notice. It is also distinctive in its characteristics, both as respects the description of goods made, and the mode of manufacture. The manufacturers of Carpetings in Hartford and Lowell confine their operations, we are told, to all wool and worsted goods, made in super and extra-fines; while the manufacturers in Philadelphia not only make the better qualities, but go down to goods which are all cotton, and sell for about 20 cents per square yard. The fabrication of Cotton, and Cotton-and-Wool Carpetings, is said to be *exclusively confined to Philadelphia.*

As respects the mode of manufacture the business is distinctive, inasmuch as it is distributed among a large number of weavers; there being but one mill that employs power-looms, and only to a very limited extent. The individual manufacturers number about one hundred, who furnish employment to at least fifteen hundred hand-looms, the largest manufacturer having one hundred and fifty looms at work on his fabrics. Each loom will turn out, monthly, three pieces of 120 yards each, or 4,320 yards Carpetings yearly; consequently, the annual production for 1,500 looms would be 6,480,000 yards. The prices of Ingrain Carpetings range from 20 to 85 cents—a low average being 40 cents, which would give an annual value of $2,592,000.* The persons employed are, weavers 1,500; and all others, winders, spoolers, warpers, assistants and dyers, say 1,000 more—in all 2,500 persons. The average price for weaving Carpets is 9 cents; and the

---

* An excellent article of supers, worth 80 to 85, and extra-fines, 65 to 70 cents per yard, is made for New York, and for Chestnut street retailers. J. Bromley & Son are particularly noted for the weight and excellence of their extra-fines. Creagmile & Brother make Damask Venitians, from $1.05 to $1.15, which are fully equal to the imported, and unequaled in this country. Other makers will be subsequently mentioned.

average earnings of weavers $6 a week, or $300 a year. The whole amount paid to weavers and others, for labor, will reach $695,000 per annum.

The "Glen-Echo" Mills, at Germantown, A. McCallum & Co. proprietors, have one hundred looms in operation, a few being power-looms, employ two hundred hands, and produce an average annual product of over $200,000. This firm, and James Lord, spin and dye their own yarns, and are thus exceptional in conducting all the processes of manufacture from the raw material.

Rag and *List Carpets* are also produced to the extent of 1,680,000 yards annually, yielding, at 30 cents per yard, $504,000. The weavers employed in this branch have frequently but one loom each, and rarely over eight. The principal manufacturer has only about twenty looms. Weavers, when they supply the chain, receive about 20 cents a yard; or for weaving alone, from 6 to 10 cents, according to quality. The cotton chain for the better qualities is obtained of yarn dealers, costing about 20 cents per lb., dyed. Carpet balls, for filling, cost from 6 to 7 cents per lb. This description of Carpet sells from 25 to 50 cents per yard.

The entire production of Carpetings, in Philadelphia, we state as follows :—

| | No. of Looms. | Earnings of Weavers, &c. | Production. | |
			Yards.	Value.
Ingrain,	1,500	$695,000	6,480,000	$2,592,000
Rag,	560	126,000	1,680,000	504,000
Total,	2060	$821,000	8,160,000	$3,096,000

The persons employed in the Carpet manufacture are English, Irish, Scotch, and German ; but very few Americans, as we are informed, are known to be engaged either in weaving or spinning. The economy in manufacturing would be greatly promoted, it is supposed, if there were larger mills in the city for spinning and dyeing yarns.

4. WOOLEN HOSIERY—FANCY KNIT WORK.

The importance of this branch of industry, and the success of the Philadelphia manufacturers, entitle it to separate notice.

For more than two hundred and fifty years Nottingham and Leicester were the chief seats of the Hosiery manufacture in Europe and America. The Knitting trade had its origin in Nottingham, through the invention of the Stocking-frame, by the Rev. Mr. Lee of that place, in 1589. At the present time, it is estimated that there are at least 50,000 Stocking-frames in operation in Great Britain, employing 100,000 persons, and producing an annual value of $18,000,000. So diversified are the articles produced in color, shape, and adaptation to markets, that one Leicester manufacturer thought he could not fairly represent his production at the Great Exhibition, in 1851, except by sending 12,500 specimens and prices. Until within the last fifteen or twenty years, America looked exclusively to foreign sources for her supply of the various articles designated as Fancy Woolen goods or Woolen Knitwork. Within that period, however, the manufacture has taken such deep root in Philadelphia—particularly in Germantown and Kensington—that the Nottingham articles no longer find any considerable sale in the American markets, or even in the Canadas. The term "Germantown Woolen Goods," is now as familiar to most dealers as Nottingham Hosiery; while the quality of the American product is really far superior to that of the foreign. The Philadelphia manufacturers have such special and important advantages over the English in the price of wool—being able, therefore, to use much finer grades in the production of articles costing the same price—that they may reasonably anticipate a period not remote when their goods of this class will find a sale, as they certainly will receive a preference, in the English market. A few large establishments, well managed, and combining all economies, it is the opinion of competent judges, could even now export these commodities to England with profit.

The manufacture, as at present conducted, is essentially a domestic one. In Germantown, in which the production is so large as to give its name to the goods produced, there are a few extensive mills employing steam-power; but the distinctive feature of the business is its hand-looms and domesticity. Fully one half of the persons engaged in the production have no practical concern with the ten-hour system, or the factory system, or even the solar system. They work at such hours as they choose in their own

19*

homes, and their industry is mainly regulated by the state of the
larder. But the inherent, natural industry of this class of operatives, who are largely Leicester and Nottingham men, will be
inferred from a visit to Germantown, and practical observation
of the neatness of the dwellings, and the air of comfort that per-
vades all its street and avenues.

In the city proper, there is one large factory engaged in producing Hosiery, Opera-hoods, Comforters, Scarfs, &c., employing
five hundred hands, and consuming annually upward of 250,000
lbs. of American wool.* The hand-frames and machines it is

---

* The factory alluded to is that of MARTIN LANDENBERGER, and we extract from the "Ledger" the following description:—

"The factory has a fine front of thirty-eight feet, is over two hundred
feet deep, and with the basement is five stories high. It is an attractive
brick structure, its external neatness vieing with its internal arrangements
in every respect.

"In the basement we find woolsacks upon which the Lord Chancellor of
England never took his seat; for they contain American wool, woolen yarn
of every describable shade and color, and goods generally, which are
packed and ready for dispatch. Here is also a steam-engine of fifteen
horse power. Near the engine is a large Wool-Scouring Machine. The
wool is then passed into a drying-room, heated by steam; after drying,
it is submitted to the services of the (devil) picker; it then passes through
two sets of cards, which may be termed breakers and finishers. This process of carding prepares the collected and straightened fibres for twisting.
In the twisting department there are ten machines constantly at work, consisting of four hundred spindles. In the spinning department eight sets
of mules are engaged, consisting of twenty-five hundred and sixty spindles.
The yarn is then warped and reeled; subsequently it is bleached, dyed, and
printed according to certain designs. From the warps the yarn is then arranged upon beams for the loom, and from the reeled yarn large spools are
filled by a hand-winding process performed by small boys. The yarn is
then ready for weaving, in which process upward of fifteen different kinds
of looms are at work. We noticed particularly a new loom, the invention
of the proprietor, for weaving neck-comforts. This loom, after much labor
and thought expended in its construction, was started some months ago
This loom weaves four neck-comforts of a double fabric, and each of a different pattern. The Jacquard principle is about to be applied to this loom,
so that by control of the Jacquard index, almost any design will be produced by it. There is another loom of a different construction now in preparation, and will soon be put to work. The other looms used are of vari-

almost impossible to ascertain with accuracy ; but they exceed seven hundred, of which about five hundred are employed on Woolen Hosiery. The average product for each frame exceeds $1,650 annually ; and the whole Hosiery and fancy Woolen goods production in Philadelphia, in 1857, was about as follows :—

500 Knitting Frames, averaging $1,657.50 each,	- -	$828,750
7 Factories in Germantown and Kensington,	- -	800,000
Total value of Woolen Hosiery,	- - -	$1,628,750
200 Knitting Frames on Cotton Hosiery, $897 each,	-	179,400
Total, - - - - - - - -		$1,808,150

The foundations of the American Woolen Hosiery and Fancy

ous kinds and calibre. All the new machinery used in the establishment is made on the premises, upon such a principle that it is impossible for outsiders to copy the construction or mode of operation. Every new style demands some action upon the machinery, which calls out some new demonstration of inventive genius on the part of the proprietor. Here are manufactured hoods, talmas, opera-cloaks, neck-comforts, scarfs, and hosiery of every conceivable description and variety. Every room is set apart for some particular branch in the process of manufacture, and the regulations prevent any laxity of morals on the part of the employees; the males and females are not brought in contact with each other at all. Gladness and health seemed to beam from every countenance upon the occasion of our visit. The stairs and floors are kept thoroughly clean. In the winter season the entire factory is heated by steam to a comfortable degree. This tends to promote the comfort of the workers, whilst it serves a good mission to the machinery.

"Fifteen years ago Mr. Landenberger commenced operations with about twelve hands, and had then to compete with the foreign manufacturers, so that he had to work to get along; but being determined to overcome the importation of woolen hosiery, he laid himself out for the task, and has succeeded admirably." He gives employment to nearly *five hundred* hands, and manufactures every year upward of 250,000 lbs. of American wool which, through his agent, Mr. L. purchases from the grower. He consumes about 2,500 gallons of lard oil, being one gallon to every hundred pounds of wool. He manufactures eight hundred different styles of goods, of all sizes, every season. The value of the business done is about $300,000 annually.

"The majority of the men employed in the establishment are from Leicester, the principal seat of the hosiery manufacture in England. A considerable number of Germans are also employed. For cleanliness and good arrangement, Mr. Landenberger's Kensington Woolen Hosiery manufactory cannot be exceeded, and a visit to it is a *bona fide* entertainment."

Goods manufacture, it is quite evident, are laid in Philadelphia. Within ten years, by persevering and well-directed industry, Philadelphia manufacturers have succeeded in almost excluding the foreign articles from the American market; and they certainly have succeeded in enabling merchants, from all parts of the country, to obtain in Philadelphia superior goods at less than Nottingham or Leicester prices.

5. NARROW TEXTILE FABRICS—SILKS, ETC.

In England, the various manufactures included in the term Narrow Textile Fabrics, are known by the name of Small Wares; and on the continent of Europe the manufacturers of them are designated *Passamenteurs*. In this country the term usually employed is Trimmings, which represents military goods, ladies' dress trimmings, carriage laces, curtain trimmings, cords, tassels, braids, fringes, ribbons, and numerous other manufactures assimilating in character. In Switzerland, Germany, and other chief seats of these manufactures, the establishments confine themselves each to a single class of goods—one making fringes, another ribbons, and so on, but here, two or more branches are often carried on by the same parties; and in the case of one firm in this city, all the above branches are united in one establishment—the largest, beyond all doubt, in the world.

Philadelphia has long been known as the principal seat of the manufacture of Military Goods and Carriage Laces; and now, probably, one half of the whole production of the United States originates here. The branch known as "Ladies Dress Trimmings," is comparatively of modern date in this country. Up to 1839 very little was made, being principally plain fringes, a few buttons, and cords and tassels. The business, however, has become a very important item of our domestic manufactures; and, since the reduction of duties on raw silk, is rapidly expanding. Patterns the most complicated are executed with facility, from designs that are original with the manufacturers; the foreign being rarely copied, except when something very striking is imported. Importers are now timid in forwarding orders abroad until they can ascertain what styles will be introduced

Manufactory of Wᴹ H. HORSTMANN & SONS,
Cor. Fifth & Cherry Sts. PHILADELPHIA.

FORMERLY
Friends Meeting House.
Cherry St.

ST MICHAELS
German Lutheran Church.
Erected 1743.

at home, for in more than one instance they have suffered severely from temerity in this respect. The fabrics produced
here are acknowledged to be generally of better quality than
the English and German; and for several years, have com-
peted successfully with nearly all articles of French manufacture.
Fringes and fancy bindings, particularly, have been sold side by
side with French goods, and in some instances have had the preference. Jobbers certainly are interested in encouraging the
growing preference among consumers for Philadelphia goods of
this class, inasmuch as the supply of desirable goods will then
always be convenient and available, and the risks attending large
purchases, in advance, avoided.

Philadelphia is now the chief seat of the general manufacture of
Trimmings in the United States. Its growth in other places, as a
wholesale manufacture, has always been sickly, and the late financial disasters came near extinguishing it. There are now about
twenty establishments in this city engaged in the various branches,
including Carriage Laces, Regalia, and Upholstery. We shall
endeavor to mention the names of all in a list, which we append
to this article; and therefore we shall here only allude to the most
complete concern of the kind in the Union, and to one other
house, as a representative of the general trade.

The establishment of WILLIAM H. HORSTMANN & SONS is the
one alluded to, as undoubtedly the most extensive of its class in
the world. The business was established by Wm. H. Horstmann,
the father of the present proprietors, in 1815, and is consequently
the oldest established of the kind in the city, if not in this country. In the infancy of its career, the manufacture was limited to
a few patterns of coach laces and fringes; at the present time, it
embraces a wide circle of fabrics, of silk, silk and wool, silk and
cotton; and includes some that are not made elsewhere in this
country.

The manufactory of the Messrs. Horstmann is situated at the
northeast corner of Fifth and Cherry streets, formerly the burying
ground of the German Lutherans, and bought of the congregation
owning the old church, (built 1743,) on the opposite side of
Cherry street. The building forms an L, having a front of 140
feet on Fifth street, 100 feet on Cherry street, and 50 feet wide,

containing six floors. The engine-house and machine shops are in a detached building in the yard. The machinery in operation in the factory is new, much of it original, and includes

    130 Coach Lace Power Looms,
     54 Power Looms, making 600 stripes, or rows of goods,
    334 Silk Spindles,
     60 Plaiting or Braiding machines,

using about 150 Jacquard machines, ranging from 40 to 800 need les ; besides all the auxiliary machinery necessary in the business.

Adjoining the manufactory on Cherry street the firm own an additional lot bought of the Friends, containing 75 feet on the street. The engraving on the opposite page exhibits the Factory, the Old Meeting House, and the German Lutheran Church. The large meeting-house has been converted into a spacious salesroom.

Many of the most important machines, and applications of machinery that are now in use in the manufacture, are indebted to the enterprise of this firm for introduction into this country, or to their genius for their invention. The Plaiting or Braiding machines were first introduced into the United States from Germany, by Mr. W. H. Horstmann, in 1824 ; and to this introduction the Whip manufacture owes much of its success, the principle governing the operations of the Plaiting machines being applied to the covering of Whips ; thus opening up a new business requiring an increased amount of capital, and furnishing employment to many hands. In the year 1825, the same gentleman introduced the Jacquard machines. Gold Laces were made by power, in Philadelphia, several years before attempting it in the old world ; and the use of power for making Fringes may be said to have been first generally adopted here. In fact, it may be said, that this firm were the first in any country to apply power to the general manufacture.

Among the machines and improvements which are indebted to the present Messrs. Horstmann for their origin, the most recent, and probably the most valuable, is that for cutting fringes apart, patented in 1857. Fringes, it is well understood by the trade, are woven two together, and then cut apart with scissors ; but this machine does the work most accurately, and about twenty times as fast as by hand. The machine is referred to in the following paragraph, from the report of the English Commissioners upon the industry of the United States, who, after stating that Messrs. Horstmann have recently erected a very large and well-arranged factory within the city of Philadelphia, remark:

" The whole establishment presents an example of system and neatness rarely to be found in manufactories in which handicrafts so varied are carried on. Female labor is, of course, largely employed in the weaving and making-up departments, and formerly in the cutting of fringes. This, however, is now performed by a machine with a circular knife, so arranged as to cut the thread on the diagonal. The double fringe, as it leaves the loom, being either run off the beam or placed upon a roller for that purpose, is divided much more exactly than it could be by hand, and at so rapid a speed as scarcely to admit of a comparison with hand labor. Any width of fringe can be thus cut, the machine being so constructed as to be easily adapted thereto."

In another part of their report these Commissioners allude to the Clinton Company, located at Clinton, Massachusetts, long known as the largest manufacturers of Coach Lace in America; and state that " the designs of some of the best qualities in which silk is freely used are very good. The looms are of the same construction as the Brussels Power Loom." During the last year, (1857,) the entire stock of manufactured goods, materials, looms, and patent-rights of this Company, were purchased by the Messrs. Horstmann, and thus another important link was added to the chain, securing pre-eminence to Philadelphia as the greatest manufacturing city in the Union.

The Messrs. Horstmann employ four hundred hands, who receive $100,000 annually in wages; have a capital of $400,000 invested in the business; and produce an average annual product of the same amount.

The establishment that we would select as a fair and excellent representative of numerous other manufactories of Ladies Dress Trimmings in Philadelphia, is that of HENRY W. HENSEL. It employs about one hundred persons—say thirty men and boys, whose average wages is $7 per week; and seventy females, receiving $2.75 per week; or in other words, $20,000 are paid annually in wages. The looms in operation comprise twenty Jacquard looms, and twelve other looms, being thirty-two in all, and consuming annually 5,000 lbs. of silk; worsted yarn, 500 lbs.; linen do., 200 lbs.; cotton do., 3,000 lbs.; fine wire, 200 lbs.; and the total amount of sales, of goods manufactured, is about $100,000. The proprietor has been very diligent and successful in originating saleable patterns, and has thus contributed materially to elevate this class of American Textile Fabrics in the scale of popularity. It is his purpose shortly to visit Lyons, and other manufacturing districts of Europe, to examine and introduce such improved machinery as may be adapted to facilitate his general manufactures, which embrace all the usual varieties of Ladies' and Gentlemen's Silk Fringes, Bindings, Braids, Galloons, Cords, Tassels, &c. His general sales are limited, as we are informed, exclusively to jobbers.

Fly Nets are extensively made in Philadelphia; and Regalias, &c., form nearly the exclusive business of one or two manufacturers.

The manufacture of Sewing Silks is carried on by five establishments in Philadelphia, but not as an exclusive business. It is usually conjoined with the production of what is known in commerce by the terms Singles, Tram, and Organzine.* A large proportion of the raw silk imported into the United States comes from China—the Chinese silk being preferred for the pure whiteness of its color, and the strength and glossiness of its fibre. Its successful conversion into the various articles named depends largely upon the excellence of the machinery employed. In the production of Sewing Silks, our home manufacturers have been so successful, that it is supposed that the quantity now imported does not amount to five per cent. of the home production.

---

* *Singles* is formed of *one* of the reeled threads slightly twisted in order to give it strength and firmness.

Tram consists of two or more threads thrown just sufficiently together to hold, by a twist of from one to one and a half turns to the inch.

Organzine, or thrown silk, is formed of two or more singles, according to the thickness required, twisted together in a contrary direction to that of the Singles of which it is composed.

All varieties of Sewing Silk are made, spool silk, embroidery silk, saddlers' or three-corded silks ; and put up in quarter and half pound packages, or in hundred skeins, of different colors. Hundred-skein silk is so termed, because it is made up of from one to one and a half ounces of silk to the hundred, measuring about ten yards in length to the skein. This article is generally sold to peddlers and jobbers. There is another description of skein made up for retailers, which measures from twelve to twenty yards in length. It is principally used by clothing houses, who find it economical to employ the larger skeins. The capital employed in the production of Sewing and other Silks, in Philadelphia, is stated at $300,000, and the annual production at $312,000. The machinery employed for Spinning and Twisting Silk is equal to any in the world.

The oldest established and leading concern in this business, in Philadelphia, is that of B. HOOLEY & SON. The house was established nearly 20 years ago by Messrs. B. & A. Hooley of Macclesfield. The present perfection attained in the manufacture of Sewing and Fringe Silks, in this city, is largely due to the enterprise of this firm. They are now making extensive improvements in, and enlarging their mills, with the view of improving the quality of their Silk and increasing their business ; and as their standing stock of goods of every color is always large, they are enabled by their facilities, the result of experience and a large cash capital, to furnish a superior article at the lowest market rates.

6. PRINTING, DYEING, EMBOSSING, FINISHING, &c.

In the operation of Printing and Dyeing Textile Fabrics, the manufacturers of the United States have, without doubt, been greatly aided by the emigration of artisans from Europe. The attractions of Philadelphia, as a place of residence, have drawn hither the most skillful of these artisans—many of whom bring with them experience gained by almost unremitting attention to these departments of industry during the past half century, in England, France, and Germany. Moreover, the water and climate of Philadelphia are peculiarly favorable for success in dyeing. The influence of these natural agents has already been remarked upon ; but we may refer to the fact mentioned by the English Commissioners, that in Lowell it is well-known the water of the Merrimack River, though reasonably well adapted for dyeing cottons, is not at all suited for woolens. They state, "*this question of the selection of a water site for Dyeing and Printing, is a most important one in the United States, since it is quite certain that in no country is there so great a variation in this respect.*"

The principal Dye Works for Cotton and Woolen goods, in Philadelphia, are located at Frankford. The water in that locality is excellent for the purpose, and equally as well adapted for woolens as for cottons. The Messrs. Horrocks have the most extensive Dyeing Works, it is supposed, south of Providence, R. I.

In the city proper there are many *Silk Dyers* and Refinishers, who have been very successful, and are deservedly celebrated. In the introductory we alluded to one of these—a celebrated French dyer, who had experimented in various places, and found none so well adapted for producing desirable and brilliant results in dyeing as Philadelphia. De Laines, Merinos, and other French goods, are consequently now largely imported in an unfinished state, and we believe at a less rate of duty, and dyed in this city in fast and exquisite colors.

The refinishing of Silks is made an almost exclusive business by a few, and so successfully performed, that old goods are made to wear the appearance of new.

FACTORIES AND HAND-LOOMS.

The factory system of Philadelphia, as will probably be inferred from what has been already stated, is the result and offspring mainly of individual, unaided efforts. It owes but little, if any thing, to the advantages of associated capital; and has grown to a vigorous maturity in spite of foreign competition and unfriendly home legislation. The manufacturers having, from the beginning, directed their energies mainly to the production of useful fabrics, necessary to the comfort of the masses, have steadily worked on, aiming at substantial excellence in an unpretending sphere without attempting, until recently, to compete with others in the finer or more ornamental fabrics, or invoking the attention of the world by the erection of mammoth establishments. In the location of their factories, they have not generally been governed by any other than reasons of convenience and economy, peculiar to each proprietor; hence the factories are scattered throughout the city and its vicinity, the operatives forming no distinct class, the buildings attracting but little notice. In Frankford, and particularly in Manayunk, some show of aggregation is manifest; but in the latter place the exhibition is so unfavorable for a correct observation of the beauties of the system, that dispersion would be preferable.

The mills, though generally small, compare very favorably in machinery and amount of product with the medium establishments in New England. In Philadelphia, as in Lowell, several mills are often the property of one proprietor; and if we were permitted to publish statistics of individual establishments, we could enumerate one having 900 looms, 27,000 spindles, 850 operatives, and producing an annual product of 3,500,000 yards, worth $600,000; another having 432 looms, 9,774 spindles, 38 cards, 513 operatives, and producing annually $430,000; another, having 216 looms, 8,000 spindles, 50 cards, 320 operatives, producing last year 3,272,510 yards duck, Osnaburgs, &c., worth $362,162; another, having 240 looms, 300 operatives, producing yearly 2,100,000 yards ginghams, pantaloonery, &c., worth $250,000; another, having 10,716 spindles, employing 200 operatives, and producing 750,000 lbs. cotton yarn. The Washington Manufacturing Company's Mills, at Gloucester, N. J., nearly opposite our city, and of which our esteemed townsman, DAVID S. BROWN, is President, contain 36,000 spindles and 800 looms, employ 209 males and 445 females; consume 130,000 lbs. cotton per month, 225 tons of coal, 280 gallons sperm oil, &c., and produce about 6,220,000 yards per year, mostly fine printing cloths. The value, when printed, of the product of these mills, is over six hundred thousand dollars per annum. The goods are printed by the Gloucester Manuufacturing Company, another corporation, whose works are situated near the above. This corporation employs about 100 hands, mostly males. It will thus be seen, there are some factories in and near Philadelphia that will compare favorably with those of any other place; but it would be highly desirable and good policy, to erect one or more calculated, from size and arrangement, to give eclat to the manufacture.

The majority of the operatives in the factories are English or Anglo-Americans. The hours for working are usually $10\frac{1}{2}$ per day; but as operations cease early on the afternoon of Saturday, the average for the week is ten hours. In New England, and many other places, labor is extended to eleven hours or more, daily. The female operatives, though perhaps less literary than their Lowell sisters, are seemingly as attractive in appearance, skillful in manipulation, and correct in deportment. Their earnings, as weavers, are from $4 to $5; and as spinners and spoolers, who are mostly young girls, from $2 to $3 per week. The identity of interests which exists between the employer and the employed is seemingly comprehended more clearly by both, and the relations between them exhibit, on the part of the former, more paternal characteristics than is evidenced where the employers are large corporations. Some of the manufacturers are men who are distinguished for benevolent effort; and in some instances, where the factories are remote, schools and churches have been specially established by factory proprietors. The distinctive feature, however, of the Dry Goods Manufacture in Philadelphia is— HAND-LOOM WEAVING.

It is a remarkable fact, that notwithstanding the rapid substitution of power for the production of textile fabrics, and the growth of large establishments from the results of accumulated capital, there is no actual decline in the number of hand-looms in operation. There are fewer looms devoted to certain classes of goods, and in certain localities, than formerly; but the aggregate of such looms now in operation is probably fully equal to that in any former period. Philadelphia is truly the great seat of Hand-loom Manufacturing and Weaving in America. There are now, within our knowledge, 4,760 hand-looms in operation in the production of Checks and other Cotton goods; Carpetings, Hosiery, &c.; and it is probable that the true number approximates six thousand.

The material is furnished by manufacturers, and the weavers are paid by the yard. The weaving is done in the houses of operatives; or in some cases a manufacturer, as he may be termed, has ten or twelve looms in a wooden building attached to his dwelling, and employs journeymen weavers—the employed in some instances boarding and lodging in the same house as their employer. Throughout parts of the city, especially that formerly known as Kensington, the sound of these looms may be heard at all hours—in garrets, cellars, and out-houses, as well as in the weavers' apartments. Among the weavers there are many very intelligent men, and some that have been employed in weaving those magnificent damasks, and other cloths, that Europe occasionally produces to gratify the pride of her rulers. But the subject and statistics of Hand-loom Weaving are fully and well-considered by Mr. Edward Young, in the subjoined report, to which we invite the reader's attention.

REPORT.

" Sir :—In my previous report on Hand-loom Weaving, I stated that in the city there are at least 2,000 hand-looms engaged on Checks, Ginghams, Linseys, and to a small extent on Diapers. As this estimate was larger than any previously stated, you desired such evidence as should prove conclusively the correctness of my assertion, if disputed. I have therefore given much attention to the subject, but regret that longer time could not be allowed in order to investigate the subject thoroughly. As I previously stated, the manufacturers do not own the looms. Each has in operation from 20 to 100, and one has 300 looms. The greater part are situated in the Seventeenth and Nineteenth Wards, (Kensington). The following twenty-five manufacturers altogether employ 1250 looms. Four of the largest employ, on an average, 100 each.

William Beattie,	Edw. Murray,
James Beattie,	James Nolan,
John Dallas,	John Quin,
Robert Dallas,	Patrick Quin,
E. Devlin,	Arthur Rodgers,
John Elliott,	W. Rowbotham,
J. Dickey,	E. Ryan,
J. Donohoe,	D. Murphy,
James Irwin,	William Steele,
Alexander Jackson,	W. Stevenson,
James Long,	Thomas Stinson,
A. & J. Mabin,	John Whiteside.

John Scanlin, 1435 Howard street, (also diapers).

All of these, with one exception, are engaged in making Checks in connection with Linseys, Cottonades, Ginghams, &c. The exception is

20*

Patrick Quin, Master and Cadwalader streets, who makes an excellent article of Damasks, Marseilles, figured Pantaloon Stuffs, &c.

" Besides the above, there are at least 250 looms in the Northern part of the city

" In the Southern and Western parts of the city are the following:—Andrew Catherwood, James Dearie, Greer & McCreight, Thomas Dickson, James Lamb, Robert Little, Thomas Maxwell, Andrew Mitchell, Samuel Orr, Robert Paul, John Perry, and R. Selfridge. These twelve manufacturers employ altogether about 500 looms; but there are many others whom I have not seen. A very intelligent manufacturer, who has been long engaged in the business, assures me that *there are one thousand looms in the South end alone*. Anxious not to overstate the production, I place the number at 750 ; which is a low estimate, for Mr. Selfridge alone affords employment to 300 looms. The number of hand-looms employed on Checks, Ginghams, Linseys, Cottonades, Diapers, &c., I repeat, therefore, is as follows : In the Northern part of the city, 1,250 looms ; in the Southern and Western part, 750 looms—Total 2,000 looms.

" The daily production of hand-looms is as follows : On Linseys, 40 yards ; Checks and Ginghams, 30 yards—making allowance for dull seasons, the average is stated by manufacturers at 25 yards per diem. To show that this is a moderate estimate, I state the prices paid in 1857, for weaving, viz.: Linsey, 2 to $2\frac{1}{4}$ cents per yard ; Checks, $2\frac{1}{2}$ to 3 cents. During the late depression less than those prices was paid. At these rates, a weaver who makes 25 yards the year round, will earn from $3 to $4\frac{1}{2}$ a week ; a sum seemingly inadequate to support a family.

" The production of 2,000 looms, at 25 yards, is 50,000 yards daily ; and counting 300 working days in a year, is 15,000,000 yards yearly— which, at an average of 11 cents per yard, amounts to $1,650,000. The number of hands employed are—weavers, 2,000 ; winders, spoolers, &c., 1,000—Total, 3,000. The amount paid yearly to operatives, is $650,000.

" Now as to Hosiery. With the reputation of Germantown Hosiery and Woolen goods, you no doubt are familiar. If you speak favorably of it, 'Uncle Sam' will confirm your remarks ; for not only has he the Shoes and Clothing, but the Stockings for his Army and Navy made in Philadelphia, where experience has shown the best articles are produced. One manufacturer, T. Branson, made last year *five thousand dozen pairs* for the Government ; and is now completing a contract for 5,000 dozen more, while others are making 2,500 dozen in addition. The Hosiery business is of great economical interest, inasmuch as it affords employment to a large number of females, who sew and finish the various articles after they leave the frame; and thus at leisure hours add to the income and comforts of their families.

"There are three kinds of Knitting-frames in use, viz.: the old hand-frame, such as has been so long in use in England, and which requires a great outlay of muscular power; the lever frame, which is much easier on the operative, and will turn out nearly double the work of the old frame; and the Rotary Knitting machine, or round frame, which will do an ordinary day's work before breakfast, and at slight cost of manual labor. The last has been in use some ten or twelve years in this country. The average weekly production, either of Cotton or Woolen Hosiery on all these frames, good, bad, and indifferent, in dull and busy seasons, is 15 dozen per week, or 780 dozen per annum, which, at $1.15, the average price for Cotton Hosiery, amount to $897, the yearly production of each frame on cotton. On Woolen Hosiery the quantity is the same on the medium and larger sizes. The prices range from $1 for children's to $3 for ladies'; while a few very superior are as high as $5. Assuming the average at $2.12½, the annual production per frame on Woolen Hosiery is $1,657 50. Some manufacturers produce more of children's hose, and half-hose; but while the number of dozens will be greater the price will be less, and the aggregate value remains the same. The total number of Knitting-frames and machines in the city proper, and Germantown, exceeds 700, besides the machines in use by Landenberger and others, who have large factories. Few frames are devoted exclusively to making Cotton Hosiery; all being used for cotton or woolen, according to the demand and remuneration. As Woolen Hosiery pays better, the production, as estimated by intelligent manufacturers, is nearly three fourths of the whole.

"My summary of Hand-Loom production in Cotton and Woolen Goods, is as follows:—

	Looms.	Operatives.	Annual Earnings.	Production.	
				Yards.	Value.
Checks, Ginghams, Linseys, Diapers, &c....	2,000	3,000	$650,000	15,000,000	$1,650,000
Carpets—Ingrain and Venitian..................	1,500	2,500	695,000	6,480,000	2,592,000
" Rag, List, &c...........................	560	630	126,000	1,680,000	504,000
Hosiery—Woolen	500	750	....	....	828,750
" Cotton	200	300	....	....	179,400
Total..	4,760	7,180	....	....	$5,754,150

"The Hand-loom and Knitting-frame business in Kensington and Germantown, affording, as it does, full or partial employment to *nearly ten thousand operatives, and support to upward of thirty-five thousand persons,* is of vast importance to our city, and demands greater attention than it has ever received. A year could be profitably devoted to in vestigating it in all its operations, and in its relations to other branches of industry, and a folio volume filled with the record of such investigations. The cursory examination which I have been able to make has awakened in my mind a deep interest in that part of our city which I denominate 'The Bee-hive,' and led me to feel more sympathy with its busy operatives.

"EDWARD YOUNG.

"P. S. A manufacturer stated to me that he counted the names of 2,200 Ingrain Carpet Weavers appended to a 'strike' for higher wages, about two years ago, and that several hundred did not sign the document. He estimates them at 2,700, instead of 1,500, the number necessary to operate 1,500 looms. It will be seen that the business is really more extensive than I have stated. A celebrated English manufacturer admitted the fact, that *more yards* of Ingrain Carpeting are annually made in Philadelphia than in all Great Britain."

The following is a condensed summary of certain aggregates of production, based principally upon information derived from proprietors themselves ; but partly also from calculations of averages, and from information derived from commission-merchants and others possessing knowledge on the subject, viz. :

Woolen and Cotton Goods, by power, - - - - -		$13,163,968
" " " " " hand-looms, (exclusive of Hosiery,)		4,746,000
Hosiery and Fancy Woolen Goods : hand power, - - 1,008,150		
factories, - - 800,000		
		1,808,150
Narrow Textile Fabrics, Sewing Silks, &c., - - - -		1,600,000
Total annual product in Philadelphia of Dry Goods,		**$21,318,118**

The hands employed, including hand-loom weavers, number over 15,000 ; and the total spindles in operation exceed *two hundred thousand.* The factories are less numerous than the proprietors, for often two or more conduct their operations in the same building, and use the same power—but in other instances, one proprietor owns two or more factories. We shall not, therefore, enumerate the factories merely, but herewith subjoin—

A List of the Principal Manufacturers of Textile Fabrics in the City of Philadelphia.

Allen, William, Germantown, Hosiery.
Arbuckle, Daniel, Eagle Mill, Mixed goods.
Armstrong & Shaw, Satinets and other woolen goods.
Armstrong, John, Germantown, Hosiery.
Austin David, Globe Mill, Pantaloon stuffs.

Baird, William, Frankford, Apron checks.

Barlow, James, Haddington, Shawls.

Beattie, William, Ginghams, diapers, miners' flannels, &c.

Bechmann, G. F., Upholstery trimmings, cords and tassels.

Beaux, J. P., & Co., Silk sewing thread.

Birchell, Elias, Germantown, Hosiery.

Black, William R., & Co., Fairmount Mill, Cotton spinners.

Blundin, Richard, Cassimeres and other woolens.

Branson, T., Woolen and cotton hosiery.

Briggs' Print and Dye Works, Frankford.

Bromley, J., & Son, Fifth and Germantown Road, Carpets.

Bronson & Co., Germantown, Hosiery, &c.

Brown, David S., President Gloucester Manufacturing Company, Printing, Dyeing, Bleaching, and Finishing.

Bruner, J. P., Shawls and other woolen goods.

Burke, James, Print and Dye Works.

Button, John, & Son, Germantown, Hosiery, &c.

Callaghan, Robert, Cassimeres and jeans.

Callaghan, George, Paschalville, Heavy cassimeres.

Campbell, A. & Co., Schuylkill, Linden, and Crompton Steam Mills, Ginghams, checks, and cottonades.

Carr, Joseph, Mount Airy, Cotton yarn, wicking, and laps.

Carr, Edward, Webbing, braids, tapes.

Champrony, J. B., Ladies' dress trimmings.

Clegg, Joseph, opposite Manayunk, Woolen jeans, &c.

Clendenning, John, Aramingo Mills, Table-cloths, stair crash, &c.

Colladay & Bowers, Aramingo Mill, Checks.

Conkle, Henry, Jr., Cotton cord.

Craige, Thomas H. & Co., Star Mill, Cotton yarns.

Craige, William, Print and Dye-house.

Creagmile & Brother, Carpets.

Crowson Brothers, Germantown, Fancy knit goods and hosiery.

Dearie, J. & J., Yankee Mill, Shawls, colored checks, &c.

Derbyshire, John, Kensington Mill, Osnaburgs.

Dickson & Gans, Aramingo, Dyers and Finishers.

Divine, Wm., & Son, Kennebeck Factory, Kentucky jeans, &c.

Divine, Wm., & Son, Penn Factory, Kennebeck checks and print cloths.

22*

Divine & Tomlinson, Hosiery and fancy woolen goods.

Dobson & Co., Falls Mill, Woolen carpet yarn, &c.

Drake, Thomas, Western Mill, Print cloths and cotton yarns.

Drake, Thomas, Coaquanock Mill, Kentucky jeans.

Dudley, John, 207 Quarry st., Webbings, bindings, and bed-lace.

Erben, Peter C., Ringgold Factory, Jeans and other mixed goods.

Ervin, Alexander, Kensington, Dye-house.

Evans, George P., Fancy cassimeres.

Everett & Bohem, Sewing silk.

Ferz, Jacob, Print and Dye Works.

Finley, Thomas, Carpets.

Finley, William, Carpets.

Fleming, Joseph, Cottonades, Canton flannels, and apron checks.

Fling, Geo., & Brother, Germantown, Carpet and hosiery yarns.

Foss, G. W. & Co., Sewing silks, tram and organzine.

Foster, Israel, Pilling's Mill, Satinets, &c.

France, John & E., Germantown, Carpets.

Frazer, John, Apron checks.

Fryer, H. L., Fringes, tassels, &c.

Fullforth & Lovelage, Germantown, Hosiery.

Gadsby, John, & Sons, Hosiery.

Garsed, R. & Brother, Wingohocking Mills, Denims, ducks, Osnaburgs, bagging, &c.

Garside, Joseph, Franklin Mill, Cassimeres and wool tweeds.

Gorgas, Matthias, Wissahicon, Cotton wadding.

Graham, John, James, and Walter, Carpets.

Graham, John C., Ladies dress trimmings.

Granlees & Norris, Columbia Factory, Ginghams, plaids, linseys, and linen checks.

Greer & McCreight, Ginghams and tweeds.

Greer, Johnson, Columbia Factory, Apron checks.

Greenwood, John, Wissahicon, Carpets and carpet yarns.

Greenwood & Co., McFadden's Mill, Manayunk, Carpet yarn.

Greul, Godfrey, Coach laces.

Guy, Robert & Co., Apron checks and tweeds.

Haberstick, John J., Webbing, &c.

Haly, Robert, Wissahicon, Jeans and dyed yarns.

Harrop, Thomas, Sewing silks, &c.

Hawkyard & Whitaker, Falls Mill, Woolen yarn.

Heft, Jacob D., Wissahicon, Dye-house.

Hensel, H. W., Fringes and Ladies dress trimmings.

Henson, William, Germantown, Hosiery.

Hill, Joseph & George W., Germantown, Cotton carpet yarns.

Hill, John, Kensington, Dye Works.

Hilton, James, Flat Rock Mill, Woolen carpet yarn.

Hogg, William, Kensington, Carpets.

Hogg, James, Kensington, Carpets.

Holt, Richard, Globe Mills, Cotton yarns.

Hooley, B., & Son, Sewing silks.

Horn, Wm., & Brother, Woolen yarns.

Horrocks, J. & W., Frankford, Dyers and Finishers of Cotton goods.

Horstmann, Wm. H., & Sons, Military goods, and narrow textile fabrics of every kind.

Howorth, Israel, Dark Run, above Frankford, Woolen cloths.

Hunter, James & John, Hestonville, Print and Dye Works.

Irwin & Stinson, Montgomery Mill, Kensington, Canton flannel carders.

Jennings & Sons, Kensington, Woolen carpet yarns.

Jones, George, Hestonville, Fine cassimeres.

Jones, Thomas, Germantown, Hosiery.

Jones, Aaron, Germantown, Hosiery.

Jones & Duer, Upholstery and silk trimmings.

Kemper, J. & A., Ladies dress trimmings.

Kempton, Jas. C., Roxborough Fact'ry, Cotton checks and stripes.

Kershaw's Mill, Satinets and wool tweeds.

Kitchen, Wm. & Son, Wissahicon, Jeans and cassimeres.

Lambert & Mast, Tassels and cord.

Landenberger, Martin, Kensington, Hosiery and fancy woolen goods.

Kolmer, P., Bed coverlets.

Lafferty, M., Dyeing, &c.

Large, John, Frankford, Dyeing and Finishing.

Laycock & Holt, Pennsylvania Knitting Works, Woolen knit goods and hosiery.

Leckey, John, Kensington, Carpets.

Ledward, James, Manayunk, Woolen and carpet yarn.

Levine, A. T., Fringes, gimps, tassels, &c.

Levine, S., Girth-web fabrics.

Lodge, Fleetwood, Cotton laps and carpet yarn.

Lodge, Jonathan & Bro., Holmesburg, Cotton yarns and laps.

Long, James, Star Mill, Ginghams, checks, woolen plaids, Canton flannels, table diapers, &c.

Lord, James, Wissahicon, Woolen yarns and carpets.

Lord, Rushton, & Co., McFadden's Buildings, Fine woolen yarns.

Lucas, James, Checks, pantaloon stuffs, &c.

Marks, A. & Co., Cords, fringes, tassels, &c.

Mary-aine, A. S. & Co., Steam Dye-house.

Maxson, John, & Son, Lower Manayunk, Satinets, cassimeres, &c.

Maxwell, Thomas, Dye Works.

Maxwell, J. G. & Son, Dress trimmings.

Maynard, Henry J., Gimps, &c.

McBride, T., & Son, Franklin Mill, Checks, cottonades, linseys, &c.

McClain, Edward, Apron checks.

McCallum, A., & Co., Glen Echo Factory, Carpetings.

McCune, Clement, & Co., Ringgold Fact'y, Plaids & cottonades

McMullin, David, Kensington, Carpets.

McNutt, Bernard, Cotton and mixed cloths.

Meadowcraft & Winterbottom, Frankford, Checks, cottonades

Meves, Charles, Fringes and tassels.

Miller, James, & Son, Smith's Mill, Apron checks.

Mills, John, Refinishing, pressing, &c.

Milne, David, Ginghams, linseys, pantaloonery.

Mintzer, William G., Fringes, braids, fly nets, &c.

Mitchell, Andrew, Ginghams, &c.

Moody, Paul R., Fairhill Mill, Osnaburg stripes and checks.

Nugent, George, Falls Factory, Falls of Schuylkill, Jeans and twills, Dye-house.

Orange, W. B., Ashland Mill, Broad, bel. Coates, Sewing silks, &c.

Philadelphia Webbing Company, (J. & J. P. Steiner & Co. Agents, 9 Bank street,) Bindings, webbings, &c.

Perry, John, Mixed goods and stripes.

Preston, E. W. & J., Flat Rock Mill, Kentucky jeans.

Raby, Samuel, Ginghams and checks.

Randall, Gould & Barr, Germantown, Cotton tie yarn and twine
Reed, Thomas, Tottenham Mill, Manayunk, Kentucky-jeans.
Riggs, C., Kensington, Dye Works.
Ring & Bros., Flat Rock Mill, Manayunk, Woolen carpet yarn.
Ripka, Joseph, & Co., Manayunk Mills, Cottonades, &c.
Ripka, Joseph, & Co., City Mills, Satinets, Cassimeres, &c.
Rockord, Philip, Winpenny's Mill, Jeans, &c.
Rodgers, James B., Finishing silks and cloths.
Sacriste, Lewis, & Son, West Philadelphia, Satinets and jeans.
Scholes, Wm., Kensington, Fancy woolen hosiery, &c.
Schofield, Thos., Hill's Mill, Wissahicon, Woolen carpet yarn.
Schofield, Benjamin, opposite Manayunk, Woolen carpet yarn.
Schofield, John, & Co., Manayunk, Cotton carpet yarn.
Schofield, B. & M., Woolen yarns.
Schofield, M., McFadden's Mill, Manayunk, Cotton carpet yarns
Selfridge, Robert, Checks, ginghams, linseys, and miners' flannels.
Shaw, John, & Son, Schuylkill, above Manayunk, Woolen goods.
Siefert, L., Germantown Road, Bed coverlets.
Simons, William C., Manayunk, Cotton yarns.
Simpson, William, Thornton Works, opposite Falls of Schuylkill,
    Print and Dye-house.
Simpson, Hood, Madison Mill, Plaids, stripes, ginghams, checks,
    prints, and cotton yarns.
Solms, Sidney, Pekin Mills, Manayunk, Jeans.
Spencer, Charles, Leicester Knitting Mills, Germantown, Fancy
    knit woolens and hosiery.
Spitz, Joseph, Webbings, bindings, and diamond bed-lace.
Smith, Jesse E., Chatham Mill, Kensington, Cassimeres, &c.
Smith, John D., Marion Print Works, Satinet printing, &c.
Smith, Thomas, Belfield Print Works, above Frankford.
Smith, Thomas, Philadelphia, Silk dyer.
Smyth, James P., Washington Mill, Apron checks, &c.
Sonneboyn, Lewis, Carpets.
Stafford & Co., Manayunk, Woolen carpet yarn.
Steele, Wm., Hope Mill, Checks, ginghams, and mariners' stripes.
Stephens & Whitaker, Arkwright Mills, Manayunk, Shirtings,
    tickings, and denims.
Stone, Amasa, Quarry st., Webbing, lamp-wick, &c.

Steenson, Robert, Carpets and Dye-house.

Sutton, Geo., & Son, Perseverance Mills, Lower Manayunk, Cassimeres and twills.

Taylor, Yates & Co., Checks and plain goods.

Taylor, Robert & James, Haddington, Mixed goods.

Thompson, Andrew, Craige's Mill, Apron checks.

Thornton & Smith, Globe Mill, Kensington, Apron checks.

Walker, R. J., Kensington, Girard Finishing Works.

Watt, W. & J., Ginghams, checks, and pantaloon stuffs.

Watt, John M., Nineteenth and Pine, Mixed goods.

Watt, William, Jr., Globe Mill, Kensington, Checks.

Waters, John, Haddington, Woolen jeans, &c.

Wallace, David, Manayunk, Kentucky jeans.

Wade, Edward, Germantown, Hosiery.

Washington Manufacturing Company, (David S. Brown, Pres't.) Printing cloths, &c.

Wakefield Mills, Fisher's Lane, Germantown, Hosiery, &c.

Watson & Thorp, Chestnut Hill, Print and Dye Works.

Whitaker & Waldron, Keystone Mill, braids, cords, &c.

Whitaker, Wm., Cedar Grove, above Frankford, Cotton goods.

Winpenny, James B., Manayunk, Cotton yarns.

Wilde, Solomon, Frankford, Jeans, plaids, &c.

Willian & Hartel, Holmesburg, Pennepack Print Works.

Wilson, Charles, & Co., Summerdale Dye and Print Works.

Winterbottom & Co., Aramingo Mill, Frankford, Cotton yarns.

Wood, Henry, Marshall street, Cotton laps.

Wright, John, Craige's Mill, Checks, ginghams, and pant. stuffs.

In addition to the factories located within the limits of the city, or so close to the borders as to be called in the city, there are a great number in the adjacent counties—some of them very fine and large establishments. In the counties of Chester and Delaware there are over fifty factories for the production of Cotton and Woolen goods; in Montgomery County there are twenty-one factories not included in our statement—one of which, located in Norristown, was the largest mill, it is believed, in the United States, previous to the erection of the Pacific Mill, at Lawrence; the Harrisburg Mills; the Reading Steam Manufacturing Company; the celebrated Conestoga Mills, at Lancaster; five or six mills in and near Wilmington, Delaware; three or four near Newark, Delaware; the Exton Mill, at Extonville, New Jersey; the New Jersey Mill, at Millville; and others at Bordentown, Trenton, and other places, whose head-quarters are in Philadelphia. The production of these mills, as published recently in the *North American,* from returns received from a portion only of those known to exist, was as follows :—

Production of Delaware and Chester counties, - - -	$3,125,000
" of exterior localities and mills in Delaware, -	3,571,000
	$6,696,000
Add for Philadelphia and Gloucester, as previously given, -	21,318,118
Total, - - - - - - - - -	**$28,014,118**

This simple statement has a significance, an interest, a value to every dealer in, we may say consumer of Dry Goods throughout the Union, even to the remotest frontiers of civilization. Nearly thirty millions—probably over thirty millions of the most useful Textile Fabrics are made annually in Philadelphia and its vicinity; and found in first hands in the warehouses of Philadelphia merchants. No comments can possibly add any thing to the force of a statement, the correctness of which all subsequent investigation will confirm, or if extended more minutely, will prove to be below the truth. We need deduce no inferences from it, for the eye of self-interest, quick in its perceptions, is generally quite as correct in its conclusions as political economy. When to the fact that thirty millions of Dry Goods are produced and controled, if not monopolized by the manufacturers and merchants of Philadelphia, we add another, viz., that the manufacturers of Old England and New England, consign every season their products to be sold in this market for what they will bring, the conclusion is inevitable, that *Philadelphia is the cheapest and best market in the Union for Dry Goods;* and fairly without a rival in those Staple Goods, the bulk of every stock, which, by their intrinsic value and low price, are SPECIALLY ADAPTED TO THE WANTS OF THE PEOPLE OF THE MIDDLE, SOUTHERN, AND WESTERN STATES.

XV.

Flour, and Substances used as Food.

1. FLOUR.

Twenty years ago it was very generally believed, that good Flour could not be made except by water power. The use of steam in Flour Mills was then a novelty—there being, at the time the first Steam Mill was erected in Philadelphia, say in 1838, but few, if any others, in this country. It was objected that the mill-stones, when propelled by steam, ran too fast, and that the steam heated the flour too much, and numerous other reasons were assigned why the products of City Mills must necessarily be inferior. Since that period, however, and mainly within the last eight years, so great a revolution has been effected in popular opinion, that now City Steam Mill Flour is invariably preferred; and the products of at least one maker in this city, whose brands are designated as the "Premium" and "Red Stone," command in the Liverpool market two shillings per barrel more than any other Flour of the same grade. Not only has Genessee Flour been excluded from the Philadelphia market, and thus three quarters of a million of dollars kept at home annually, for the encouragement of Pennsylvania farmers; but the Philadelphia brands are now popular in all parts of the world to which United States Flour is shipped; while certain brands have the preference wherever they are known. The care and attention which are given by some of our manufacturers to the cleansing of the wheat, and the success attained in the production of Extra Family Flour, are unequaled in any other place.

There are now twenty-two Flour Mills in Philadelphia, with an aggregate of 90 run of stones, and having a capacity for producing 15,960 barrels of Flour per week. During the year ending July 1, 1857, the production of Flour in this city was over 400,000 barrels; averaging $7\frac{1}{2}$ each, or $3,000,000 for all, to which must be added at least $200,000 for Corn Meal, Mill Feed, Hulled Barley, &c. The Wheat consumed was 1,800,000 bushels. The following are the Flour Mills in Philadelphia, with their power and weekly production, the list having been originally prepared at the instance of the *Corn Exchange Association*, and now published with some additions and corrections:

	Horse Power.	Run of Stone.	Bbls. Flour weekly.	Bush. Wheat.
William B. Thomas, (2) - -	125	12	2000	9000
Rowland & Ervein, - - - -	100	8	2400	10800
Detwiler & Hartranft, - - -	90	6	1800	8100
Girard Mill, - - - - - -	50	4	700	3150
S. Roberts' Mills, (2) - - -	45	5	500	2250
J. C. Kern, - - - - -	40	4	900	4050
D. C. Gunckel, - - - - -	40	5	900	4050
C. Heebner, - - - - - -	40	6	750	3375
Meyers & Ervein, - - - -	40	4	600	2700
Twaddell & Smith, - - - -	40	4	500	2250
J. K. Knorr, - - - - -	40	3	400	1800
A. Comstock, - - - - -	40	4	800	3600
James Watt, - - - - - -	30	4	500	2250
E. W. Wilson, - - - - -	30	3	450	2025
Esson & Spencer, - - - -	30	3	500	2250
Keystone State, - - - -	30	3	500	2250
A. O. Boehm - - - - -	25	4	500	2250
A. Thorpe, - - - - - -	20	3	500	2250
H. W. Marshall & Co., - -	20	2	400	1800
M. B. & N. Rittenhouse, - -	20	3	360	1620
	895	90	15,960	71,820

2. BREAD, CRACKERS, AND SHIP BISCUIT.

The Baking of Domestic Bread, as at present conducted, can scarcely be called a manufacture. There are two or three of the bakers who work up daily about thirty barrels of flour each; and may therefore be said to conduct the business in a wholesale way, but the average does not probably exceed five barrels per day for each baker. The aggregate production, however, must amount to a large sum; for 600,000 persons, supposing each to consume only five cents worth of bread in a day, would expend in a year for the purpose the sum of $10,950,000.

Within the last year, however, a company was organized and incorporated as the "Pennsylvania Farina Company," having an authorized capital of $500,000, for the purpose of manufactur-

21*

ing Bread by Steam Power, on a large—in fact, a magnificent scale. A Bakery has been erected at the corner of Broad and Vine streets, provided with two of " Berdan's Automatic Ovens," and all the necessary equipments for the conversion into Bread, it is said, of eight hundred barrels of flour per diem. The theory of the construction is that all the processes, from the mixing of the dough to the final delivery of the bread, may be effected solely by mechanical agency. The kneading-machine will knead a batch of ten barrels of flour in less than twenty minutes. Another machine cuts the dough into loaves, and a self-acting register records the number. Cars, of which there are twenty-six—thirteen ascending and thirteen descending at the same time, convey the loaves into the oven, passing through the oven say in thirty minutes—the time allowed when baking common-sized loaves—but the speed varies according to the size of the loaf. The capacity of each baking-car is sixty loaves, weighing about a pound and a half each. The temperature of the ovens is regulated by self-adjusting dampers, and revealed by thermometrical indications.

The flavor of the Bread, it is claimed, is superior, and its nutritive properties increased in consequence of the retention of an alcoholic vapor, arising from the fermented dough, which in common ovens is lost. The peculiar odor observable also in home-made Bread, baked in a close oven, originates in the condensation of this vapor into a fixed oil. Theoretically, the Mechanical Bakery is a very wonderful institution ; practically, it has not as yet effected any very marked revolution in the Baking business. Its present consumption of flour is about forty-six barrels per day.

The Baking of Pies has gradually developed into a considerable business—a number of persons making it an exclusive occupation. The pies are sold wholesale, at prices ranging from three to ten cents each ; and the largest retailed again in the market-houses, or from wagons, at twelve cents ; and at Restaurants for twenty-five cents each. But the only branch of the general art, which can be said to have commercial importance, and which is properly a manufacture, is that of Baking Biscuits, Crackers, and Ship Bread.

The Crackers produced in Philadelphia have long enjoyed a

celebrity abroad, especially in the West Indies, South America, and some of the British Provinces, to which places they are exported in considerable quantities. "Wattson's Crackers" are regularly quoted in Jamaica prices current.

There are nine establishments in the city engaged in this business, having a capital invested of $250,000; consuming annually 50,000 barrels flour, 1,000,000 lbs. lard and butter, 480,000 lbs. sugar, employing 125 men, and producing about 120,000 barrels Crackers, of 80 lbs. each, of the value of $600,000.

Ship Bread is made by a few; but the principal product is Crackers, known as Water, Soda, Butter, and Sugar Crackers—three fourths of the whole being sold in the city and vicinity.

To carry on the business successfully a large capital is necessary—not only for the erection of the Ovens, Machinery, &c., but in the purchase of Flour and other materials, which are bought for cash; while the Crackers are sold in this city on time, or shipped to distant ports to await returns.

3. CURED MEATS

The capital constantly invested by citizens of Philadelphia in Beef, Pork, Lard, Hams, &c., or what is denominated Western Provisions, probably exceeds two millions of dollars. Some brands, having the highest reputation, though sold also in other markets, are controled by capitalists of Philadelphia. This market derives an additional advantage from the circumstance, that the facilities of transportation now established between Philadelphia and the West are so great that the product of the hog, for instance, as we showed in the Introductory, can be transported from Cincinnati to Philadelphia, and shipped hence half way to Liverpool, for less than the cost of transporting them to New York or Boston.

The bulk of the provisions sold in this, as in other markets, is prepared in the West, the merchants aiding with their capital; but the determination to be satisfied with nothing less than the best—characteristic of our citizens in this as in other pursuits, has impelled several firms to provide facilities for smoking Meats brought from the West and fitting them for market under their personal supervision. Establishments have been erected expressly for the purpose—some of them acknowledged to be the best arranged and most complete of their kind in the Union. Among the new constructions of this kind there are several entirely fire-proof, and capable of containing at a time 100,000 lbs. of meats—the capacity of one being 120,000 lbs. or sixty tons. A large proportion of the Meats prepared in Philadelphia being of the first quality, is consumed in the city and its vicinity; but of late years the demand from the South is regular and increasing. One firm does an extensive business in shipping to California "Clear Bacon"—that is, Bacon from which the bones have been removed, and their brand has secured almost a monopoly of that market. A considerable export trade is done directly by butchers, in addition to the regular houses, in shipping "Sheer Bacon" to Cuba.

The Curing and Packing of MESS BEEF are also largely and successfully carried on in Philadelphia. One firm has slaughtered for this purpose as many as 400 beeves per week, for several successive weeks; and the abundant supply of ice always at command, enables the packers to continue operations, without interruption, throughout the year. The cattle are generally fattened in the counties adjacent to the city; the pasturage being, it is well known, of the best description.

There are upward of thirty firms in Philadelphia engaged in the Wholesale Provision business—the annual sales of some houses amounting to $700,000 each. The value of the provisions—Beef, Hams, Shoulders, Sides, Tongues, Lard, &c., cured and prepared in Philadelphia, is estimated at $4,000,000. The Philadelphia brands have, deservedly, a high reputation in Europe, South America, West Indies, California, and wherever known. Large quantities are shipped to the South.

4. PRESERVED FOOD, SPICES, AND CONDIMENTS.

The art of Preserving Animal and Vegetable Food, in a fresh and sweet state, for an indefinite period, is a result of modern skill and ingenuity. Its practical application dates back but twenty-five years, and is intimately connected with the attempt made to explore the Arctic regions. As soon as the value of these preparations became known in cold climates, their use was extended to hot ones; and so great is their present popularity, that thousands of tons are manufactured in England and America, and used in all hot countries, and on all long voyages.

In Philadelphia, attention has been principally directed to the preservation of Fruits and Vegetables—the abundance and excellence of these articles in our markets affording superior opportunities for selection. Some idea of the extent of this trade may be found from the fact, that in one establishment, that of MILLS B. ESPY, there were put up in a single year upward of 20,000 pounds of cherries, 10,000 quarts of strawberries, 4,000 baskets of peaches, 6,000 baskets of tomatos, 3,000 bushel of plums, 100 bushels of gages, 100 barrels of quinces, 30,000 pine-apples, 1,000 bushels of gooseberries, 2,000 bushels each of corn, peas, and beans, besides 300 hogsheads of pickles, &c. Although a comparatively small quantity of oysters are put up here, nearly 12,000 cans were prepared in this house, as well as thousands of cans of fresh beef, mutton, veal, milk, and other articles. The sealing process adopted is so perfect that fruits will keep for years in any climate, without losing their natural flavor, or in any manner impairing their beauty of appearance. It is quite common for New England families to send to Philadelphia for these articles.

The grinding of Spices and the preparation of Chocolate and Mustards, occupy the attention of several firms. The Messrs. Fell, who have the oldest and probably the best arranged Spice Mills in the country, have attained great celebrity in this manufacture.*

* THE FAULKLAND SPICE MILLS, C. J. Fell & Brothers, proprietors, were established more than three quarters of a century ago, by Jonathan Fell, and have since that time increased from a single-horse mill, to an establishment possessing all the new improvements in mill machinery, and using a steam-engine and water power, equal to one hundred horses. The principal mill is located near Wilmington, and runs (speaking technically,) nine pairs of stones. These are devoted to the manufacture of Mustard, all the different preparations of Cocoa, the grinding of Spices, and the making of Hominy. The last is so prepared by a new process, that it resists the effects of any climate, and keeps sweet and good for years. The requirements in tin and wooden boxes, kegs, &c., for packing Spices, furnish employment to a large number of persons. For this purpose, the Messrs. Fell have also a machine, propelled by steam, which *weighs* accurately, and

Within the last few years, the extending popularity of the *Essence of Coffee*, in connection with the preparation of Vermicelli, Maccaroni, Baking Powders, Turkish Coffee, Spices, &c., provides business for a half dozen firms. The oldest preparation in the market is known as Hummels, made by BOHLER, TOMSON & WEIKEL; but the original has been much improved upon since its introduction, and the demand is increasing. The Essence of Coffee is extensively used in private families, and first-class hotels and boarding-houses; for besides being more economical, it is said to make, in connection with a portion of real coffee, a decidedly finer flavored and more pleasant drink than the best Java. One firm in Philadelphia manufacture about forty thousand dollars worth yearly. Philadelphia doubtless exceeds any other place in the extent of the manufacture, as well as in the quality. New York makes very little; New England, little or none at all.

5. VINEGAR.

The manufacture of Vinegar is carried on in this, and all our principal cities, as well as in the country, to a much greater extent than is generally supposed, or its apparently limited culinary use would seem to warrant. But in addition to consumption in this way, and in the preparation of preserved food, this article is indispensable in several branches of manufacture, as in the dressing of Morocco Leather—an extensive business in Philadelphia—and in Dye and Print Works. One manufacturer, Mr. J. G. Peale, informs us that he has supplied one establishment in the latter business with about ten thousand gallons annually. One

packs the Spices neatly in bundles. Its ingenuity and speed are remarkable.

The Faulkland Mills are, we believe, the oldest Mustard, Chocolate, and Spice Mills in the country; and the advantages of long experience, the best machinery, together with the business integrity of the proprietors, are realized in the celebrity of these mills for the purity and extent of their production The motto of this house, for three generations, has been, "never to sell an article otherwise than as represented;" and by adhering to this rule, and avoiding all adulterations, they have given a high reputation to the Spices prepared in the Philadelphia market, while they have attained a fortune for themselves.

Vinegar maker alone, we are also informed, produces daily about ninety barrels, much of which he exports to other parts of this country, as well as to the West Indies and the British Provinces, There are some twelve or fifteen Vinegar manufacturers—some of whom, as well as these referred to, make large quantities; the whole business amounting to at least $300,000.

The process of manufacture, as at present conducted in these establishments, is much more expeditious than that still in use in the country for making vinegar on a small scale.

The latter method consists in placing the cider or other vinous liquid in casks, with open bung-holes, in the sun, and the slow action of the atmosphere upon their contents requires nearly two years to perfect the acidifying process. By the improved mode, the liquor employed, is, by the addition of saccharine or other matter, and a suitable temperature, so managed as to induce its fermentation; after which, it is slowly filtered by a kind of percolation, through tall cisterns or tubs packed with shavings, &c., which minutely divide the liquor, and thus expose nearly every drop separately to be acted upon by the air, which has free access from beneath. The liquor thus absorbs oxygen from the atmosphere, and being drawn off by a pipe near the bottom of the butt, and the same process repeated as often as may be necessary, the acetification is complete in a very short time. The Vinegar in this state is set away to clarify, a process which may also be artificially hastened, and in one or two months is ready for use.

This is a brief outline of the process, though other minor precautions are taken to regulate it; and we believe that proper manipulation and care may even still more abridge the time, as well as modify the color and strength of the product. Cider, whisky, wine, infusions of malt and ale, liquids capable of the acetous fermentation, will make vinegar: but the first three are chiefly used here.

XVI.

Furniture, Chairs, and Upholstery.

Sir: In compliance with your request, I furnish, as far as I am able, a brief abstract of the Furniture business in this city.

In reply to your first question I can answer, that the Cabinet-making business has very much progressed, both in point of taste and extent of production, the last few years. In 1840 there were but few Furniture stores in Philadelphia, and they mostly small ones; keeping samples of the styles of goods, but relying mainly on orders from their customers to supply work for their employees. A Spring-seat Sofa was then a luxury—almost a novelty. The art of Veneering was just beginning to be understood. Previous to this period a crotch of Mahogany wood, (which was then mostly used for furniture,) was cut into Veneers by a narrow blade saw, drawn laterally by two men. They could not get more than four Veneers out of an inch thickness. This was a great waste of the finest class of material, and the Veneers could only be applied to flat work or very slight curves. About this time Circular Saws, some of which were seven to eight feet diameter, were introduced, and gradually improvements were made, so that at the present time it is not uncommon to produce sixteen Veneers to the inch. Mahogany, Rosewood, Walnut, and all the finer woods, are now used in Veneering with such skill, that elliptic ogees, or oval surfaces of common wood, are covered with a thin coating of fine wood, thus reducing the consumption, comparatively, of the finer woods. In the course of time, Mahogany became scarce; and growing in mountain fastnesses, it was procured only at a great expense. Rosewood has always been equally difficult to obtain. To supply the deficiency, the merits of American Walnut were examined, and on trial it was found equally suitable for fine Furniture. The grain of the wood, and the feathery character of the curl, (where two main branches separate from the trunk,) are similar to Mahogany, except in color; the Walnut being of dark purple shade, though varying in color according to the latitude and nature of the soil. Walnut is now used more than all other woods combined. The supply on the rich bottom lands of Indiana, and the Western States generally, is enormous, and the quality so superior that some is shipped to Europe.

All varieties of these woods—Mahogany, Rosewood, Walnut, and others, are used by the Cabinet-makers of Philadelphia. There are nearly one hundred employers in the business, and at least ten large warehouses, where the most fastidious tastes may be satisfied from goods already made. Philadelphia has a well-merited reputation for the production of fine Furniture; the carved work is really superb; and the less elaborate, known as Cottage Furniture, is distinguished for excellent workmanship, high polish, tasteful painting, and moderate price. An oak Sideboard, carved by a Philadelphia sculptor, I notice, was recently regarded by the visitors to the American Institute, in the New York Crystal Palace, as one of the most remarkable specimens of skill in the exhibition. The Southern demand, which is proverbially fastidious and luxurious in the choice of Furniture, is almost entirely supplied from this city. With the increasing demand for fine Furniture, there has been a corresponding improvement of taste in design; and it may be well doubted whether France can, at this time, exhibit more magnificent displays than can be seen in the Cabinet Warehouses of Philadelphia.

In respect to novelties, about which you inquire, I had not the good fortune to discover any very remarkable. The trade are generally satisfied with substantial excellence, without aspiring to any very striking effects. In Mr. I. Lutz's establishment, on Eleventh street, my attention was attracted to an ingenious method adopted by him, to prevent the liability of carved Mahogany to break. In carved Chair work, for instance, he divides the Mahogany into several lateral parts, and joins them by glue in such a manner that the grain of the wood runs in different directions. The strength of the wood is, by this method, increased in proportion to the number of times it is divided; and in the manufacture of Sofas, large Arm-chairs, &c., its advantages are especially apparent. Mr. Lutz employs fifty hands, and has supplied Furniture for some of the finest mansions in this city. Two Sofas, furnished to order, at a cost of $175 each, then on exhibition at his warerooms, were remarkable specimens of elegant workmanship.

In George J. Henkel's establishment, I was particularly struck with the immense quantity of finished Furniture on hand as well as the richness and fine effect produced by its arrangement. The rooms then occupied by him were 175 feet long by 27 feet wide, four floors in number. The leading purpose of this establishment is to supply a complete assortment of first-class Furniture for an entire house; by which all the articles from the attic to the kitchen correspond in style, modified, of course, by their situation. In the construction of Extension Tables, Mr. Henkels is deservedly pre-eminent—the extension being formed by cross-arms working at right angles on metal hinges, which preserve it from swelling or shrinking in a variable climate.

In MOORE & CAMPION'S, WHITE'S, KLAUDER, DEGINTHER &. Co.'s, W. & J. ALLEN'S, and other warerooms, the display of elegant carved Furniture is truly magnificent.

Church and *Library Furniture* constitute a special depart ment of both the carving and furniture business. In Philadelphia there is at least one—perhaps many others—who has attained de served distinction in this branch. For nearly a quarter of a cen-tury, Mr. JOHN HARE OTTON has devoted a large share of his atten tion to Carving and making Pulpits, Lecturns, Book Cases, &c., and his collection of designs now embraces the best examples in every known style. In so long an experience, he has executed a large number of the most elaborate carvings; and in all his recent work, especially, has manifested excellent taste, and an apprecia-tion of appropriateness in ornament that is rarely seen in Amer-ican decorative art. Mr. Otton has also executed some masterly patterns in *Iron* and in *Stucco*, which constitute a branch of his business.

Besides those who are engaged in the wholesale manufacture of Furniture, there is a large number occupied in making special articles. At least twenty-five establishments in the city— some of them of considerable extent—make *Cane-seat* and *Windsor Chairs*. One manufacturer has substituted Whalebone for Cane, which is an evident improvement. Chair findings are largely supplied by Mr. McCullough, and a new establishment is about being opened solely for the supply of chair bottoms.

There are several manufactories and warerooms of *Office* and *Counting-house Furniture* exclusively. Articles of this descrip-tion are both supplied to order, and kept on hand in large quan-tities. Several articles belonging to this category were remark-able as novelties; but among those which seemed to me to com-bine novelty and usefulness in an eminent degree, I was particu-larly attracted by the Patent Elevating and Graduating Top

Tables, which are truly a business luxury. The top can easily be raised or lowered to suit any attitude; placed upon a horizontal plane or inclined as the lid of a desk. The construction is firm, and all the appendages of drawers and boxes are complete. The Office Furniture manufacturers are entitled to very great credit for the specimens of workmanship that they exhibit.

Billiard Tables have been made in Philadelphia since 1809—the date when Mr. THOMAS DAVIS, still a leading manufacturer—commenced business. These tables are now made at four or five establishments; but the business in this line has been a good deal crippled by the preference given to the Patented Tables and Cushions, made in other cities. The deficiency in this respect, however, is compensated for by superiority in another and more important branch, viz., the manufacture of VENITIAN BLINDS. It is believed, by persons professing knowledge on the subject, that this business is larger in this city than in all the rest of the United States. They are sent to almost every part of the Union, and to the British Provinces. The lightness and beauty of the work could not be too highly praised, and the cornices and trimmings are adapted to the furniture of the room in which they are hung, with remarkable taste.

Upholstering is carried on in connection with the manufacture of Furniture, and also as a separate business. It embraces the manufacture of Curtains, Pew and other Cushions, and the making up of Carpets, Hair Mattresses, Buff Window Shades, &c. There are about twenty principal concerns engaged exclusively in this business, besides a vast number of small ones. The fitting up of churches furnishes considerable employment for the Upholsterers; an average bill for a modern fashionable church of medium size being $1,500, the pews alone costing $10 each. The West Arch Presbyterian Church paid $3,000 for Upholstery; and the Academy of Music a much larger sum. The entire business of the city, in this branch, is about a half million of dollars annually.

It is extremely difficult to arrive at the statistics of the Furniture manufactured in Philadelphia; but my opinion, after laborious investigation is, that including all the above-named branches, the annual business will reach two and a half million of dollars. Some manufacturers state it at three and a half millions.   S.

XVII.

Glass Manufactures.

Intelligent foreigners have repeatedly complimented the manufacturers of Glass in the United States—not only for excellence in the production of useful articles, to which they have hitherto given their attention principally, but also for various successful attempts that have been made in producing those rich and decorative works which belong to luxury rather than to utility. The imitations of Bohemian Glass and Opal Glass, made in several establishments throughout the Union, are considered better than a great portion of those produced in Europe. In Philadelphia, the Glass manufacture, though surpassed by many others in amount of production, is nevertheless sufficiently extensive to be called a leading pursuit. The locality, by reason of the facilities for procuring the raw materials, is one of the best in the Union. The finest qualities of sand are obtained from the adjacent State of New Jersey, and the alkali are supplied by the Chemical factories in the city.

There are at least thirteen manufacturers of Glass, whose headquarters are in this city, though the factories of some are located in New Jersey, and outside of the city limits, viz.: WHITALL, TATEM & CO., WHITNEY & BROTHERS, BODINE & BROTHERS, BURGIN & SONS, PHILADELPHIA GLASS COMPANY, JOHN H. MOORE, BENNERS & BROTHERS, SHEETS & DUFFY, JOSEPH PORTER & SONS, HAY & CO., RICHARDS & BROTHERS, JOHN CAPEWELL, President of the United States Glass Company, and THOMAS MILLS.

The leading business is the manufacture of Green and Crown Glass Bottles, including all kinds of Druggists' Vials, Jars, Demijohns, Carboys, &c. This kind of Glass is made of ordinary materials—generally sand with lime, and sometimes clay, and alkaline ashes of any kind; but great care and considerable experience are required, particularly in making bottles that are to contain effervescing fluids. The materials must be carefully and thoroughly fused, and the thickness uniform throughout, to resist the pressure of the contained carbonic acid. The loss of bottles by bursting, in the Champagne trade, is from twenty to thirty per cent. A machine has been contrived to test their strength, which should bear the pressure of from twenty-five to thirty-five "atmospheres." In bottles which are to contain acids, the alkali and the lime must be chemically united to prevent action of the acid. The green color is said to be owing to impurities in the ashes, generally to oxide of iron.

Window Glass is made in several establishments; and in addition to the various sizes and qualities, most, if not all in this business, make double-thick and cylinder Plate Glass, suitable for coaches, pictures, and extra-large windows; some of which is quite equal in quality to the English and French Cylinder Plate Glass. At some establishments, white and colored, plain and figured Enameled Glass is made.

One firm, Messrs. BURGIN & SONS, have, in addition to their furnaces for the manufacture of Black and Green Glassware, one devoted exclusively to the manufacture of a kind of Glass, new in this country, which they designate "German Flint Glass;" and although not as beautiful in appearance as Flint Glass containing lead, it is preferable to it for many purposes, particularly for holding acids and alkalies, as they have no effect upon it. It is a very strong variety of Glass, and is much used by Chemists, Apothecaries, and Perfumers; it can be colored, moulded, and pressed into all the various patterns and forms of Flint Glass, and is sold at intermediate prices between Green and Flint Glass.

The Philadelphia Glass Company was established for the manufacture of Rough Plate Glass, particularly rolled or hammered Glass for green-houses, &c., and flooring Glass—articles which previously had not been made in this country. The excellence of their product so effectually alarmed foreign manufacturers, that they reduced the price at once, from $2.25 per square foot to 75 cents, and are now actually losing money on their sales, in order to crush an American competitor. The advantages of this locality for this manufacture, however, are so great, that with proper encouragement, this Company believe they can continue business even at the reduced price. They are now manufacturing a Glass Furnace, which they consider equal to any in the world.

Besides the manufacturers of Glass above enumerated, there are several whose attention is devoted to supplying orders for special

22*

kinds of Glass, particularly Tubes for Philosophical Apparatus, Syringes, &c., for druggists. Of Glass-cutters there are several in the city; while the Glass mould and press makers are entitled to a compliment for their success in originating novel designs and skill in their profession, particularly for being able to make a glass bottle precisely similar to another in size and appearance, but which will contain considerably less in quantity!

STAINED GLASS.

The origin of this beautiful art is lost in the dimness of antiquity. The process employed in modern times is described as follows. After the figure to be put upon the plate is drawn upon paper, and painted as desired, it is transferred to the glass, which has been prepared to receive it. This has to be done with artistic skill, equal to that employed upon an oil painting, and requires much more care in its execution. In transferring fruits and flower pieces, all the delicate tints of the objects must be copied with the greatest nicety. The glass is then put into a kiln, and submitted to a heat almost sufficient to fuse it, which not only has the effect to add greatly to the beauty of the painting, but makes it a part of the glass itself, no power being able to remove it.

There are two principal manufacturers of Stained Glass in Philadelphia, Messrs. J. & G. H. GIBSON, and FRANKLIN SMITH. The former firm has just completed the magnificent glass ceilings for the House of Representatives at Washington, composed of plates having the appearance of enameled work; the Coats of Arms of the United States are done in rich colors, giving the effect of Mosaics set in silver. They have also been engaged to furnish the Senate ceiling in a similar manner. The Stained Glass made in this city is considered quite equal to that of European manufacture.

XVIII.

Hats, Caps, and Furs.

The Hat, which is regarded by some as more indicative of the social position of its wearer than any other garment, affords a wide field for research, a theme for many speculations, and could be aptly used in illustrating the mutability and instability of earthly things. We, however, have no leisure for any further reflections than to express gratification that the heavy Fur and Wool Hats, whose heat and weight muddled the brains of our ancestors, are superseded by light and more handsome styles. Much of the progress that has been made, particularly in diminishing the weight, it is proper to state, is due to American enterprise, the most important improvement being that of "waterproofing" the bodies previous to their being napped. The elastic properties of the gums used in this process, when dissolved in pure alcohol or naphtha, impart a body to the materials which enables the maker to reduce a considerable proportion of their weight. As an illustration of the value of this improvement, we may mention that, about twenty years since, ninety-six ounces of stuff were worked up into one dozen ordinary-sized hats for gentlemen, while at present from thirty-three to thirty-four ounces only are required to complete the same quantity. It is therefore scarcely surprising, as we learn from a late traveler, that American Hats are superseding the use of the Turban in Turkey!

In Philadelphia there are extensive concerns engaged in the Hat manufacture, though the number of those that make an entire Hat is quite limited. The furs and other materials used are for the most part prepared abroad, on the continent of Europe, where children are largely employed in the various operations; but three fourths of the Hatters' materials used are imported direct by houses in this city. The mode of manufacturing is partly a domestic one, the materials being given out to workmen who shape them in their own houses, though the principal portion of the work is done in large manufactories, where several hundred hands are employed. In one establishment, which commenced operations within the last two years, machines are largely employed in all the various processes of making soft Fur Hats, and Hat bodies. By means of a machine, known as Wells' Patent, the shell or skeleton body is made so expeditiously, that two men and a boy, with its aid, can form three hundred Hats in less time than ten Hats could be produced by the old method. In this manufactory there are seven machines in constant operation, capable of producing over two thousand Hat bodies per day. There are nine other machines for the separation of the hair from the fur. Pickers, propelled by steam, are employed for mixing the Furs ; and even the Hats are washed by machinery—all these operations being performed better, and more cheaply, than they can be done by hand. The proprietor, Mr. Wm. O. Beard, has an engine of sixty-horse power, and employs one hundred and eighty hands.

The branch of the general business, in which Philadelphia Hatters claim to excel all others, is in the production of Silk and White Fur Hats. For producing the Pearl White and Light Colored Hats, it is claimed that, in the water of this city, the makers have peculiar advantages ; while for the manufacture of Silk Hats, they have an advantage in being able to command at all times the most skilled workmen.* The importance of this will be understood when we state, that the Silk Hat passes through six distinct departments before its completion ; each department requiring hands who generally serve an apprenticeship but to one branch of the business. The fashions, as a general rule, are not imported, but originate with the leading houses, with only a slight reference to those prevailing in Paris.

Journeymen Hat-makers may be said generally to command good wages ; though their earnings, inasmuch as they work by

* The senior partner of one of the most extensive of the Hat manufacturing concerns of Philadelphia—that of P. Herst & Co.—was one of the pioneers in the manufacture of Silk Hats. He commenced business here some fourteen years ago in a very small way, but has gone on increasing and extending his operations, until now the firm employ one hundred and fifty persons in making Silk Hats, and supply to some extent nearly all the markets of the United States. Mr. Herst also claims to have been the inventor and introducer of the *Satin Under Brim*, now so much admired for its beauty and durability. It was first presented to the public about five years ago, and has superseded nearly every other material previously in use. The firm of P. Herst & Co. are probably more extensively engaged in the manufacture of Silk Hats than any others, and are now annually producing many thousands, mostly of the first qualities, and specially adapted to the fine retail trade. They also make Beaver and Cassimere Hats, of all shades, for summer and winter wear. For softening the brim previous to shaping, this firm use *Billing's Brim Heater*, said to be an admirable invention. By taste in the modeling of styles, and fidelity in workmanship, they have secured a pre-eminence among the fashionable trade, creditable alike to themselves and to Philadelphia workmen.

the piece, depend very much upon the state and prosperity of the country. Body-makers often earn only six dollars a week; but at other times they make thirty. Finishers make from ten to twenty; and shapers and curlers, from fifteen to thirty dollars per week. It is a peculiarity of this trade, that a workman wishing employment in an establishment never applies to its proprietor, but to the foreman, who possesses the chief power to employ or to discharge men.

STRAW HATS are made to a large extent to supply the Southern demand, which continues throughout the year; and the Northern market during the spring and summer seasons. The work is done in work-rooms provided by the employers, or at the houses of the operatives; whose average weekly wages are, for men, $7.50; for women, $4.50. The Straw Braid is chiefly imported from England, Switzerland, and Tuscany. Panama Hats are obtained from Panama, Maracaibo, and other parts of South America, while coarse Straw Hats are brought in large quantities from Canada. As these goods are generally imported ready-shaped, the principal preparation for the market is trimming, and adapting the Hats to the prevailing fashion. The value of the labor performed on those imported, and the production of Straw Hats, will amount to $350,000.

2. CAPS.

The manufacture of Caps is a business distinct from that of Hats. There are a large number of concerns occupied exclusively in making Caps; those of Cloth constituting the chief part of the business, though Plush, Silk, Glazed, and other Caps, are also made. The Caps made in Philadelphia are distinguished for durability and excellence of quality, rather than for fanciful decoration, and command the market wherever these qualities are appreciated. Some have been exported to Russia; and exports are made regularly to the West Indies, South America, and to California.

The Cap manufacture furnishes employment to a large number of females, whose wages in the business will average about $4 per week. Sewing-machines are largely employed; being, in fact, indispensable in consequence of the expansion of the trade. The annual production is about $400,000.

3. FURS.

FURS are prepared by at least twenty establishments, either as a distinct business, or in connection with Hats. It is the object of the Furrier, by dyeing the inferior skins, to imitate the more perfect kinds; and so successful are many, that the permanence of the color of the dyed Sable, for instance, is equally durable with the natural color. Philadelphia Furs are more tastefully made than those of New York; and are considered equal to the Boston Furs, which have a very high reputation. A difference of opinion, wide as the Atlantic, exists as to the comparative value of Furs—the Americans preferring those of Europe, while Europe seems to prefer the American Furs.

The following List will exhibit the demand for American Furs in Europe, and the kinds which this country principally contributes.

Import into London of Furs and Skins from the United States, and Hudson's Bay Company in British America, for one year, from Sept. 1856, to Sept. 1857.

Totals.	Names of Skins.	Hudson's Bay Company.	United States.
1,164,461	Muskrat	302,131	862,330
99,198	Beaver	90,604	8,594
15,941	Otter	11,573	4,368
9,586	Fisher	5,561	4,025
1,548	Silver Fox	7,071	477
4,751	Cross Fox	3,143	1,608
55,086	Red Fox	10,498	44,588
6,597	White Fox	4,940	1,657
11,142	Kitt Fox	5,776	5,366
186,355	Marten	170,956	15,399
123,601	Mink	45,091	78,510
355	Sea Otter	188	167
24,165	Lynx	23,341	824
10,796	Black Bear	7,483	3,313
1,058	Brown Bear	942	116
769	Gray and White Bear	769	....
477,916	Raccoon	1,894	476,022
9,872	Wolf	9,831	41
941	Wolverine	916	25
7,949	Skunk	7,740	209
7,157	Wild Cat	184	6,973
52,672	Opossum	....	52,672
2,271,916			

Besides these Furs of American origin, the principal ones are the *Russian Sable*, everywhere esteemed as the most beautiful, costly, and useful Fur the Arctic zone produces; the *Baum* or *Pine Marten;* the *Stone Marten*, more valuable for the excellent qualities of its skin than the beauty of its fur; *Ermine*, a Siberian and Norwegian Fur, the whitest known, though in summer the animal is a dingy brown; the *European Fitch,* or *Polecat,* a Fur remarkable for durability, and smell, which it is difficult to counteract; the *Tartar Sable,* of which the tail is used exclusively for artists' best pencils; *Nutria,* a Fur used extensively in making hats, and having considerable resemblance to Beaver; *Hamster,* a German Fur; European *Gray Hare,* and the *Chinchilla,* a South American Rabbit.

The Skin that is probably the most extensively used is that of the *Siberian Squirrel.* Of these little animals, not much larger than our common red squirrel, 15,000,000 are every year captured in Russia; the color varies from a pearl gray to a dark blue gray.

The business done in the preparation of Furs, in this city, is estimated by a principal manufacturer at $350,000; and when we remember that Capes alone are sometimes sold at $800 to $1,000, the amount is not probably overstated.

Recapitulation:

Silk and Soft Hats,	$800,000
Straw Hats,	350,000
Caps,	400,000
Furs,	350,000
Total,	$1,900,000

An increased amount of capital could be profitably invested in Philadelphia, in all branches of the Hat and Cap manufacture.

XIX.

Iron and its Manufactures.

It is probable that in no branch of the general manufactures of Philadelphia, is her superiority so widely known and generally conceded as in the fabrication of Metals. The abundance of Iron produced in the vicinity of the city, and its consequent cheapness, have naturally concentrated attention upon its manufactures, as well as extended its uses; while the fame of our Engineers and Machinists attracts from abroad a large and constantly increasing patronage. It is not necessary, therefore, for us to prove what is already admitted, nor to exhibit in much detail and minuteness what is neither doubtful nor disputed, but the subject is too important to be very summarily dismissed. In our introductory remarks we gave some statistics of the Iron production of Pennsylvania, and stated, that of 782,958 tons of Iron produced in the United States in 1856, Pennsylvania furnished 448,515 tons. We also showed that Philadelphia is *situated in the district which is entitled to be called the centre of the Iron production in the United States.* We shall therefore limit our present remarks to a brief outline of the processes employed in the manufacture of Iron, for the benefit of the general reader, besides exhibiting, so far as we can in a limited space, the present development of its manufactures in Philadelphia, particularly with reference to the manufacture of Hardware and Tools, and the construction of Machinery.

Iron, we may remark, exists naturally as an ore—in the form of a rusty, metallic stone. The ores are found both on the surface of the earth and in deep underground veins. Within the limits of Philadelphia we believe there are neither ore beds nor opened mines; though just beyond the city limits, in Montgomery County, ore is dug in considerable quantities; and near Phœnixville, Chester County, there is an extensive Iron mine, which is supposed to be the oldest in the United States. It was opened a few years before the Revolution, and is yet worked with much success. It is 150 feet deep, and has been mined over sixteen acres of surface. The great Rail Mills of the Phœnix Iron Company, successors to Reeves, Buck & Co., obtain a considerable portion of the ore used by them from this mine, known as the Warick Mine.

The ore, after being dug from or raised to the surface, is generally broken and washed in water. It is then most commonly roasted, to drive out the sulphur which exists in many ores. The roasting is done in large kilns or stacks, heated with coal. Many Iron-works, however, do not practice roasting their ores. The great primary process—the first step in the long course of the Iron manufacture—is "smelting." This is the expulsion of the water and oxygen of the ore,—the driving off, by heat, of the natural impurities which enclose and are mixed with the pure iron. This is effected by means of a "Blast Furnace," using as fuel either Anthracite coal, coke or charcoal. The furnace is kept "in blast" night and day, until some vital part is destroyed by the heat. The hearth is tapped at regular intervals, and the iron drawn off and run into "pigs," moulded in the sand-floor in front of the furnace. Fresh materials are as regularly added at the top. The largest class of furnaces produce from 120 to 160 tons weekly, and even as much as 200 tons have been produced, in a few cases, in a single week. The product of the Blast Furnace, or rather the Iron, after being drawn from the furnace and moulded, is called by the familiar term—Pig Iron.

Having passed the first stage of its manufacture—or in other words, been separated from the clay, sand, and other impurities with which it was mixed in the ore, it is now fusible and ready for conversion into Wrought or into Cast Iron. The conversion into Wrought Iron is effected simply by an additional heating, which heat is prolonged for some time at just above the melting point, and during which the iron is stirred up until every particle has been brought under the cementing action of the heat.

There are two kinds of furnaces in use, either of which produce Wrought Iron from Pig. In either case the iron is only melted, and stirred stoutly for a considerable time in that condition.

The forge fire is employed for converting pig into the better kinds of Wrought Iron. A large open forge fire, with the tweer in one corner, is used; a trough or pit is hollowed out beneath the tweer, and the broken pig or coal brought together to a melting heat. The action of the blast from the tweer drives the coarse and lighter impurities to the opposite side of the trough, leaving the melted metal to settle in the trough, to be stirred and turned until it becomes Wrought Iron. When the metal acquires a sufficient consistency to admit of being removed, it is taken out, and the impure end cut off. In this state it is called a "Bloom" or "Loop," and it is ready to be reheated in the heating furnace, and to be brought under the hammer.

The forge fire is used only for the best and choicest kinds of iron, as it is too expensive in coal and labor for making the cheaper kinds.

The Puddling Furnace is the most common of all means of reducing Pig Iron to Wrought Iron. This is a covered furnace like an oven, a grate being placed at one end, and a pit or trough being made in the centre. The chimney or stack is at the opposite side from the grate. The puddling furnace may be worked either with or without blast. Coal and wood are used alike for fuel. Mills in which rail-road iron is manufactured generally work

23

either Anthracite or raw Bituminous coal, with blast, for puddling furnaces. The pig iron is placed in the puddling furnace, and melted in about three quarters of an hour. It is then stirred with a suitable hook or poker, worked by a "puddler," having charge of the furnace. The stirring goes on until every particle of the puddle has been thoroughly exposed to the fire, and until the iron adheres in a spongy mass. It is then divided, while in the furnace, into four or five balls or lumps. These are taken successively to a stout hammer, called a shingling hammer, or else to a machine called a squeezer, either of which acts by compression, to get rid of the coarse cinder contained in the iron. This runs off in a melted condition, leaving the bulk malleable, and possessing the distinctive qualities of Wrought Iron.

It is then, while still retaining a great portion of its original heat, shaped, by rolling or hammering, into such forms as are found to be most saleable in the market.

In the vicinity of Philadelphia, Forges and Rolling Mills are generally separate establishments—the former considerably outnumbering the latter. In 1856 there were 116 Forges in Eastern Pennsylvania, and 63 Rolling Mills, including those in the city ; working about 175 hammers, and having about 500 forge fires, with the heating and puddling furnaces, and turning out an aggregate product, for the district tributary to the city, of about five millions of dollars annually.

In Philadelphia, Forges are usually combined with Rolling Mills, there being but one exception, viz., the Fairhill Forge, of which Patterson, Morgan, and Caskey, are proprietors. The Rolling Mills are as follows :—

Kensington Iron Works and Rolling Mill, James Rowland & Co., proprietors.

Penn Rolling Mill, Kensington, Verree & Mitchell, proprietors.

Treaty Rolling Mill, Kensington, Marshall, Plunkett & Co.

Robins' Rolling Mill, Kensington, Stevens Robins, proprietor.

Oxford Rolling Mill, Twenty-third Ward, W. & H. Rowland.

Fairmount Rolling Mill, Fairmount, Charles E. Smith & Co.

Fountain Green Rolling Mill, two miles above Fairmount, Strickland Kneass, proprietor.

Pencoyd Rolling Mill, below Manayunk, west side of Schuylkill, A. & P. Roberts, proprietor.

Flatrock Rolling Mill, Manayunk, A. P. Buckley & Son, proprietors.

Cheltenham Rolling Mill, one mile below Shoemakertown, Rowland & Hunt, proprietors.

In these establishments, over 700 men are employed, and receive annually in wages about $250,000. The aggregates of production were recently made up and published in the *United States Gazette*,* as follows:

	Tons.	Value.
Spring and Cast Steel, - - - - - -	2,100	$283,500
Bar, Rod, and Band Iron, - - - - - -	13,310	880,500
Boiler and other Plate, - - - - - -	1,660	150,000
Aggregate, inclusive of other items, for the nine Rolling Mills of the city, - - - - - - - - - - -		1,455,000
Distinctive production of Rolling Mills, simply, - - - -		1,206,500
Total for Forges and Rolling Mills within the city, - - -		1,801,150

The products, besides those above enumerated, include for the Pencoyd Rolling Mill, Rolled and Hammered Car and Locomotive Axles, and for the Fairmount Iron Works, Charles E. Smith & Co., proprietors, Rail-road Chair Iron, Marble and Stone Saws, and Bands and Bars of extra sizes.

Passing from the manufacture of Wrought Iron and Steel to that of CASTINGS, we are led to the consideration of Foundries and Cupola Furnaces, in which the smelting is usually accomplished by a process somewhat similar to that employed in the reduction of the ore in the blast furnace. The metal is mingled with coal in a capacious receptacle lined with fire-brick, and subjected to a furious blast of air from tweers beneath. The height

* A very careful statistical investigation of the Manufactures of Iron was recently made by the present indefatigable Secretary of the Board of Trade, and published in the *United States Gazette*—a leading commercial journal of this city. At the request of several Iron workers, who decline to furnish additional individual particulars, we shall, in many instances, adopt the results as published in that journal. It will also be observed by those who are familiar with the volume edited by myself, and published two years ago, that several paragraphs in this article, descriptive of processes, are extracts from that work.

of the cupola furnace, however, rarely exceeds ten or twelve feet, while that of the blast furnace approaches forty or even fifty, and the pressure required to force the air upward through the sinking mass of materials is of course proportionably less. No lime or other flux is employed in the cupola furnace; and the temperature required is presumed to be considerably less than in the blast furnace, the heat necessary to melt iron being generally assumed at 2300 to 2800 degrees of Fahrenheit's scale, or some fifteen times hotter than boiling water; while the temperature, in the hottest portion of the blast furnace, is supposed to reach 5000 degrees.

When drawn from the cupola furnace, the iron is poured into moulds of tightly compacted earth, the varieties employed being clay or loam and fine sand carefully mixed; the "moulding sand," as it is termed, being in most cases enclosed in boxes called flasks. If the desired casting is a column, an exact model or pattern is embedded, one half in each of two flasks of sand. The sand having sufficient cohesion to retain any impression given it, the pattern is then withdrawn, and the flasks fitted accurately together, as before, leaving a cavity to give just the required shape to the metal, which is afterward poured in. If the casting is to be made hollow in any part, a "core," or solid mass of sand, is previously baked in a suitable box, of the shape of the desired cavity. This core of sand, being placed in the flask, and adjusted suitably to the mould, leaves a cavity of its own shape in the casting. For, while the core assists in confining the melted iron to the desired limits of size and form, it can be punctured and removed, after the casting has cooled, with the same ease that the mould itself may be broken up, and the sand be again used for another mould.

The interior surface of the moulds is generally dusted or rubbed with finely-powdered coal or other material, technically termed "blackening," the object of which is to induce a smooth, perfect surface on the casting. The astonishing smoothness and delicacy of the small statuettes, known as "Berlin Castings," it is believed, are the result of some secret with regard to blackening the moulds.

The exterior of a large casting is invariably harder than the interior. This effect is probably due to the rapid cooling of the parts in contact with the sand, as the hardness is found to depend very much upon the rapidity of cooling. This fact has induced many experiments, and the quite general adoption of several different processes, according to the quality of the work required.

Iron required to withstand wear, as hammer faces, car-wheels, gudgeons, &c., is cast in close contact with a large mass of cold iron ; and iron in which a great uniformity of strength and a general softness is required, as small portions of machinery which are to be drilled, planed, &c., is cast in moulds previously heated to a tolerably high temperature. The former are called "chilled," and the second "dry-sand castings," as distinguished from the first described, or "green-sand castings." Very large moulds are built up with brickwork and lined with clay ; and the products are termed "loam castings." There are other processes for rapid cooling besides that above mentioned, one of which consists in a rapid circulation of water through pipes in the vicinity of the part to be chilled, but all act in a substantially similar manner, and with the same result.

There is a process of "annealing" metals, by heating, and then gradually cooling under favorable circumstances, which we will refer to when we come to speak of Car wheels. A species of cast iron, produced by a modification of this process, is called " Malleable Iron," and combines in a high degree the tenacity of wrought with the cheapness of cast-iron shapes. An immense number of Locks, and other articles in the Hardware trade, is produced by this process, which may again be alluded to in a separate division of our subject.

During the last few years the demand for Castings of great size has severely tested the skill of founders, but they have invariably responded to the calls by producing specimens more remarkable than any heretofore attempted. Cylinders, in which the tallest man could stand upright, have repeatedly been cast at Foundries in Philadelphia ; and those of the "Erricson," cast at I. P. Morris & Co.'s Foundry, were eleven feet five inches in diameter. The boring was executed by their great Vertical Boring Mill, which was in use in this city before introduced into New York. The cylinders for the blowing machinery of the Lackawanna Iron

25*

Works, at Scranton, Penna., cast at the same establishment, are one hundred and ten inches in diameter, and ten feet stroke. But the heaviest casting ever made in this country, and probably in the world, was the bed-plate for the Baltic, which weighed 130,148 pounds. The bed-plate for the Arctic weighed forty-five tons, and that for the Atlantic thirty-seven tons.

The products of Foundries, disconnected from Machine-shops, consist principally of Stoves, Hollow-ware, Iron Building work, and Railings, Safes, &c., to each of which we shall briefly refer.

1. STOVES AND HOLLOW-WARE.

Five large Foundries in this city are devoted exclusively to the manufacture of Stoves; while two others make Stoves, together with miscellaneous castings. The capital invested in the manufacture is about $600,000, and the annual product about 12,500 tons, worth $1,250,000. The designs, in many instances, are remarkable for their elegance, and the establishments are not surpassed in facilities or in extent by any others. The moulding-room of one firm is three hundred and sixty feet long and sixty feet wide, being the largest moulding-room, with the exception of one also in Philadelphia, in the United States. The Foundry of another firm has facilities and capacity for turning out 30,000 Stoves per annum. The cheapness of the raw material, and mildness of the winters, enabling the manufacturers to continue operations without cessation throughout the year, are marked advantages, and the fineness of the castings induces professed manufacturers in other places to obtain their supplies from this city. The varieties made here embrace almost every description, from the old Franklin Stove, and the *Ten-plate* Wood Stove, down to the most modern styles and patterns, including Gas Cooking Stoves. In originating patterns and beautiful styles, the Philadelphia manufacturers and Stove pattern-makers have been remarkably successful; and Stoves from this city have been shipped to Oregon, California, Australia, and Europe; while in our own markets no others can compete with them.

In addition to the establishments devoted, either entirely or in part, to the production of Stoves, there are about fifty Stove-makers who get their castings from founders, and finish them in their own shops. As some Cooking Stoves have sheet-iron ovens, and many Parlor and Office Stoves are chiefly composed of Russia Sheet Iron, the value of the castings, in some instances, is increased from two to three times. About nineteen establishments, besides the above, are engaged in the manufacture of *Hot-air Furnaces*, (or Heaters,) and *Cooking Ranges*. They usually originate or purchase the patterns, and get the castings executed at the regular foundries. The varieties made, embrace the most complete, convenient, and economical, as well as a fac-simile of the article so long used in New York; and the Summer Range, or Gas Oven, which originated here, and is said to be unknown elsewhere. The above establishments furnish employment to at least six hundred metal workers, and consume a large amount of Russia, English, and American Sheet Iron, besides Tin-plate, Fire-brick, &c., &c. *Ornamental Iron Parlor Grates*, for which we have long been dependent upon New York, are now made here of great elegance, and in various styles, by at least one firm, who has recently erected ovens for baking on the enamel.

Three of the Foundries in Philadelphia are occupied almost exclusively in casting Hollow-ware and Hardware Goods, which are subsequently enameled or tinned. The establishment of one of these, that of Messrs. STUART & PETERSON, is probably more extensive than any other of the kind in the Union. In this manufacture great care is necessary in the selection and commixing of the different brands of Iron, in order to obtain castings of proper tenacity; and after such are obtained, the inside surface of the ware must be made smooth and bright to protect the enamel. In England this is effected by turning the article in an ordinary foot-lathe, the tool being guided by hand; but the inhalation of particles of Iron proved most destructive to the lives of the operatives. The firm above alluded to, employ for this purpose self-acting tools or lathes, the invention of their master-machinist; and so admirably do they conform to the irregularities of the surface to be turned, that they seem to be endowed with almost human intelligence.

The products of this establishment embrace a great variety of

Culinary and Household articles—Pots, Kettles, Stew-Pans, and other articles, from the smallest to the largest, as Caldrons, &c.*

The other Hollow-ware Foundries in the branch, are those of Messrs. SAVERY & Co., and LEIBRANDT, McDOWELL & Co., late Finley & Co. The former has been established about twenty years, and has produced an immense number of Pots, Pans, Kettles, &c., besides Plows, and other agricultural implements of great variety and acknowledged excellence. This firm employ about one hundred hands, and have been very successful in producing castings remarkable for their size, as Caldrons, Sugar Boilers, &c., capable of holding hundreds of gallons. The establishment of FINLEY & Co. is new, and has recently changed owners. In its present hands it will no doubt soon take rank with the others.

2. BUILDING AND ORNAMENTAL IRON-WORK.

The use of Iron, as a material for building purposes, must be ranked among the modern applications of this wonderful metal. The gentleman who erected the first Iron Building in the United States is, we believe, still prepared to receive orders. The oldest Foundry in Philadelphia, devoted to the production of Building castings, was erected in 1804 ; and its proprietor, Mr.

---

* We extract the following remarks from a circular of Messrs. STUART & PETERSON, who are certainly entitled to very great credit for their successful efforts in competing with foreign manufacturers.

"We now *anneal* and *turn out bright* the *inside surface* of all the ware we enamel or tin, the annealing making it *less liable to break by sudden exposure to heat*, and turning off the casting surface makes it retain the enamel more perfectly ; and even after long use, if the enamel should come off, the surface left will be smooth and easily kept clean, altogether making it more serviceable than the ware of those manufacturers who do not prepare their ware thus. *We wish it to be known particularly* that we do not put into the mixture, *or use in any way in the preparation of our enamel*, any lead or other metallic oxides.

"We desire to call particular attention to our Tinned (usually called PATENT METAL) Ware. We prepare it for tinning in the same way, use the same quality and quantity of tin on each piece, use the same quality of iron, and finish it in the same way, making it in all respects the same quality as *English Patent Metal Ware*. It will stand as much heat and use, and many of our customers have been pleased to say, is more bright and perfect than any they have ever seen imported."

James Yocom, was one of the first in this country to make Iron fronts for buildings. The business now employs six Foundries, almost exclusively; and as the advantages of Iron for this purpose, combining as it does strength and durability, with cheapness and facility of elaborate ornamentation, become more manifest, the architectural popularity of the metal will extend.

At the present time, the firms engaged in producing Building Castings, may be said to execute work for the whole country. During the last year, Messrs. H. C. ORAM & Co., made and put up a five-story Iron front in New Orleans; an Iron front in Savannah; another in Nashville; supplied Ornamental Castings for the Town Hall in Wilmington, N. C.; a Cast Iron frame for the New Orleans Gas Company; besides putting up a large number of fronts in Philadelphia and its vicinity; casting thirty-six bowstring girders from sixteen to forty-three feet long, six hundred and fourteen columns from eight to twenty-two feet, and the magnificent Cast Iron dome and ceiling of the Bank of Pennsylvania. Another firm, Messrs. HAGAR, SANSON & FARRAND, has executed extensive orders for Galveston, Texas; and supplied numerous places in the South and West. This firm make a *Revolving Iron Shutter*, which is extensively used and highly appreciated. Its ability alike to resist the assaults of fire and of burglars, as well as its durability and convenience, has increased its popularity and induced an extensive demand. These Shutters are known as Mettam's Patent, and are corrugated, which gives the slats increased strength. Mr. Sanson has invented a machine which cuts and corrugates the slats at the same time. This cheapens the production, and enables this firm to supply Shutters at a reduced price.

It may be safely said that the firms now engaged in producing Building work have a most complete and extensive stock of patterns, and every facility for the execution of orders, however difficult may be the design or configuration desired.

The manufacture of *Roofing* is made a distinct branch of the Iron Building work, in this city. One firm, Messrs. R. S. HARRIS & Co., is very extensively engaged in the manufacture of *Corrugated Iron Roofing*, an article introduced here some years ago by Asa Whitney, and found peculiarly well adapted for covering buildings of great size or span, as Rail-road Depots, Foun·dries, Banks, &c., while also well adapted for smaller buildings It is well-known that a wooden roof, if the span be great, say sixty or eighty feet, requires a very heavy frame; but by the pro cess adopted by this firm, a roof superior in durability is obtained with less weight. The material used is generally American Galvanized Iron, (unless common iron painted, of which the first cost is less, be preferred;) and is supported on a peculiar Patent Independent Truss, supplied by this firm. The corrugating so strengthens the material, that Iron No. 22, possesses all requisite strength for the largest building. The works of this firm, situated at Prime and Eleventh streets, are well provided with machinery for making every part of a roof on their premises; but probably the most remarkable of their machines is one for punching, by which the sheets of a roof are so accurately and uniformly punched, that the proper place of each can be known, and the entire roof put up by others than the manufacturers.*

Messrs. Harris & Co. also make Patent Galvanized Cornices, which are cheaper and lighter, and more ornamental than stone, and more durable than wood. Specimens can be seen at the Academy of Music, in Philadelphia, and at Nassau Hall, Princeton, N. J. It is hoped that Roofing and Cornices, of the description which this firm manufacture, will come into more extensive use than heretofore; for, being entirely fire-proof, they are a protection to a city.

Ornamental Iron Work, and especially the manufacture of *Iron Railings*, constitute to some extent a distinct business, though generally associated with Architectural Iron-work in some of its forms. The Iron Railings made in this city are of a very superior character, both as regards the construction and decorative arrangement of the parts; no expense being spared by the

---

* Among the numerous buildings covered with Corrugated Iron Roofing, we might mention the United States Mint, the Masonic Hall, John Grigg's Fireproof building, the Depot of the West Philadelphia Passenger Railway, the Phœnixville Iron Works, the Gas Works in Cincinnati, the Custom House in Mobile, the Charlotte Branch Mint, the very extensive buildings of the Georgia Central Rail-road Company at Savannah, the Gas Works at Richmond, and at Winchester, Va., and many others in the chief cities of the South and West; besides several in Havana, Cuba.

leading manufacturers to obtain beautiful and tasteful designs. Most of the Cemeteries and Public Squares throughout the whole country are adorned by work executed in Philadelphia; and every city, probably every town in the Union, contains some specimen of our manufacturers' skill and taste. The English Commissioners, in their report on the Industry of the United States, refer, in terms of high commendation, to the Ornamental Cast Iron-work of Philadelphia, and mention with appreciation the cast-iron statue of the late Henry Clay, fifteen feet high, recently furnished from Mr. Wood's establishment, and erected in Pottsville.*

In the manufacture of Railings, though Cast Iron is principally employed, Wrought Iron is used in considerable quantities. The latter is. considered superior to Cast Iron in the power of resisting strains or concussions; and since the discovery of the process of *weaving bars* of any size, recently introduced into Philadelphia, it is possible to attain equal strength in the construction of Window Guards, Gratings, Railings, &c., with much less weight of material than formerly.

Within the last few years the applications of Cast Iron-work have been greatly extended. Iron Bedsteads, of all sizes, are made largely by at least two firms, Messrs. WALKER & SONS, and SAMUEL MACFERRAN. The Ornamental Iron Bedstead of Mr.

---

* We extract from our Reporter's notes the following remarks with respect to another of the Philadelphia manufactories.

"At the establishment of W. P. HOOD, I saw samples of finished work that certainly cannot be excelled for taste and beauty. This gentleman is successor to Messrs. Moore & Gallagher, a well-known firm, who founded the establishment in 1848. The enclosure of Lafayette Square in Washington, and the Iron-work of the Patent Office, were both executed at this establishment. The assortment of designs is, I think, larger than that of any other establishment in the Union, and I have seen the most of them. A specimen of the late production of this establishment may be seen in the heavy and rich railing around Lodge No. 2, A. Y. M., at Mount Moriah Cemetery. The work executed embraces, in addition to Railings for public and private purposes, Settees, Chairs, Tables, Fountains, Garden Urns, Brackets for Porticos, Mantels, Winding Stairs, Fire proof Doors, Window Shutters, and Iron Vaults for burial purposes in place of marble. The superintendent of this establishment has had fifteen years' experience in the business; and Mr. Hood, the proprietor, is justified in claiming that he can please the most fastidious taste in regard to beauty of design and well-finished work."

Macferran is highly recommended by the Franklin Institute, as an article combining neatness, and light weight, with sufficient strengh.   His manufactures include a great variety of Ornamental Iron Castings, Hat Racks, Umbrella Stands, Water Coolers, Washstands, Sinks, Fountains, Settees, Dogs, Lions, Tables, Chairs, Towel Racks, with a great variety of Brackets, Hitching Posts, Spout Castings, &c., &c.

This gentleman is noted for his taste in designs, and ingenuity in originating desirable patterns.   He also manufactures the celebrated Champion Hot-air Furnace, and Ranges, Gas Ovens, &c.

3. SAFES.

The manufacturers of these articles, almost indispensable among a mercantile people, have so effectually "cried aloud and spared not," that the public are probably more familiar with their relative merits than we are.   The metal portion of the Safes consists of stout and tough Wrought Bar and Plate Iron; and the space between the outer and inner surfaces is filled with a chemical preparation, which is a good non-conductor of heat.   The interior is rendered wholly impervious to damp; and books, papers, and jewelry, may be preserved in them any length of time without blemish from mould or mildew.   Rival makers have manifested a very determined disposition to burn up each other's Safes; and if none have succeeded in doing this, we must infer that all are equally proof against fire.   The annual production in this city is about $150,000.

The principal restriction hitherto to the more extended use of Iron has been its tendency to oxdyation or rust, but happily mechanical ingenuity has overcome this difficulty.   Iron is now coated with another metal, forming a combination impervious to atmospherical influences, and known as GALVANIZED IRON.

The process of effecting this great change in this useful material, and forming Galvanized Sheet Iron, is described to us by a leading firm in the business, as follows : The Iron is first rolled into sheets as ordinary Sheet Iron ; but for the purpose of galvanizing, a selection is necessary, for experience has proved that Iron, though of good quality, will not in all cases combine with the zinc which is used in coating.   The sheets selected are rolled very smooth and well trimmed to the size required, and cleansed from all impurities by a weak acid. The effects of the acid are in turn removed by immersion in a tank of clear water, and then the sheets are dried in an oven. The iron thus prepared is placed in contact with the zinc, and the two metals being brought to the same temperature combine and fuse, and form a material impervious to rust, and requiring neither paint nor any preservative agent. The proper regulation of the temperature of the zinc and the iron is a point of great nicety, requiring in the manufacturer much previous experience.

The firm to whom this material is indebted for much of its present popularity and even intrinsic value, and who, we understand, were the first to introduce the manufacture of Galvanized Sheet Iron into the United States, are Messrs. McCullough & Co., of Philadelphia. Their works, it is believed, are the most extensive of the kind in the Union. Their mills for rolling Sheet Iron, of which they own three, are located in Cecil County, Maryland—two at North East, and one at Rowlandsville. They are driven by water-power—the former by the feeders of the North East Creek, and the latter by the waters of the Octorara; and are capable of producing from 1,500 to 2,000 tons of Sheet and Flue Iron annually. The firm employ at their various works from 200 to 250 men, use 2,000 to 2,500 tons of Pig Iron and Blooms, and consume about 1,500 tons of Anthracite, and some 2,000 tons Bituminous coals. The works for Galvanizing are located in the city, at the corner of Eleventh and Prime streets, and all the Iron made is brought from the mills, by way of the Philadelphia, Wilmington and Baltimore Rail-road and the River, and delivered at the warehouse in connection with these works.

The Galvanized Iron of this firm has been tested by the eminent chemist of the Mint, Professor Booth, who pronounced it equal to that of English manufacture; and in certain tests by sulphuric and other acids, it proved superior. Its applications are necessarily almost as numerous as Iron itself, being available wherever exposed to corrosive influences, and specially adapted for Roofing, Iron-work for ships, Water and Gas Tubing, Window Shutters, Telegraph Wire, &c.

In the northern part of the city there is another establishment for Galvanizing Sheet Iron, Wire, &c., MARSHALL, GRIFFIN & CO.

24

proprietors.   At these works sixty men are employed, and twelve miles of Telegraphic Wire are galvanized in a day, at a cost of about $10 per mile.

We have thus briefly traced and narrated the processes employed in the production of the various kinds of Iron—Pig, Wrought, and Cast, from the period of its extraction from the earth in the form of Ore, down to its introduction to the market. A more comprehensive and connected view of the whole, however, may be obtained from an examination of the subjoined table, or ANALYTICAL VIEW OF THE MANUFACTURE AND USE OF IRON

Natural *Iron Mines* and *Iron Beds* contain

- Magnetic Ore, or Black Oxide of Iron, used in U. S., Russia, Sweden, India, &c. The richest ore known.
- Specular Ore, or Red Oxide of Iron, used in U. S.
- Hematite Ore, or Brown Oxide of Iron, used in U. S.
- Clay Iron Stone, used in England, Scotland, &c.
- Black Band Ore, a compact Carbonate of Iron, used in England, Scotland, &c.
- Sparry Carbonate of Iron, used in Germany.

These undergo in the *Blast Furnace*

Digging or Mining, Roasting,* SMELTING.

and assume the form of *Pig Iron,* rated as

White Iron, Mottled Iron, Bright Iron, Gray Iron.

- Foundry Iron, No. 1, flows very readily, worth $32 per ton.
- Foundry Iron, No. 2, worth $28 per ton.
- Foundry Iron, No. 3, worth $27 per ton.
- Forge Iron, No. 1, worth $26 per ton.
- Forge Iron, No. 2, worth $25 per ton.
- Forge Iron, No. 3, worth $23 per ton.

They next in *Iron Works* and *Foundries* undergo

Refining* PUDDLING, Squeezing or Hammering, Rolling.   Melting, MOULDING, Cleaning.

And are put in *use* as

- Anchors, Wire, Shafting, Rail-road, Tee, Angle, Scroll, Sheet, Band, Boiler, Round Bar, " Flat Bar Iron, " " " " " " "
- Pavements, Coffins, Cradles, Bridges, Buildings, Girders, Ornaments, Statues, Fences, Gratings, Furnaces, Stoves, Hollow Ware,

And when *worn out* become

Wrought Scrap Iron.   Cast Scrap Iron.

* Those processes marked by a Star are sometimes omitted.

XX.

Iron and its Manufactures Continued. Machinery.

The manufacturers of Machinery, considered with reference to the nature of their occupation, are divided into two classes, who may be styled *special* and *general* Machinists—the former being those who confine their operations to a special and particular class of Machines and Tools, and the latter being those who have the disposition and facilities to execute orders for almost all kinds of Machinery, heavy and light. The Machine-makers of Philadelphia are, in this view, principally general Machinists; but each of the following classes—*Cotton* and *Woolen Machinery ; Railway Machinery ; Machinists' Tools ; Paper-makers', Printers'*, and *Bookbinders' Machines; Fire Engines; Gas* and *Water Apparatus*, and probably some others,—has extensive establishments devoted expressly to its production. We shall briefly advert to the most important of these special classes, commencing with

1. COTTON AND WOOLEN MACHINERY.

It is stated, in apparently authentic records, that the manufacture of some parts of the machinery necessary in the production of Textile fabrics, was carried on in Philadelphia in the time of the Revolution. As early as 1778, we learn from Scott's Gazetteer of the United States, published in 1805, that the eminent Philadelphia machinist, OLIVER EVANS, manufactured

" Wire from American Bar-iron, which he made of excellent quality, on the most improved plan carried on in this country ; also wrought it into wire for cards, in the way described by those who had seen them made in Europe. But thinking the process too tedious, he invented a machine by which he could work the wire into card teeth, at the rate of nearly three thousand per minute, by the simple motion of turning a winch, or wrench, by hand ; also, a machine for punching the holes in the leather for the teeth, by which he could prick by the motion of his hand one hundred and fifty pair of cards per day. He also planned a wire mill, with machinery to make the wire into card teeth as fast as drawn. This he has often declared was one of the greatest productions of his mind. He applied to the Legislature of Pennsylvania for aid to carry it into effect ; but this was not granted, and this was lost. When peace was established he declined this business, and in the year

1783 commenced the building of a merchant flour mill, which led him to the study of the improvement of the art of manufacturing flour; and invented the machines which he has denominated the Elevator, the Hopper-boy, the Conveyor, and Drill, by means of which, when properly applied, the greatest part of the manufacture and labor which were before necessary is now saved."

But the first regular manufactory of Cotton Machinery was established at Holmesburg, in 1810, by ALFRED JENKS, who had been a pupil and colaborer with the celebrated Samuel Slater, and who brought with him from New England drawings of every variety of Cotton Machinery, as far as it had then advanced in the line of improvement. He supplied the first mill started in this portion of the State of Pennsylvania, with the requisite machinery; and subsequently the Keating Mill, at Manayunk, now owned by J. G. Kempton. In 1816 he built for Joseph Ripka, a number of Looms for weaving Cottonades. A record now before us states:

"Under the universal impetus given to home manufactures during the last war, Mr. Jenks greatly extended his business operations, and in 1819 or 1820 removed to his present desirable location in Bridesburg, the increased growth of which is owing in no small degree to the personal efforts and enterprise of himself and the importance of his establishment. Here, where he possessed the necessary facilities for shipping to his more distant patrons, he conveyed his old frame building from Holmesburg on rollers, which yet stands amid the more substantial and excellent structures beside it. This, however, was found too small for his increased business, and was extended by the erection of a stone building thirty feet long, now forming the north end of the present main building, which is four hundred feet in length. When the demand first arose for Woolen Machinery in Pennsylvania, Mr. Jenks answered it, and at once commenced its manufacture, and furnished the first Woolen Mill erected in the State, by Bethel Moore, at Conshohocken, with all the machinery necessary for this manufacture.

"In 1830, Mr. Jenks, impressed with the idea that the labor of manipulation was insufficient to supply the wants of the population, or to meet the commercial demands, invented a Power-loom for Weaving Checks, and introduced it into the Kempton Mill at Manayunk, where its success produced such excitement among hand-weavers, and others opposed to labor-saving machinery, as to cause a large number of them to go to the mill, with the avowed purpose of destroying it, from doing which they were only prevented by the presence of an armed force. This, and other improved machinery made by Mr. Jenks, soon acquired an extended reputation, and induced the erection of larger buildings and the introduction of increased facilities. The numerous valuable improvements made by Mr. Jenks, from 1819, when he obtained his first patent, to the present time, and those of his equally ingenious and skillful son, are embraced in such a vast number of patents, and are so various in their nature and construction, as to prevent us from even enumerating their titles and objects in this limited notice. They are, however, well-known to manufacturers."

The present works of Messrs. ALFRED JENKS & SON are unquestionably among the most extensive and important for the manufacture of Cotton and Woolen Machinery in the Union. The present sole manager, BARTON H. JENKS, Esq., has been untiring in his efforts to improve and perfect the general system of manufacturing; and so successful in this, and in originating improvements with reference to special articles, that, at the present time, we do not believe there are any other works in the entire Union that can be compared with these for the purposes for which they are designed. The development which we have shown has been attained in the manufacture of Textile fabrics in Philadelphia, and its vicinity, is no doubt in part due to the excellence of the machinery supplied from this establishment; but its benefits are by no means local, for its products are as regularly shipped to New England as to Manayunk, and to the South as to Gloucester. We, however, shall reserve a description of these Works for the APPENDIX.

Cotton and Woolen Machinery is made at several other establishments in Philadelphia; but in these the scope of operations is either restricted to certain particular Machines, or is so extended as to embrace general Machinery. J. & T. WOOD, proprietors of the Fairmount Machine Works, for instance, do

24*

an extensive business in the construction of Looms, for which they are provided with all the requisite facilities ; and in 1856, they turned out four hundred and eighty Power Looms, or forty per month. Their Looms, we believe, are so constructed as to be adapted for use either in Cotton or Woolen Factories. But the business of this firm takes a much more comprehensive range, embracing the construction of *Embossed Calenders, Lard Oil Presses*, of which they can make twenty per month, and all kinds of *Shafting, Pulleys, Hangers, Couplings,* &c., and machine work in general. They employ regularly about seventy hands. The Messrs. Wood have an excellent reputation for doing thoroughly and well whatever they undertake.

Messrs. HINDLE & SONS, in West Philadelphia, employ about fifty hands, principally in the manufacture of Woolen Machinery ; and Messrs. ECCLES & SON make Looms, and a variety of Machinery for Cotton Factories.

Hepworth's picking stop-motion for Drop-box Power Looms is made by J. J. Hepworth ; and Lead Wire for Looms is made at the Lead Pipe Works of Tathem & Brothers, which are believed to be the most extensive of the kind in the world.

Card Clothing is made very extensively at one establishment, where fifty-three of those wonderful and ingenious machines, which Webster is reported to have said seemed to be endowed with human intelligence, are in constant operation. The original Machine was patented in 1810, by Thomas Whittemore, though the real inventor was Elizur Smith, of Walpole, Mass. Various improvements have been made from time to time ; and now so perfect and automatic are its operations, that only three men are required to tend fifty-three machines. It seizes the wire in its iron fingers, bends it, punches holes for it in the leather, then inserts it; and if the slightest derangement take place, or the least imperfection is manifested in the manufactured product, it stops and waits until the difficulty is remedied.

The proprietors of this establishment, Messrs. JAMES SMITH & Co., are experimenting with reference to the substitution of Cloth for Leather in their manufacture. The average price of the Card Clothing made in Philadelphia, is $1 per square foot, and the quality superior to that made elsewhere in the country.

One firm, Messrs. W. P. UHLINGER & Co., are extensively engaged in the manufacture of Ribbon Looms, Jacquard Machines, and Rotary Knitting Machines. This excellent establishment employs from forty to seventy mechanics, and does an annual business of over $50,000. The most ingenious and complicated Machinery is made here—Ribbon Looms, for instance, being self-acting, and combined with the Jacquard Machine, to be propelled by power or hand. These are supplied largely to New York, Connecticut, and Massachusetts—the extensive Manufactory of Ribbons and Trimmings, at West Newton, in the last-named State, being wholly supplied with Machines by this firm.

Rotary Knitting Machines for Stockings, Jackets, Shirts, &c., are made at this establishment. For this Machine, Mr. Uhlinger received a first-class premium from the *Franklin Institute;* and its practical value is shown in the patronage bestowed upon it, both by power and by hand-loom weavers.

Mr. Uhlinger's establishment was founded in 1850; and though its transition from insignificance to importance has been rapid, its present equipments, perfection in machinery, and quality of its manufactures, entitle it to rank among the important ones of Philadelphia. The demand for Sewing Machines has induced the proprietor to provide himself with superior facilities for their manufacture; and hereafter these important Machines, so largely sold in this market, will also be extensively made here.

The common Knitting Frames are made by two persons in Germantown; and Looms, &c., by several manufacturers in a small way throughout the city.

Shuttles are an exclusive article of manufacture in at least two concerns—those of Mr. H. SERGESON, and E. JACKSON, each of whom make annually 20,000 Shuttles of all sizes, from the small ones for Silk or Lace, and Hand-looms, to the largest sized ones, with wheels, used in weaving Broad-cloth. The prices range from $4.50 to $22 per dozen. Philadelphia has peculiar advantages for the production of an excellent Shuttle, at a moderate cost, from the fact that a better quality of wood, used for this purpose, is here attainable than elsewhere. This is a fine quality of Dog-wood, which grows upon the Isthmus, between the Chesapeake and Delaware Bays, and for this manufacture it is nearly equal to the best Turkey Box-wood. The varieties grown further South, or more inland, are softer, and of inferior quality, the sea air apparently conducing to the perfection of the wood.

Small Bone and Ivory Shuttles, for ladies' use, are also made, and both common and fancy Shuttles are sent from Philadelphia to all parts of the Union.

Reeds and *Heddles* constitute a distinct business for several parties. One firm has employed as many as sixty hands; and the annual product has attained a value of $150,000.

2. RAILWAY MACHINERY.

The activity of the American people in constructing Railways, already extending, as they do, to more than 24,000 miles, or a distance as great as the circumference of the globe, has necessarily called into existence immense establishments, exclusively devoted to supplying a demand for Railway equipments. Four years ago, it was estimated that the capital then invested in Locomotive building was $3,000,000, employing over 6,000 hands, who received $2,700,000 yearly for labor, and turned out $8,000,000 in value of manufactured products.

Twenty years ago, it is believed, there were not six Locomotive establishments in the Union. A story is told of a gentleman who, about that time, received an offer from a capitalist of New York to furnish him the necessary capital to engage in the manufacture of Locomotives, if he thought it would pay, and, as such offers were rare, was quite desirous to accept of it; but, after visiting the principal shops, reported to the capitalist that the business would not pay, "for with three hundred men and Baldwin's shop, in Philadelphia, he could build all the Locomotives the country would need for twenty years." This gentleman is probably now a wiser, as well as an older man; for, by examining the late report of the *Pennsylvania Central Rail-road*, he could learn that this Company alone has in use 216 Locomotives.

The concentration of Rail-roads, and the advantages for economical construction, and especially the experience and eminence of her Engineers, have made Philadelphia a great centre for the construction of Railway Machinery. The establishments which are principally occupied in this pursuit, are among the most extensive, important, and interesting in the city; and this remark applies not merely to those which are employed in producing complete Machinery, but also those occupied in making parts, as Wheels, Axles, Tubes, Turn-Tables, &c. We shall advert to the more prominent of these, commencing with LOCOMOTIVES.

It is a somewhat singular fact, that the same eminent Philadelphia Engineer, to whom we referred as a pioneer in the construction of Cotton Machinery, is also credited with having built the first Locomotive Steam Engine, taking the word locomotive in its derivative signification as "self-acting."*

It may also be claimed, that the first entirely successful American Locomotive was built by a Philadelphia mechanic; while it is conceded that here many of the most important improvements in its construction and capabilities had their origin. The workshops of this city have sent forth nearly 1,800 Locomotives to perform their part in extending civilization, some of which are now thundering up mountain grades, on the long lines of the Pennsylvania Central and Baltimore and Ohio roads, while others are extending the fame of American genius in Continental Europe. The establishments, of which there are two in this city, date from the organization of the manufacture into a distinct business, and a brief outline of their history will not be deemed inappropriate.

---

* Scott's Gazetteer, published in 1805, speaking of Oliver Evans, says:— "He is now just finishing a machine called the *Orukter Amphibolis*, or Amphibious Digger, for the purpose of digging either by land or water, and deepening the docks of the city of Philadelphia. It consists of a steam-engine on board of a flat-bottomed boat, to work a chain of hooks to break up the ground, with buckets to raise it above water, and deposit it in another boat to be carried off. This principle he can no doubt apply to dig canals to make great dispatch. *Orukter Amphibolis* is built a mile from the water; and although very heavy, he means to move it to the water by the power of the engine. *Its first state will then be, a Land Carriage moved by steam.*"

M. W. BALDWIN & CO.'S LOCOMOTIVE WORKS.

The founder of these works, Mr. M. W. Baldwin, is a native of New Jersey, but has been a resident of Philadelphia for over forty years. He commenced his mechanical career as an apprentice to the Jewelry manufacture; but, on attaining his majority, saw proper to apply the knowledge so obtained to the production and improvement of Bookbinders' Tools, which at that time— thirty or thirty-five years ago—were generally imported. In partnership with David H. Mason, he prosecuted this manufacture with success; and, by the introduction of new designs, largely extended and improved Ornamental Bookbinding. To this business was added in 1822, that of engraving rolls for printing cotton goods, which became the source of large profits. They were the originators of this business in this country, and pursued it without competition until they had brought it to a degree of perfection that defied foreign competition. Subsequently, Banknote engraving was attempted with fair success. These pursuits required the invention and manufacture of a variety of tools and machinery adapted to particular uses, the getting up of which gradually introduced the Machine business, and the manufacture of Hydraulic Presses, Rolls for Calendering Paper, Stationary Engines, and finally the Locomotive. In 1830, at the request of Mr. Peale, the proprietor of the Philadelphia Museum, Mr. Baldwin constructed a model Locomotive Engine for exhibition, which was put in use in 1831, hauling five or six passengers in a train of cars, and attracting crowds to the then novel sight. This led to an order for an engine from the Philadelphia and Germantown Railroad Company; it was completed in 1832, and placed on the road in January, 1833. This was, undoubtedly, the first successful American Locomotive Engine; and, from the records in the newspapers of that day, its performance was not exceeded for years after, having made a mile in less than a minute. The business was now commenced, and extended as rapidly as. the necessary tools, patterns, and fixtures, could be obtained. During the years 1833–34, five engines were built, and the large shops on Broad, above Callowhill street, now occupied as their works, were commenced and completed. In 1835, fourteen Locomotives were manufactured; in 1836, forty; and in 1837, between forty-five and fifty. The financial revulsions of the period reduced the number, in 1838, to twenty-four. The leading features of the engines built by Mr. Baldwin, and which established his reputation upon a permanent basis, were their simplicity, strength, and durability. The greater portion are yet in use; and, within the limit of their power, are still doing duty profitably to their owners, and creditably to the skill of the builder. The plan of attaching the cylinders to the outside of the smoke-box, now almost universally adopted, originated with Mr. Baldwin; and also the metallic ground joints, and various minor improvements, upon which the present perfection of the Locomotive Engine depends.

In 1842, Mr. Baldwin introduced the six and eight-wheel connected engine, with an arrangement of truck for adaptation to the curves and undulations of the road. The superintendent of the largest coal freighting road in the United States says of these: "They are saving us thirty per cent. in every trip on the former cost of Motive or Engine Power."

In 1854, Mr. MATTHEW BAIRD became associated with Mr. Baldwin, under the present firm style of M. W. Baldwin & Co. Mr. Baird is a practical mechanic, who is familiar both with the details of the Locomotive business since its commencement, and with other mechanical pursuits, and is a gentleman of much and deserved popularity. Contributing to the concern capital, energy, and practical knowledge, it has, with his accession, taken a new lease of prosperous activity.

The proprietors of these works have for years been engaged in perfecting a system of engines, by means of which they could be adapted to economical working on almost any grade or curve. Several distinct kinds, and numerous sizes of each kind, from three to thirty-five tons weight, are manufactured with from two to eight driving-wheels. The system of adaptation, and its advantages, are seen in its results. On the Pennsylvania Rail-road, Eastern Division, where the grades are moderate, a passenger engine, has been running over eighteen months 133 miles per day without the loss of a trip for repairs.

The success with which difficulties are overcome by engines of his firm's construction, is specially illustrated in a pamphlet published by Charles Ellet, Civil Engineer, describing their working on a mountain top, over the Blue Ridge. He says:

"We should not regard mountainous regions as necessarily excluded from participation in all the comforts and conveniences due to the railroad, because they can only be reached by lines of very steep grade or very abrupt curvature. The American Locomotive can penetrate into the most retired valleys of Switzerland, and bring forth the products of their industry. Wherever men can go to cultivate the earth with profit, there the locomotive can follow to take away the produce of their soil. In fact, the engines daily running on this road, and drawing after them regular trains of forty or fifty tons of freight and passengers up grades rising at the rate of 296 feet per mile, and swinging their trains of eight-wheel cars around curves of less than 300 feet radii, are capable of carrying the artillery and supplies of an army up the steepest slopes of the present road over the Simplon, and offering facilities to an invader that would have been deemed impossible a very short time ago.

"This road was opened to the public in the spring of 1854, and it has now, in the autumn of 1857, been in constant use for a period of more than three and a half years. In all that time the admirable engines relied on to perform the extraordinary duties imposed upon them in the passage of this summit, have failed *but once* to make their regular passage.

"The locomotives for this severe duty were designed and constructed by the firm of M. W. Baldwin and Company, of Philadelphia. The slight modifications introduced at the instance of the writer, to adapt them better to the particular service to be performed in crossing the Blue Ridge, did not touch the working proportions or principles of the engines, the merits of which are due to the patentee, M. W. Baldwin, Esq. During the severe winter of 1855–56, when the travel upon all the Railways of Virginia, and the Northern and Western States, was interrupted, and on many lines for days in succession, the engines upon this mountain track, with the exception of the single day already specified, moved regularly forward and did their appointed work. In fact, during the space of three and a half years that the road has been in use, they have only failed to take the mail through in a single instance, when the train was caught in a snow-drift near the summit of the mountain.

"These results are due, in a great degree, certainly, to the admirable adaptation of the engines employed to the service to be performed; * * * the difficulties overcome in the location and working of the line, very much exceed those which have made the Austrian road over the Söemmering famous throughout Europe, while they have confirmed the claim of the American Locomotive, in climbing steep grades, to unrivaled pre-eminence."

The present extent of the works of M. W. Baldwin & Co., will be best illustrated by the following items of materials consumed during the year 1857, viz.:

Bar Iron	1,294,257 pounds.	Sheet Copper	103,692 pounds.
Boiler and Flue Iron	646,177 "	Ingot Copper	55,492 "
Sheet Iron	35,831 "	Banca Tin	14,536 "
Tire Iron	292,235 "	Springs and Steel	114,868 "
Pig Iron	1,901,536 "	Anthracite Coal	2,000 tons.
Axles and Forgings	315,981 "	Bituminous Coal	25,300 bushels.

Value $223,766 69.

Iron Flues	value $17,027	Lumber	value $9,017
Files and Hardware	"......11,745	Oil, Paints, Glass, &c	"...... 7,322

In addition to the above, Sheet Brass, Spelter, Charcoal, Belts, Hose, Locomotive Lamps, Steam Gauges, Moulding Sand, Fire-Brick, Clay, Boiler Rivets, &c., &c., were purchased to the amount of $30,000. Over 600 hands were employed, producing machinery equal to seventy-two Locomotive Engines, during the year.

THE NORRIS LOCOMOTIVE WORKS.

The Norris Locomotive Works originated in 1834, in a small shop, employing but six men, whose united wages was but thirty-six dollars per week. The power was furnished from an adjoining wheelwright shop, by a connecting shaft through a hole in the wall. Previous to this, in 1831, Mr. Wm. Norris, in connection with Colonel Stephen H. Long, General Parker, George D. Wetherell, and Dr. Richard Harlan, had formed a company for building "Locomotors," (as they were then called,) intended for the use of Anthracite coal as fuel. The first Engine was built under the immediate supervision of Colonel Long, at the Phœnix Foundry, Kensington. On the 4th of July, 1832, steam was raised, and it was tested on the New Castle and Frenchtown Rail-road. The trial proved their first attempt a failure, in consequence of the limited grate and fire surface. The Locomotive would run a mile at fair speed; but would then stop short, until a fresh supply of steam was generated.

At these works it is said an Engine was first constructed, capable of ascending heavy grades with loaded cars. This feat was

25

performed by the "George Washington," in 1836. This success excited attention everywhere to the superiority of Philadelphia Locomotives, and orders from Europe were received. In 1837, the Gloucester and Birmingham Railway, England, was supplied with seventeen Locomotives from these works, some of which are still in use.

The present works are very extensive, embracing numerous buildings situated on Hamilton, Fairview, Morris, and Seventeenth streets, on the locality formerly known as Bush Hill. In the year 1853 over one thousand hands were employed in them; and with the improvements in buildings, tools, &c., made since 1853, they can now accommodate over fifteen hundred hands.

There are several leading principles observed in the administration of these works, which appear calculated to insure their highest efficiency, and the best quality in their productions. One is the manufacture, upon the spot, not only of the Engines, but as far as possible, of the materials also of which they are composed. All the forged work—Tires, Tubes, Springs, Brass and Iron Castings, Chilled Wheels, and other parts, are here made in the best manner, and with the aid of every fixture to be found in establishments supplying separately each of these items. Another is the greatest possible substitution of machinery for manual labor. The tools are adapted, in a special manner, to the execution of each portion of the work; and each class of tools is specially appropriated to distinct portions of the work. Another is the entire independence of the different departments of the works from each other. Hardly any two distinct branches of labor are carried on together in the same apartment; but, at the same time, there is the utmost facility for all necessary communication between the separate departments. In the materials used for the Engines, wrought iron is used wherever practicable, and to the exclusion of cast iron. Hammered charcoal iron is used for the boilers; thick brazier's copper is used exclusively for the tubes; and tough scrap is used for all important forgings.

Up to the present period nine hundred and thirty-seven Locomotives have been constructed at the Norris Works; the average for the last ten years being about forty Locomotives per year. Of this number, one hundred and fifty-six were on foreign account, having been shipped to England, France, Austria, Prussia, Italy, South America, Cuba, &c.

The cost of a Locomotive, complete, varies between $6,000 and $12,000, although the price is somewhat confused, from the practice of taking stock or bonds of a road in total or part payment, and often at some nominal price, without reference to their real value. The weight of a large first-class Locomotive, whether for freight or passengers, reaches as high as from twenty to thirty tons, exclusive of the tender. It is expedient in practice to use large Locomotives and haul heavy trains, in preference to the reverse, as the expense of attendance, and, to a certain extent, of repairs, is no greater for a large than for a small engine.

The workmen employed in the Locomotive establishments of Philadelphia are a very superior order of mechanics, of whom the citizens of Philadelphia may justly be proud. The greatness of their mechanical creation is, in some respects, a prototype of their physical and mental characteristics.

Cars are made in Philadelphia at two establishments, which in excellence of production, if not in extent, rank among the first in the country. The Philadelphia builders have constructed Cars for more than fifty of the Rail-roads in the United States; and for beauty of finish, thorough workmanship, strength, and durability, their Cars have no superiors. Nearly all the Passenger and Freight Cars of the Pennsylvania Central, and all for the North Pennsylvania Rail-road were built by them, as well as large numbers for Rail-roads in the Southern and Western States, and in Cuba and the British Provinces.

The locality is one of the best for this manufacture in the country; for, connected as Philadelphia is by Rail-roads with every part of the United States, and being on tide-water, builders have every facility for convenient and cheap transportation to any part of the world; and, with iron and coal—the two heaviest items in their business—cheaper here than in any other shipping port in the Union, they necessarily possess unrivaled advantages for manufacturing Cars with the greatest economy.

For the manufacture of CAR WHEELS, A. WHITNEY & SONS have an immense establishment, the buildings of which cover 8,000 square feet of ground. The moulding room is four hundred feet by sixty feet—probably the largest in this country ; having two Railways extending its entire length, on which carriage Cranes are propelled, and used for removing the molten iron from the furnaces to the moulds, and the wheels from the moulds to the cooling pits. There are five large furnaces in all—three of which communicate by tubes with an immense caldron for containing melted iron. There are thirty-six cooling pits, having a capacity for holding at a time two hundred and fifty wheels.

The wheels are taken from the moulds as soon after they are cast as they can bear moving, without changing their form, and before they have become strained while cooling. In this state they are put into a circular furnace or chamber, which has been previously heated to a temperature about as high as that of the wheels when taken from the mould ; as soon as they are deposited in this furnace or chamber, the opening through which they are passed is covered, and the temperature of the furnace and its contents is gradually raised to a point a little below that at which fusion commences. All the avenues to and from the interior of the furnace are then closed, and the whole mass is left to cool gradually, as the heat permeates through the exterior wall, which is composed of fire-brick four and a half inches thick, inclosed in a circular case of sheet iron, one eighth inch thick.

By this process the wheel is raised to one temperature throughout before it begins to cool in the furnace ; and, as the heat can only pass off through the medium of the wall, all parts of each wheel cool and contract simultaneously. The time required to cool a furnace full of wheels is about four days ; and thus, in this manner, wheels of any form, and of almost any proportions, can be made with a solid nave.

The wheels while hot are then removed from the pits, and the centre part is placed in a hole communicating by means of a flue with a chimney one hundred and twenty feet high, and the edge is packed round with sand. A draught is thus created, which cools the mass of iron near the centre of the wheel, and in some measure prevents it from contracting unequally during the operation.

These works were established in 1847, and since that period they have turned out 164,000 wheels.

The manufacture of the minor parts of Railway Machinery constitutes in the aggregate an important business, but it is carried on usually in combination with other machinery.

CAR AXLES are made at several establishments, but principally at the Pencoyd Rolling Mill, of which A. & P. ROBERTS are proprietors. These works are almost exclusively engaged in the manufacture of both Rolled and Hammered Car and Locomotive Axles. Their products are in use on most of the leading Railroads of the United States, Canada, Cuba, and South America, and deservedly enjoy a high reputation for quality and finish. Since they have been in operation, until January 1, 1858, they produced 12,982 Hammered Car and Locomotive Axles, and 16,410 Rolled Car Axles, of various diameters and lengths. Their Axles are all stamped in the "*centre*" with the name of the works, year and day of month on which made, and are all centered ready for the lathe before leaving the works.

They have three Heating Furnaces, and one Trip and one Steam Hammer, and one train of Rollers ; and employ, when in full operation, about seventy-five hands.

CAR SPRINGS, of every description, are made in the city; and large quantities, of sizes and patterns most in demand, are usually stored in anticipation of orders. The best of material and work are availed of, and much pains are taken to secure the best form and construction of every detail appertaining to the business. About four hundred tons of Springs are manufactured annually in a single establishment.

Railway *Turning* and *Sliding Tables* and *Pivot Bridges* are made upon a new and economical plan, and of any required length. Messrs. WILLIAM SELLERS & Co., who, by the engineering ability they have displayed, are entitled to a rank among the most eminent of European and American Engineers, make a Turn-Table of peculiar construction—the largest size being fifty-four feet in diameter. It consists of a quadrangular centre-piece or box, upon which the arms for carrying the rails are keyed in a very substantial manner. At the outer end of the arms are placed two cross-girths, carrying four truck wheels, which are intended to take the weight when the load is going on or off. The centre rests upon *Parry's Patent Anti-Friction Box;* and the power of

25*

one man is sufficient to turn the table and its load, easily, without the intervention of any gearing. They are so constructed, that water in the pit, within eighteen inches of the top of the rail on the road, will not impair their efficiency or durability.

Lap-welded Boiler Flues, for Locomotives and other Engines, are made by Messrs. MORRIS, TASKER & Co., and of various sizes, from one and a quarter to eight inches, outside diameter, cut to a specific length. The reputation of this firm will be esteemed, by those who know them, as a guarantee for the excellence of every article they produce.

Car and *Locomotive Lamps* are made in several establishments ; and Mr. H. W. HOOK, to whom we referred in connection with type-metal, makes a very superior *Anti-Friction Metal* for bearings, which is apparently indestructible.

3. MACHINE TOOLS.

The excellence of the Machine Tools, made in Philadelphia, was referred to at some length in our Introductory, as contributing to the manufacturing advantages of this city. Since those remarks were written, a gentleman, who has a practical and technical acquaintance with the subject, has testified that Philadelphia Tools unquestionably surpass those made elsewhere in this country, in *strength, proportion*, and *workmanship*, and assigns practical, satisfactory, and technical reasons for this superiority. Their strength, he says, is insured both by the amount and quality of their material. Machine Tools require great solidity of parts— much inertia, to prevent injurious vibration under work—any jar being incompatible with accurate work, besides injuring the tool itself. The Beds and other important parts of the Tools, made by Bement & Dougherty, and by Wm. Sellers & Co., average nearly or quite double the weight allowed by other American builders. The distribution of cast iron is extremely stiff. The iron itself is selected from the best qualities known in the manufacture ; very little used in Philadelphia Tools having a tensile strength of less than 22,000 pounds, while much of it stands 28,000 pounds per square inch. The Castings, especially, are of a quality peculiar to this city, being of singular perfection. The lathe-spindles are made of *cast* steel, hammered to shape in Sheffield, and costing no less than sixteen cents per pound. The Boxes are of gun-brass, nine parts copper to one of tin, a most expensive combination, worth forty-four cents a pound. The Wrought-iron work is made from best Pennsylvania Charcoal Iron, or other equally good bar; and all parts which can be properly so treated are carefully case-hardened.

The strength and quality of material are applied to the best advantage through good proportion. In this point, the Tools under notice have everywhere met the highest approval. The excellent distribution of metal relieves the castings from any appearance of clumsiness. The bearings are of ample sizes, the cone-pulleys of such width as will bear a belt equal in power to the strength of the machine; and the gearing, screws, and all other parts, are in corresponding proportion.

While the finish of Philadelphia Tools, though chaste, is severely plain, the entire workmanship is of the best character. All parts are made to standard gauges, whereby each will fit its corresponding parts in a hundred tools. The wearing surfaces are *scraped* together,—a slow and patient process, which insures the highest accuracy of fit,—absolute contact at every point. The bolt-holes are all reamed and the bolts turned and driven home. The gearing is cut to a perfect form of tooth in each case. The screw-cutting cannot be surpassed.

A Lathe or Planing Machine, made with such care and accuracy, will accomplish *double* the work of a tool of ordinary construction, of the same nominal capacity. This has been fully proved in the various manufactories and rail-road shops where both have been tested in comparison with each other. The Philadelphia Car Wheel Works has lathes which can turn regularly ten ordinary axles in a day of ten hours. The same Works has Boring Mills, in which seventy car wheels can be bored and squared up in the same time. The Pencoyd Works has Lathes which, by doing double the work of ordinary Lathes formerly used, have saved in attendance, in one year, the extra amount of their first cost. The Camden and Amboy Company has a Lathe at Bordentown, which has turned off four flanged locomotive-tires in six and a half hours! Commercially, a tool that will "stand up to the work," in this manner, is worth more than double the price of an ordinary machine of the same nominal capacity, for it does double the work with the same attendance,—saving hands and shop room,—while the work is also much better done. This saving and advantage are so great that the leading rail-roads, when securing new equipments, cannot hesitate in their selection of Tools. The Pennsylvania, the Camden and Amboy, Virginia Central, North Carolina, Georgia Central, Memphis and Charleston, and many others, have stocked their shops chiefly with Philadelphia Tools. The United States Government has purchased, and is still purchasing them for the Navy Yards; and large quantities have been furnished the Russian government.

The two principal Machine Tool-making firms in Philadelphia are WILLIAM SELLERS & Co., and BEMENT & DOUGHERTY. They employ about three hundred hands, and turn out an average annual product of $350,000. Both of these firms have produced machines that may fairly be regarded as mechanical triumphs; and have given a permanent reputation to the manufacture which will make Philadelphia, if not already fairly entitled to be so called, the great seat of this business in the United States.

4. STEAM AND FIRE ENGINES.

Steam Engines are a leading article of manufacture in nearly all the machine shops of Philadelphia. There are more than a dozen establishments in the city, provided with facilities for constructing any size or description of Stationary and Portable Engines; but there are none in which the Steam Engine is an exclusive article of manufacture, or none which keep a large stock of finished Engines constantly on hand. The necessity for anticipating orders has not hitherto been felt by the makers, their facilities being such as enable them to meet the demand, as it arises, with sufficient expedition.

Within the last few years the attention of ingenious men in this, as well as in other places, has been directed largely to simplifying the Steam Engine, removing all essentially unnecessary parts, cheapening its price, and diminishing its size. At least one firm in Philadelphia has been remarkably successful in all these points, and now construct a Portable Engine, with a vertical cylinder, peculiarly adapted for confined situations.

Messrs. LIST & DAVIS, of West Philadelphia, the firm referred to, are now constructing a Ten-horse power Engine for a Pill Manufacturer, that will not occupy more space than six feet square. The water, before entering the boiler, passes through heated tubes, and consequently no cold water is at any time admitted into the boiler. The cost of an Engine and boiler of this description, of ten-horse power, will not exceed $1,050.

One Engine builder and machinist, Mr. A. L. ARCHAMBAULT, devotes a large part of his attention to the manufacture of a peculiar *Portable Steam Hoisting and Pumping Engine*. This Machine requires but one man to keep up steam and attend to the brakes ; and by its aid pig iron can be discharged from a vessel at the rate of twenty-five tons per hour, and still more expeditiously if it can be got ready. This Engine is also arranged for driving Portable Saw Mills ; and wherever it has been tested, it has excited attention, given entire satisfaction, and elicited much commendation. First premiums have been awarded the manufacturer, on several occasions, by the Franklin Institute, and by Agricultural Societies in various States ; and there is no doubt whatever that both labor and time can be greatly economized in Hoisting, Pumping, &c., by the use of Archambault's Portable Engine.

Mr. JOHN L. KITE also makes Portable Engines for Plantations, Hoisting, Pile-driving, &c.

Propeller Engines are a leading article of manufacture in the great establishment of REANEY, NEAFIE & Co., who have built and put in successful operation a greater number than any other firm in the United States. For the last fourteen years they have made this subject almost their entire study ; and, with an experience derived from having built over two hundred Engines of this class, may be not inaptly called "*the* Propeller builders." It would be tedious to name a fractional part of the vessels constructed by them ; but the "Pampero," whose stanchness deserved a better service than her Cuban expedition, the "Granite State," "Martin White," "Mount Vernon," the "Baltimore," "J. K. Hammitt," "Lancaster," and others, are worthy and lasting monuments to the fame of their eminent builders.*

* The firm of Reaney, Neafie & Co., is composed of Thomas Reaney, Jacob G. Neafie, and John P. Levy. Both Mr. Reaney and Mr. Neafie have

Messrs. Reaney, Neafie & Co. are also proprietors of the patent right for the " Curved Propeller," which has attained deserved popularity ; and the demand for their peculiar wheel has been so great from the Canadas and on the Lakes, that they have found it necessary to connect themselves with several extensive establishments on the Lakes. They are now building the engines for the Government sloop " Lancaster."

Quite recently, as we have elsewhere stated, this firm constructed several *Steam Fire Engines*, which have operated so successfully as to revolutionize popular opinion with respect to the availibility of steam for extinguishing fires.

The building of *Hand Fire Engines* is inseparably associated with the name of one Philadelphia maker—JOHN AGNEW. He has been engaged in the business for thirty-five years, and has constructed, up to the present time, 606 Fire Engines. Several of his Engines are in service in New York, three in California, a half dozen in Cuba and the other West India Islands. He employs generally about thirty-four hands. The average cost of a Fire Engine, finished in ordinary style, is $1500.

5. MILL GEARING AND SHAFTING.

The manufacture of Gearing, Shafting, Couplings, &c., ordinarily constitutes a branch of the general Machine Business, but about eleven years ago it was taken up as a specialty by Messrs. Bancroft & Sellers, then located in the District of Kensington, and has been continued by them and their successors, Messrs.

had a long and practical experience in machine shops—the former having served his apprenticeship with Mr. Holloway, the first Marine Engine builder in Philadelphia; while Captain Levy, the financial partner, is a practical seaman and shipwright, possessing a familiar knowledge of the hulls, rigging, and engines of steamers. The result of this union is, that the firm are prepared to build any description of steam vessel outright, and owners have but one contract to make, and that with very responsible parties. In the construction of *iron boats* of all classes, both side-wheels and propellers, this firm do a large business, having at least two on their stocks at all times. The also make all kinds of engines and boilers, high and low-pressure, heavy and light forgings, and iron and brass castings of all sizes and patterns. Having made it a rule to preserve all patterns, their stock at present is very large.

Wm. Sellers & Co., to whom we have already referred more than once, and who have now probably a larger assortment of modern patterns than any other house in the country. The journals or bearings on which the shafting runs, was one of the first points to which they directed their attention; and the article manufactured by them, and known as the *Ball-and-Socket Hanger*, has attained an enviable reputation. Since the introduction of the Hanger, and their method of connecting gear-wheels, the attention of this firm has been chiefly directed to improving the method of Coupling the Shafts together, so as to render them perfectly firm and rigid whilst at work, and at the same time to allow of detaching, at any particular point, without driving out keys or using sledges or screw-presses, as heretofore necessary. This object they have at last accomplished by means of their Double-cone Adjustable Couplings, which can be released from the shaft by slackening two small nuts; the whole coupling being smaller and less expensive than any other in common use.

The thorough system introduced into this branch of business, enables these manufacturers to employ the same patterns in a great variety of ways, and to provide special tools particularly adapted to the work, thereby decreasing the cost to consumers, who, to a great extent, are the machinists themselves in various parts of the country.

6. PAPER-MAKERS', PRINTERS', AND BOOKBINDERS' MACHINERY.

Philadelphia now contains one of the most complete establishments in the Union for making Paper Machinery, being provided with facilities for equipping at least twenty-five Paper Mills annually. The proprietor, Mr. Nelson Gavit, is well known as an ingenious mechanic, and has been very successful in turning out good machines, both of the ordinary cylinder and the celebrated Fourdrinier machine, which cost the Messrs. Fourdrinier $300,000 to invent, and caused their bankruptcy. The cost of one of these machines is now from $3,400 to $6,000, and of a cylinder machine from $1,800 to $3,400. One peculiarity noticeable in the machinery of this establishment is, that the shafting, of which there is 500 feet, turns upon *glass journals* inserted in the ordinary cast iron box, thereby avoiding a great deal of friction, and runs with much less noise, and requires oil only once, say, in two months. About sixty-five hands are constantly employed.

Bookbinders' Machines are made in part at a number of establishments, but particularly and largely at the shop of Mr. H. Howard, and Mr M. Riehl, who have been in operation about four years and have turned out a large quantity of very superior Machines, which have been shipped to all parts of this country and to Cuba. While making all kinds of Machines desired and ordered by Bookbinders, this firm is exceptional, inasmuch as they control and are sole manufacturers of a great number of superior patented machines. Mr. MICHAEL RIEHL is well known to the trade as an inventor, extremely successful in originating practical and valuable improvements. His patented *Book-cutting Machine* is in use in the principal binderies in Philadelphia, New York, and Boston, and in the Government Book Binderies in Washington. The motion of the knife is a diagonal forward one, which "draws" less and does more execution with less power, than any other similar machine, while the knife will preserve its temper and edge for a longer time. Riehl's Patent Cutter is adapted for either hand or steam power, and the price ranges from $150 to $600, according to size. Mr. Howard also makes Riehl's improved *Embossing and Mashing Machines*, weight 3,800 lbs., price $600, his *Improved Stamping Press*, $150, a *Paper-cutting Machine* for Printers, price $150, and a variety of other similar machinery. Bookbinders and Printers throughout the country, we think, would consult their interests by communicating with an establishment that is fully equipped and prepared to supply them with improved machinery on the most favorable terms.

The manufacture of *Printing Presses* is limited, we believe, to the well-known Ramage Press, made by Mr. Bronstrup, and to Dow & Co.'s Card and Job Press, which is said to possess great merit. The large Power Presses, most popular, are generally patented, and made in other cities. But minor articles for Printers' use are made here largely; as Printers' Furniture, Chases, Rollers, made by Mr. Cosfeldt and others, and Brass Galleys, Rules, Stereotype Blocks, and Rules cut to Pica ems, with various faces, made by L. JOHNSON & Co., who furnish Printers' supplies generally.

Lithographic Ruling Machines, equal to the imported, are manufactured by Mr. Saxe; and Lithographic and Copper Plate Presses, Geometric Lathes, Hydraulic and Transfer Presses, and Engravers' Machines and Tools generally, by GEORGE C. HOWARD,* and several others.

7. GAS AND WATER APPARATUS.

Philadelphia, it is generally known, is the chief seat of Gas-making Machinery in the United States. Nearly all the principal Gas Works, particularly in the South and West, besides Brooklyn, Buffalo, Newport, and New Bedford, were constructed or enlarged by Philadelphia machinists; and larger Gas Castings have been executed in foundries in this city than in any other place—the Gasholder frame of the Philadelphia Works, made by Merrick & Sons, being, it is said, the largest in the world. The eminence that has been attained in this branch is, no doubt, due largely to two circumstances: first, the advantages of Philadelphia for executing heavy castings economically, because of the abundance and cheapness of Coal and Iron; and secondly, because there are establishments in this city better provided with patterns, tools, and facilities specially adapted for the manufacture of Gas Apparatus than any others in the United States. Some branches of the manufacture, which are now of great importance, had their origin here; as, for instance, the manufacture of *Wrought Iron Tubes* and *Fittings,* first undertaken in 1836 by Morris, Tasker & Morris, the predecessors of the present firm of MORRIS, TASKER & Co., who are undoubtedly the leading

---

* Mr. GEORGE C. HOWARD ranks among the most ingenious, reliable, and honorable of Philadelphia machinists. For some years he has made all the *Bonnet and Hat Pressing Machines* used in Philadelphia. His establishment is well-provided with tools and facilities for constructing any machinery of moderate size, including Machinists' Tools, Stationary Engines, Millwright Work, &c.; and his reputation is a guarantee that whatever he undertakes to do will be well done.

Mr. Howard is known as the inventor's friend; he having been of peculiar and essential service to inventors in making their drawings, patterns, &c., and adapting their ideas to practical results.

26

manufacturers in this country of these articles and Gas Fitters' Tools, while they make also Cast-iron Gas Pipes, Gas-works Castings, Retorts, etc. Since this house commenced the manufacture of Tubes, to the expiration of the first quarter of the present year, they made 30,788,000 feet of Tubes—the smallest amount made in any one year being 60,000 feet, in 1836; and the largest, 3,647,273 feet, in 1855.*

Gas Pipes are also made at another establishment in Philadel-

---

* The house of MORRIS, TASKER & Co. was founded in 1821, by Stephen P. Morris, who commenced the manufacture of Coal Grates, Stoves, and Smith-work in general, at the corner of Market and Schuylkill Seventh sts., Philadelphia. In 1828 he removed to the corner of Walnut and Third sts., where a foundry was put in operation, and the business greatly extended. Not long afterward he was joined by Henry Morris and Thomas T. Tasker, under the firm style of Morris, Tasker & Morris. In 1836 they commenced the erection of their present works, known as the "Pascal Iron Works," on South Fifth and Franklin sts., and also commenced the manufacture of Wrought-iron Tubes and Fittings for gas, steam, and water, being, as we stated above, the first of the kind in this country. Subsequently, they added to these the manufacture of Cast-iron Gas and Water Mains, Lap-welded Flues for Boilers, Gas and Steam-fitters' Tools, &c. On the 1st of January, 1856, the firm style of Morris, Tasker & Morris, was changed to Morris, Tasker & Co., and is now composed of Stephen Morris, Thomas T. Tasker, Jr., Charles Wheeler, Jr., and Stephen P. M. Tasker. In their works, which have been extended by additions until they cover an area of nearly four acres, about 400 men are usually employed, and over 6,000 tons of Anthracite coal alone annually consumed. The tools used, as well as those made, are subjects of wonder and admiration; and the machinery, as may be supposed, is of the most perfect description. Water is furnished for the boilers of the five steam engines from two Artesian wells; and by means of hose and pipes, communicating with tanks, it is conveyed to every room for immediate use in case of fire. A portion of the iron for the tubes and flues made by this firm is prepared in a mill of their own near Fairmount, and all the pipes are tested by an hydraulic pressure of at least 300 lbs. to the square inch.

Messrs. Morris, Tasker & Co. are also extensively engaged in the manufacture of Apparatus for warming public and private buildings, both by hot water and by steam. One of the partners, Mr. Tasker, is the inventor of a Self-regulating Hot-water Furnace, by which the temperature in a house can be maintained at any required point for an indefinite period of time, without further attention than an occasional supply of coal. The Committee on Science and the Arts, constituted by the Franklin Institute, have

phia, viz. : at the Girard Tube Works, of which MURPHY & ALLI-SON are the proprietors.

Gas Meters are made by five firms or persons in Philadelphia ; the principal manufacturers, however, being CODE, HOPPER & Co., who claim the credit of having been the first to introduce into this country the manufacture of this very ingenious instrument. This house is the oldest and most extensive in the United States, having made, up to the present time, over one hundred thousand Gas Meters, wet and dry. They employ constantly more than one hundred persons directly in their manufactory, besides a large number indirectly outside their walls. In addition to every variety of Gas Meters, they make all kinds of Gas Apparatus, such as Photometers, Minute Clocks, Pressure Registers, Indicators and Gauges, Exhausters, Governors, Meter Provers, Centre Seals, &c., &c. Every part of the meter is made on the premises ; to which is attached a foundry, with steam power and the most approved and perfect machinery,

Portable Gas Works for generating Gas from resin or oil are a comparatively recent invention ; but the manufacture is increasing in importance. The apparatus of Messrs. Stratton & Brother is said to be capable of making 100 cubic feet of Gas per hour from resin : one pound of resin making nine cubic feet of Gas, and one burner consuming two cubic feet per hour. One and a half bushels of Anthracite coal will supply the requisite heat, it is said, to make 500 cubic feet of Gas.

In the manufacture of *Water Works* Apparatus, Philadelphia firms have been as successful as in Gas-making Machinery. A number of Cornish Pumping Engines, of the largest size, have been constructed by I. P. Morris & Co., but we shall defer a mention of these until we speak of Heavy Machinery.

reported at length upon its peculiarities, and state that it is free from any of the disadvantages to which Hot-air Furnaces ordinarily are subject.

Within the last few years they have also introduced *Galvanized Iron Pipe* for water, as a substitute for Lead Pipe, over which it possesses many advantages—such as strength, durability, and economy, and is rapidly coming into use.

This firm has contributed materially to advancing the good name as well as fame of Philadelphia Iron Workers throughout the Union.

The minor articles for Gas Works, as Stop-cocks, Valves, Drip Pumps, &c., are made at several establishments; and also Cocks, Fire Plugs, &c., for Water Works, Valves and Pipes for Tanks at Rail-road Stations, and a great variety of articles included in the term Water Apparatus.

So much for Special Machinery. If we were to elaborate the subject, and, besides referring to the special classes of machinery which are represented by special establishments, we were to enumerate also the articles which are made prominent and leading by certain general machinists, our task would be far from complete. *Fan Blowers* are a leading article of manufacture in two establishments, that of KISTERBOCK & SON, who announce Dempfel's Fan, and of MANOAH ALDEN, who, for more than thirty years, has been engaged in business as general Machinist. The Blower of which Mr. Alden is inventor (it has been established by repeated tests, as we are informed) will produce a stronger blast, with less power, and less noise, and less liability to derangement than those of any other construction. His Blowers are now in use in nearly all the large foundries, machine shops, and iron-working establishments in Philadelphia. Mr. Alden, during the last year, has also made for the inventor, Mr. Ager, several *Rice-cleaning Machines*, which, it is supposed, are superior to any others ever constructed for the purposes of cleaning rice perfectly without damage to the kernel. *Mint Machinery* is a leading article of manufacture with the firm of MORGAN, ORR & CO., who supplied with machinery the Branch Mint in California, and also a Mint for the Peruvian Government. They have all the requisite patterns, tools, etc., for doing this kind of work successfully, and are noted for their fidelity in executing orders. They employ on an average seventy-five hands. *Mining Machinery* is made at several establishments, but at one, of which THOMAS J. CHUBB is proprietor, it is a leading article. Chubb's Patent Pneumatic Ore Separators, Crushers, Drying Cylinders, Sifting Machines, Elevators, etc., it is expected, will supersede all machinery now in use for the same purposes; and, if so, its manufacture would engross the capacity of all the machine-shops at present in Philadelphia. *Bakehouse Machinery* is made principally by R. J. HOLLINGSWORTH, who has patterns for cracker-making by hand, which, it is supposed, are not possessed by any one else in the United States. Brick Machines constitute the exclusive business of two shops; and Braid, Cord, and Whip-plaiting Machines are made at the establishments of P. GOSFELDT and CHARLES DIEDRICHS. The first-named has had thirty-three years' experience in the manufacture of these curious machines. JAMES FLINN & Co., in connection with Agricultural Implements, make Wood-boring, Chamfering, and Wooden Pin Machines, of which they are said to be sole manufacturers. At the present time, Messrs. HUNSWORTH, EAKINS & Co. are making what it is believed will prove to be a very successful machine for driving *Steam Plows.* But these establishments, and the others that we have noticed, with but few exceptions, are provided with facilities for constructing a great variety of machinery other than that we have designated; and therefore they are properly classified among the shops for the manufacture of General Machinery.

We had hoped to conclude this chapter with a list of all the Machines which have been constructed in Philadelphia since 1850, and had expended considerable time and money in preparing it, with the quantities of each, when we discovered that circumstances which we could not control—principally apathy on the part of the manufacturers—would inevitably defeat its completeness. Enough, however, was done to inform us that such a list would prove incontestibly that the machine-shops of Philadelphia can construct almost any machine which the genius of man has invented or can invent. It would contain a number of machines, mysterious and almost awe-inspiring in the seeming intelligence, concealed in arms of wood and fingers of steel, which directs their automatic movements.* It would demonstrate, moreover,

---

* A chapter, describing the curious machines that are in operation in the various manufacturing establishments of Philadelphia, would be one of the most interesting we could insert, if our space were not preoccupied. We can only say, see the Card Clothing Machines of James Smith & Co.; the Paper Bag Machines of the North American Paper Bag and Envelope Manufacturing Company; the Independent Straight Line, made by P. R. Receveur, and in operation in the Watch Case establishment of T. Esmonde Harper; the Planing and Moulding Machines in the Wood-working estab-

26*

that there have been constructed in Philadelphia some of the largest Engines and Machines, as well as largest Castings, ever made in this country. But this point—the capability of Philadelphia to construct Machine-work of extraordinary dimensions—can be successfully established without any very extended enumeration of particulars, and by referring to a few of the products of only two establishments—those of I. P. MORRIS & Co. and MERRICK & SONS.

At the "Port Richmond Iron-Works," of which the former firm are proprietors, were constructed the large engines of the U. S. Mint; the large Steam Engine of the Lake Erie Steamer *Mississippi*, being a beam-engine, with a cylinder 81 inches diameter by 12 feet stroke of piston; two Cornish Bull Pumping Engines for Buffalo Water-Works, each having steam cylinders 50 inches diameter by 10 feet stroke; the lever-beam Cornish Pumping Engine, steam cylinder 60 inches diameter, 10 feet stroke, at the Schuylkill Water-Works; the Bull Cornish Pumping Engine, cylinder 40 inches diameter, 8 feet stroke, at Camden, N. J., Water-Works; and the Iron Light-House for Ship Shoal in the Gulf of Mexico, to be put up on screw piles in water 15 feet deep, and at a distance of 12 miles from land. The whole height of this structure, from the water to the top of the spire, was 122 feet, and from water to focal plain, 107½ feet. The structure above the foundation to the deck, a height of 93 feet, was erected in their yard, complete in all its parts, before shipping. The Blowing Machinery for the Lackawanna Coal Co. at Scranton, probably the largest ever constructed, the dimensions of which were given on a previous page, was built at these Works. This firm also constructed the large Blowing Machine for the Lehigh Crane Iron Co., a lever-beam condensing engine, having a steam cylinder 58 inches diameter, 10 feet stroke of piston, and a blowing cylinder 93 inches diameter, 10 feet stroke. The beam of this engine works on a column of cast iron 30 feet high, and the whole is set upon a heavy cast iron bed-plate. They also

lishments; the machines of Alfred Jenks & Son; the Uhlinger Machines; the Lathes, Planers, and Borers in the establishment of I. P. Morris & Co., Merrick & Sons, J. T. Sutton & Co., William Sellers & Co., and Bement and Dougherty.

made the Direct-acting High-pressure Blowing Machine for Sey fert, McManus & Co.'s furnace at Reading, steam cylinder 40 inches diameter, blowing cylinder 102 inches diameter, both 7 feet stroke of piston. This firm, it will be seen, have built the largest Engines ever constructed for making Iron with *Anthracite coal;* besides a large quantity of less capacity, but which rate among first-class machines.

At the "Southwark Foundry," of which MERRICK & SONS are proprietors, were constructed the great Iron Pile Light-Houses illuminating the Florida coast, stationed at Sand Key, Cary's Fort Reef, Coffin's Patches, Rebecca Shoal, N. W. Channel, Dry Tortugas, as also those on Brandywine Shoal (Delaware Bay) and the harbor of Chicago, besides iron lanterns for Cape Hatteras, Cape Florida, &c., and beacons for other points. The first three are among the largest in the world, being respectively 120 feet, 112 feet, and 137 feet high (water to focal plane), and 50 feet square, 50 feet diameter, 56 feet diameter at the base respectively, and weighing from 250 to 300 tons each. This firm made the great Gasholder frame for the Philadelphia Works (the largest in the world), being used for a gasometer 160 feet in diameter; it weighs about one thousand tons, consisting of twelve Gothic pentagonal iron towers, 90 feet high, braced apart by girders 36 feet long and 8 feet deep, ornamented Gothic, and weighing eighteen tons each in one piece. At this shop were made the 140 feet Gasometer and framing for the same Works.

This firm constructed almost all the machinery for the steamers of the U. S. Navy; among which may be specified the *Mississippi,* paddle, two side-lever engines of 500 horse power; *Princeton,* screw, two oscillating-piston engines of 300 horse power; *San Jacinto,* screw, two geared-engines of 450 horse power; *Wabash,* screw, two direct-acting engines of 800 horse power. Of these, the former is too generally known to need any comment, and the latter is confessedly the finest of her class in the world. Here also were made the boilers of the U. S. Steamers *Susquehanna* and *Saranac,* 800 horse power each; the machinery for the surveying steamer *Corwin;* machinery and hull (iron) for the surveying steamer *Search;* and, for private parties, the machinery of the *Keystone State,* paddle, 400 horse power, *State of Georgia,* paddle, 350 horse power, *Quaker City*, paddle, 450 horse power, *Phineas Sprague*, screw, 250 horse power, *Alfonso* and *Cardenas*, screws, 275 horse power each, and others.

For pumping purposes, the same firm constructed the great iron Elevating Wheel at Chesapeake City, Md., for feeding the canal. This wheel is 38 feet diameter, 12 feet wide, driven by two condensing beam-engines of 200 horse power, and raising two millions of gallons sixteen feet high every twenty-four hours. More recently, for the Midlothian Coal Mining Company in Virginia, they made a sixty-inch beam Cornish Engine, with one " drawing lift" and three forcing lifts, which pumps one million of gallons in twenty-four hours from the pit, which is 770 feet deep. Both these firms, and others, have made Sugar Machinery for the West Indies of the largest size; but the examples enumerated demonstrate conclusively that *Philadelphia has the ability to do the* HEAVIEST ENGINEERING WORK.

The machine work executed in the leading establishments of Philadelphia, we may remark in conclusion, is distinguished by certain characteristics, which enable a competent judge to pronounce with confidence upon the source of its production, or, in other words, to detect a Philadelphia-made machine by the "ear-marks." Excellence of material, solidity, an admirable fitting of the joints, a just proportion and arrangement of the parts, and a certain appearance of thoroughness and genuineness, are qualities that pervade the machine work executed in Philadelphia, and distinguish it from all other American-made Machinery.

We pass to the third and last division of the Manufactures of Iron, viz., Hardware and Tools.

XXI.

Iron Manufactures Concluded—Hardware and Tools.

The term *Hardware* is one of those indefinite comprehensive nouns of multitude, of which it may be said that it almost includes, as its name imports, every ware that is hard. Popularly, it is understood to embrace all the unclassified manufactures of Iron and Steel, including all the appendages of the mechanic arts, from a file to a mill saw; many of the details of common life, from a rat-trap to a coach-spring, articles as various in appearance, size, and uses as can well be conceived—in fact, whatever is sold by a Hardware dealer. In view of the almost infinite variety of articles included in the term, almost all of which are made in Philadelphia, the utmost that we can hope to accomplish is to exhibit the state of the business in its leading branches, as in the manufacture of *Saws, Forks, Shovels, Files, Locks, Bolts, Rivets, Scales* and *Balances, Edge Tools* and *Cutlery*.

If we were to consider the classes of articles designated as Hardware, in the order of their relative reputation abroad, we would come first to Saws. Every country merchant, as well as every wood-worker, is familiar with the excellence of Rowland's Saws, Cresson's Saws, Disston's Saws, and Conaway's Saws—all Philadelphia makers. The works now known as Rowland's Saw Works were founded by William Rowland in 1802, and are believed to be the oldest established of the kind in this country. They have supplied, and continue to supply, a large proportion of the large-sized Saws in use, Mill and Cross-cut, varying in length from six to eighteen feet. About fifty hands are engaged in this establishment, and two hundred Saws are produced daily.

The works of WALTER CRESSON are located beyond the city limits, at Conshohocken, in Montgomery County, but the business is transacted exclusively at the warehouse on Commerce street.* The material employed is mostly cast-steel, manufactured in England expressly for Mr. Cresson, and rolled under his orders to proper thickness, and cut into sheets of convenient size. For making Circular Saws, Mill Saws, and Cross-cut Saws, it is imported trimmed to the size and shape required. Among the productions of this establishment are a great variety of Hand Saws, Circular Saws, Back Saws, Wood Saws, Mill and Cross-cut Saws, Hay Knives, Loom Springs, and other Tools. His Saws have

---

* The factory of Mr. Cresson was destroyed by fire on August 17th, 1854, but was immediately rebuilt on an enlarged scale. The energy of the proprietor was peculiarly illustrated on that occasion, for even while the fire was still burning, he rented an old foundry, fitted it up, and on the next day set his men to work in it. In three months the new factory was completed, and in full operation. The whole of the machinery of this establishment is driven by water obtained from the Schuylkill Canal, and applied by one of Journal's Turbine Wheels.

repeatedly received the highest commendation from competent judges for their finish and make.

The "Keystone Works," of which Mr. HENRY DISSTON is proprietor, are probably the largest of the kind in the country. They consist of four buildings—three of them three stories high, and cover an area of over 20,000 square feet. The Machinery is of the most complete description, and driven by an engine of seventy-horse power. The Saws made at this establishment comprise nearly every variety, though principally Cast-steel Circular, Hand, and Panel Saws, all of which are tempered by *Sylvester's Patent Tempering Machine*, and the Circular Saws are ground by Southwell's Patent Grinding Machine. Among the novelties produced in this establishment might be mentioned, the *Patent Combination Saw*, comprising a perfect twenty-four inch square, straight edge, twenty-four inch rule and scratch-awl, and a handsaw, with a patent attachment for gauging any required depth. Mr. Disston employs one hundred and fifty hands.

The "Union Saw and Tool Manufactory," 402 Cherry street, of which Mr. WM. CONAWAY is proprietor, is also a well-known establishment, and produces, besides, all kinds of Saws, Trowels, Curriers' Knives, Carpenters' Gauges, etc. BRINGHURST & VERREE also manufacture Saws extensively.

Besides these celebrated makers, there are five others who make Saws to some extent, viz.: THOMAS GAMBLE, J. HUGEL, CHARLES LAME, JOSEPH NICHOLLS, and JAMES TURNER.

2. FORKS. There are four principal establishments in Philadelphia for the manufacture of Forks, viz.: SHEBLE, LAWSON & FISHER, MYERS & ERVEIN, RIDGWAY & RUFE, and HARPER & HOLT. The oldest of these establishments is the one represented by the last-named firm. Messrs. RIDGWAY & RUFE have a factory at Germantown, where they also make Coffee-mills, Shutter Bolts, &c. Messrs. MYERS & ERVEIN commenced the manufacture of Forks in 1847. Their establishment is well equipped with machinery, and their manufactures, which comprise all the usual varieties of Hay and Manure Forks, including Sluice Forks, a peculiar article for miners' use, have a good reputation. The youngest of these establishments is that of SHEBLE, LAWSON &

FISHER; but its annual production is now supposed to exceed that of any other Fork Factory in the Union. Previous to its establishment in 1851, the Eastern manufacturers almost monopolized the Philadelphia market; but, at the present time, city Forks are so generally preferred, that special inducements alone can effect a sale of Eastern, or even New York made goods. In less than five years, the demand for Forks of this firm's manufacture so far exceeded the ability of their works at Fairmount to supply it, that, in 1856, they erected new and more extensive works on the Frankford Creek. They have now five Trip Hammers, run eleven Polishing Machines, and employ forty-five hands. Their manufactures embrace a greater variety of Forks than any similar establishment known to us in the Union, and include every kind or style—Hay, Manure, Spading Forks and Spading Hoes—Forks for manufacturing purposes, as Glue, White Lead, Bone, Coke, &c.—in a word, Forks from two to twelve tines from the solid piece, and varying in price from $2 to $50 per dozen. Spading Forks and Hoes are new articles in this country; but the improvements made in them by the above firm, have brought them into use in place of the old style Spade and Hoe for agricultural purposes.

Forks are now exported from Philadelphia to England.

3. Of *Shovels* and *Spades* there are six principal manufacturers, having a capital invested exceeding a quarter of million of dollars, and producing annually about 85,000 dozen of Shovels, worth, at $6 per dozen, $510,000. The machinery employed in these establishments is of the most perfect description, and the manufactured product is equal in quality to the best made in the country, and far better than the Shovels imported. The raw material is mainly American iron and steel, though a share of the English metal is used.

4. FILES. These articles, which are of the first importance to all workers in metal, are made at nine establishments in this city; and it is said another, upon a large scale, will soon go into operation. Rasps are made here more cheaply than in Europe, except the large sizes, where the price of steel may affect the cost of the manufactured article. The Files made include every variety,

from those used by jewelers, dentists, and watchmakers, to those required by metal-workers in heavy operations.*

5. LOCKS. It is claimed and, we believe, conceded, that the ingenuity and enterprise of American mechanics have placed this department in advance of the efforts of all other nations. In the finer and more expensive class of Locks, it has been abundantly proved that the American production is superior to any other; while in Locks, Latches, &c., adapted to the wants of the builder or to commerce, the American mechanic supplies the home demand; and, to some extent, orders from abroad.

In Philadelphia there are a large number of very ingenious Locksmiths, who provide every variety of fastening essential for the security of dwellings, &c. The business is usually carried on in connection with Bell-hanging, Silver Plating, &c.; but the demand for special and patented Locks furnishes occupation to several special Lock-making establishments. One manufacturer, Mr. CONRAD LEIBRICH, employs forty-five hands, and makes twenty-one different Spring Locks of his own invention, besides a great variety of Pad and Door Locks, Night Latches, &c. The Scandinavian Padlock for stores, &c., it is claimed, is perfect protection against burglars. But probably the most remarkable Lock of modern times is the Permutation Bank and Safe Lock made by L. YALE, JR., & Co., at 248 Front Street. The manufacturer, Mr. Yale, it will be remembered, picked the "Parautoptic," or great Hobbs' Lock, which the English Locksmiths had unsuccessfully attempted; he having previously picked most of the celebrated Locks of the day. His Permutation Lock,

---

* The largest File-making establishment in the city, and we believe in the State, is that of J. B. SMITH, on New street. The goods manufactured by him embrace every description of Files and Rasps, from the smallest to the largest, and from the finest to the coarsest, a large portion consisting of kinds not generally imported. The popularity of his *Shoe Rasp* has induced unscrupulous persons to import from Europe a worthless imitation, which is strong evidence of the superiority of his manufacture. These Rasps of his manufacture are extensively used in New England, as well as throughout the Union. He is likewise manufacturing an improved description of *Saw File*, which is said to be of unrivaled excellence. Mr. S. is a practical File-maker of thirty-five years experience, and is therefore thoroughly acquainted with every department. Some of the best workmen in the United States have graduated at this establishment.

which may now justly be regarded as more secure than any other in the world, is distinguished by several important improvements. It is entirely without *springs*, those most fruitful sources of failure in fine locks. The tumblers, or security parts, are so placed in the Lock that they never can be seen, or felt by any picking tool, nor can powder ever affect them. The key is small, and can be conveniently carried in the wallet; yet its permutations are so numerous, that a lifetime is not long enough to ring all its changes.

Their latest improvement is their *Double Treasury Lock*, in which two Locks are so combined as to control the same main bolt. On locking both sides, the bolt is securely locked out; but on unlocking either side, the door can be opened. Hence, in daily use, one side only is used; but, in case of losing the key, or other accident, instead of being compelled to cut down the door, as in ordinary cases, the reserved key, on being applied to the other side, will open the door at once. They offer $3,000 to any one who can pick or force this Lock, which, for strength of material and beauty of workmanship, is unsurpassed. This firm also manufacture Bank Vaults and Safes, which are worthy of the Lock that secures them. Constructed of hard, chilled iron, cast around a basket-work of wrought iron rods, they are proof against drills, cutters, or sledge hammers, and are seemingly burglar-proof. The Treasury Department has adopted their work, and their customers are found in every State in the Union.

Excellence of workmanship, lowness of price, and adequate security characterize the Lock Manufactures of Philadelphia.*

† The Locksmiths and Bell Hangers of Philadelphia include some of her most ingenious mechanics. Mr. HOCHSTRASSER is called upon in all desperate cases. Mr. P. RODGERS has made a beautiful, rabbeted mortice Front-door Lock, which was ordered from Paris. Mr. J. B. SHANNON, on Sixth street, has attained considerable distinction in this branch of business, making largely for first-class dwellings and public institutions. He has constructed Locks for a large number of Insane Asylums, where they are wanted in sets, to differ from each other, and yet to be passed by a master key by the superintendents, watchmen, and others. Similar Locks are also applied to dwellings; and one member of a family may have a key which will open all the chambers and closets, and yet the respective keys of the doors will not open any except the one to which it belongs.

Bells are so arranged in dwellings, by Mr. Shannon, that a person taken sick may communicate the need of help to any other part of the house. Invalids can thus dispense with constant attendance.

6. BOLTS, NUTS, &c. The manufacture of these articles has always constituted an item in the business of blacksmiths, machinists, and others, and large quantities continue to be made by hand. But within a few years machinery has been brought into requisition for the purpose, by which an article is produced in all respects superior to the hand-made in workmanship, quality, uniformity, and price. There are now two very extensive establishments in Philadelphia for making Bolts and Nuts by machinery, viz.: HOOPES & TOWNSEND and E. & P. COLEMAN ; and so perfect are their equipments, so varied and extensive their dies, etc., that we believe this city is now the chief seat of the manufacture in the United States. The former firm make Machinery Bolts principally, the latter Carriage Bolts. They employ about one hundred and seventy hands, and produce annually a value of $175,000. In addition, large quantities of all kinds of Bolts, particularly Carriage Bolts, are made by other parties. A firm of Bridge builders consume three hundred tons of square and rolled Iron, and produce annually Bridge Bolts of the value of $45,000.

7. RIVETS. These articles may be said to occupy the same relative position to iron that nails do to wood. To a person not familiar with their uses, the quantities consumed in various constructions must seem incredible. In the Iron Light-House made for the Ship Shoal Reef, Florida, some six tons of Rivets were used, all of which were furnished by the " Philadelphia Rivet Works." In a first-class freight engine, the boilers and fire-boxes will use about 2,000 Rivets, weighing about 500 lbs. ; and the tender will require about 150 lbs. of Rivets of smaller diameter. The Point Breeze Gasholder, the largest in the world, used about one million of Rivets—the gasometer alone having consumed about six tons of Button-head Rivets. The principal firm in this manufacture in Philadelphia is PHILLIPS & ALLEN, proprietors of the " Philadelphia Rivet Works," and they can produce about five tons of Boiler Rivets per day, and about 3,000 lbs. of smaller sizes.*

---

* Messrs. PHILLIPS & ALLEN, we may remark, are fully sensible of the very great importance of using only the best materials in the construction of Rivets, particularly Boiler Rivets. They employ the severest tests to prove the quality of all iron used by them. The Rivet being made

8. SCALES and BALANCES. Eminence in the manufacture of
Scales and Balances, we are quite sensible, presupposes, in the
manufacturer, very considerable mechanical skill, fidelity in execution, and taste and accuracy in workmanship. It is a branch,
moreover, in which Philadelphia makers hold, and have always held,
the leading position. Some of the finest specimens of Balances
in use, every one will admit, are the product of the establishments of F. Meyer & Co., and their former partner and present
successor, HENRY TROEMNER. Mr. Troemner constructed all the
Balances, Weights, etc., required for the U. S. Mint, Custom
Houses, and Repositories, and several Scales for the Mexican
Mint. Some of the Balances made for the Assay Office in New
York, and for the Branch Mint of San Francisco, cost as much
as $1,000, and one made several years ago cost $1,250. Besides
Balances like these, which must turn with the thousandth part of
a grain, Mr. Troemner constructs Patent Balances that will weigh
twelve tons. His manufactures comprise Mint Balances, Bankers'
Scales, Jewelers', Druggists', Grocers', Confectioners' Scales, &c.
—in fact, any kind required for weighing purposes. Nearly all
the Banks in this city, New York, and other places, have his
Scales in use.

red hot, is hammered equally when it is hot and cold; and should the quality of the iron change from fibrous to a crystalline character, the iron is
regarded as worthless, as the force of contraction upon cooling would pull
off the head of the rivet.

For Boiler Rivets this firm employ four machines, which convert the bar
iron, purposely rolled for them (heated in coke, in order not to impair its
quality) into Rivets. These machines are of their own make—and cut off
the lengths, and shape and head them at one operation. For smaller Rivets
the best charcoal iron-wire is used, as the lengths and diameters must be
exceedingly accurate. The machines are most carefully built and managed.
After the Wire Rivets are made they are annealed, which allows them to be
driven cold, or headed with ease. Their customers are the principal machine
shops in the United States, among them Messrs. I. P. Morris & Co. Messrs.
S. V. Merrick & Sons, Reaney, Neafie & Co., Baldwin & Co., and Norris, Philadelphia; Messrs. Harlan & Hollingsworth, of Wilmington, &c., &c.; and
the well-known reputation of these establishments would seem to be a sufficient guarantee for the excellence of their work. They furnish every kind
of rivet, from eight inches long by any diameter, to the smallest wire
sizes.

There are two firms engaged almost exclusively in making *Plat-form-Scales*, viz.:—ABBOTT & CO. and A. B. DAVIS & CO. Mr. Thomas Ellicott, a partner in the firm of Ellicott & Abbott, of which Abbott & Co. are successors, is accredited with the distinction of having made the first Platform-Scale in the world. It was made in 1825, for the New York Coal Company, and continued in use until 1850, when it was removed to the factory of Abbott & Co., where it now is. This firm recently made a Track-Scale that will weigh one hundred tons, entirely of iron, for the Little Schuylkill Railroad. But the largest Scale probably ever made in the country was made, not long since, by A. B. DAVIS & CO. of this city, for the Pennsylvania Rail-road, at Columbia. It is one hundred and thirty-five feet in length, and will weigh two hundred tons. Three others, constructed for the same firm by the same road, were eighty feet each in length. For the Mine Hill Rail-road, Messrs. Davis & Co. constructed a Scale of one hundred and twelve feet long; and for the Baltimore and Wilmington, and Reading and other Rail-roads, they have made some Scales remarkable both for size and accuracy. Their achievements entitle them to rank among the leading Platform-Scale makers in this country.*

9. EDGE TOOLS AND CUTLERY. For the manufacture of *Edge* and *Hand Tools* there are two principal establishments in this city, and four others in the vicinity of the city whose products are sold here exclusively. The number of men employed is stated at one hundred and twenty-five, and the annual product at $127,000. The Edge Tools made in this city are of good material, well finished, and believed to be fully equal to any made.

The branch in which Philadelphia has attained peculiar distinction is the manufacture of *Braces and Bitts* and *Carpenters' Tools.* In nearly every report of articles exhibited at fairs within the last few years, one may find terms like the following, which is taken from the Franklin Institute Report for 1856 :—*"Braces and Bitts, Saw Pads, Spoke Shaves, and other light tools, by* BOOTH & MILLS, *Philadelphia. Excellent quality, good workmanship, and reason-able in price.—A First Class Premium."*

* Mr. Davis is also an inventor, having just patented a *Corn Sheller*, having an endless picker chain working over an angular bar-grate. This machine has been highly recommended as well adapted for the West and South, and other corn-growing districts.

This firm, BOOTH & MILLS, are the successors of T. E. Wells & Co., who received the prize medal at the London World's Fair; and the workmanship which was awarded that high distinction over Sheffield and Birmingham, was executed, in part at least, by the members of the present firm. In 1857, the American Institute in New York awarded them a prize medal; and it would seem that, by the consent of experts and good judges, both in England and in the United States, they are unsurpassed, if equaled, in the production of these important tools. They claim that the best or first quality of their manufactures is both cheaper and better than any imported. The mode of tempering adopted by them is said to be specially remarkable for its durability; and the stocks of their tools are certainly noteworthy for their elegance. Their manufactures comprise Braces and Cast-steel Bitts, Squares, Bevils, Spokeshaves, Turnscrews, Saw-pads, Pricker-pads, Cast-steel Gimlets, &c. They are sole manufacturers of T. E. Wells & Co.'s Braces, Bitts, &c.; also, Patent Anti-friction Braces. They employ about twenty hands.

Cutlery is made by several firms, and of a very excellent quality. Gilchrist's Razors, Clarenbach & Herder's Shears and Scissors, and Richardson's Table Cutlery, are all celebrated makes. Cutlery is also made by the manufacturers of Surgical Instruments, to whom we will subsequently refer. At Beverly, N. J., there was a company lately engaged in the manufacture of knives and forks, and their products were sold principally in this city.

10. RIFLES and PISTOLS are made quite extensively, and in quality equal to the best in the world. The barrels are either imported or made at Reading, and in Lancaster County, Pennsylvania—a considerable trade being done in barrels alone. One manufacturer claims that he can make a Pistol for sixty dollars, that will hit a horse at nine hundred yards. Generals Henningsen and Walker provided themselves with a pair each, previous to embarking on their memorable expedition to Nicaragua. Orders are now being executed for Rifles with solid gold and silver mountings, chased stocks, &c., the cost to the purchasers being two hundred dollars each. During the last year Messrs. Sharp & Co. erected a very extensive and complete rifle factory at Fairmount, for the manufacture of their celebrated Rifle and Pistol. (For a description see

27*

Appendix.) At the Arsenal at Bridesburg, under the superintendence of a very ingenious and courteous gentleman, Major Hagner, a
great variety of very remarkable and ingenious machinery is employed in converting Muskets into Rifles, making Percussion-caps, &c.

11. Wire Work of all kinds, plain and fancy, particularly
Wire Sieves, Wire Cloth, Screens, &c., is made very extensively
at about a dozen establishments. The iron wire is principally obtained from Easton, Pa., and from Trenton, N. J., and woven, in
this city, into every article into which wire can be twisted. The
Philadelphia goods are of a most substantial character, the average quality being acknowledged by all to be superior to the
New England make, and command readily a higher price.*

* The first Wire-working establishment in this country was founded by
John Sellers, in the year 1750. His two sons continued the business at the
corner of St. James and Sixth streets, and were so employed at the close of
the Revolution. The house thus established is continued by Sellers &
Brothers, the great-grandsons of the original founder, and ranks among
the largest in the business in this country. Their list of manufactures in-
cludes Wire-work for buildings; Fourdrinier Cloths, Brass and Copper Wire
Cloth, &c., for *Paper-makers;* heavy Twilled Wire-work for Spark Catchers,
Sieves of all kinds, Circular and Standing *Screens;* in a word, *Wire-work*,
Wire Cloth, and *Sieves*, of every description.

Watson, Cox & Co., are another prominent firm of Wire-workers. Of
Brass Wire Cloth they make all numbers from two meshes to one hundred
meshes to the inch; Iron woven Wire, from one mesh to sixty meshes to the
lineal inch; or in other words, Wire for Coal Screens to Wire for superfine
Flour. They also make a very strong Wire for cleaning rice; Sieves and
Riddles of the best kind for Iron Founders; and besides, a great variety of
Plain and Fancy Work, Covered Cylinders and Dandy Rolls for Paper-
makers. This firm is probably more extensively engaged in making *Brass*
Wire-work than any other. Their manufactory occupies five rooms, in
which five Looms and ten operatives are constantly employed.

Baylis & Darby are a young firm, but composed of men who have had
a long experience in Wire-working in the old-established houses of this
city. Their standing among Paper-makers, for Brass and Copper Wire
Cloth, is highly spoken of; also, their unusually heavy Coal Sieves for Coal
Dealers. Their stock of Sieves, of Silk, Hair, Brass and Copper Wire
Cloth is of every mesh; and their heavy Founders' Sieves and Fancy Wire-
work have been repeatedly commended. They received the Diploma at the
Pennsylvania State Fair in 1854, and since then they have applied Woven
Wire to Plastering purposes; the ceiling and dome of the Academy of

Besides the articles and branches of the Hardware manufacture I have thus specially alluded to, there is an immense number of articles made in this city, and which come within the category of Hardware stock. *Wrought Nails* are made largely; and *Cut Nails* are produced at the Quaker City Nail Works, near Fairmount, and at the Cumberland Nail Works, Bridgeton, N. J. The latter Company has about one hundred and twenty nail-machines in operation, capable of producing over 125,000 kegs of Nails per annum. The supplies of this establishment are furnished from this city, and the products disposed of in it. *Horse Nails*, chiefly for local consumption, but of superior quality, are made by four or five persons. Plasterers' and Brick *Trowels* are made in several establishments in this city, and these goods are sent to all parts of the United States. *Clock* and *Sash Weights* employ one or two foundries. The metal used in this manufacture is of the coarsest kind, 400 tons being about the annual consumption. The first *Screw Auger* made in this country was made in West Philadelphia, and the manufacture is now carried on by the successors of the originator, producing an article which is acknowledged to have no superior. *Hammers* of all kinds, from those used in watch-making to sledges and trip-hammers, are made, comparing favorably in prices with either foreign or domestic manufactures, and superior in quality and finish. *Cast Iron Butt Hinges* are made; and the Franklin Institute has pronounced the pivot-butts of one firm equal in finish to any American, and the principle better. A manufactory of *Strap* and *Reveal Hinges*, of every size and variety, has been in operation here for several years. *Piano Forte Hardware*, including gimlet-pointed screws, is made very extensively by one firm, established in 1822, and said to be the oldest in the business in the United States. They have ten machines in operation, and make eighty gross of Screws per day. *Shoemakers' Tools* are made by five or more establishments, who make their tools of steel, thus securing a large sale in the Eastern markets, where cast iron is principally employed. *Awl Blades* are made in every variety by Mr.

Music being a sample of this production. Their list of manufactures includes Wire Railing for Cemetery lots, Piazzas, Trellis-work for vines, of beautiful patterns; also, Wire Furniture and Iron Bedsteads.

Partridge, who is the son of the largest manufacturer of these articles in England.   He was himself formerly a manufacturer in Birmingham, and has recently sent to England for workmen. To this branch belongs a great variety of similar articles, besides binders', saddlers', printers' and shoemakers' sewing and trenching awls, mattress, sail and collar needles, lasting and other tacks, &c., which are made by him.   *Coffee Mills* of high reputation are made by Selsor, Cook & Co., and Ridgway & Rufe, in Germantown; and also *Shutter Bolts, &c.*   A *Rotary Knife Cleaner* or *Polisher* is made at one establishment, and by means of it, the labor that formerly required hours may be performed in a few minutes.   Of small tools the variety is infinite.   *Jewellers'* Tools, Rollers, &c., of the most delicate description; Oyster and Butchers' Knives, and Garden Tools in great variety, are made by several persons; and Patent Curry-Combs, of iron and brass, open back and covered; Patent Bake Pans and Meat Mauls, all of improved styles, are largely made by one manufacturer.   Of Beach's Curry-Combs it is worthy of remark that, while they are greatly superior in durability and neatness of finish to most of the imported, they can be furnished at considerably less cost.   Steel *Stamps*, *Brands*, and *Punches*, Stone-cutters' Tools and Mill Picks, Curriers' and Tanners' Knives and Tools of all kinds, Saddlers' Tools, Binders' Tools, Coopers' Tools of excellent quality, Ice Tools, and Umbrella-makers' Tools and Furniture, each employ one or more, and some of them several establishments.   *Skates* are made of improved construction, being fastened to the foot by springs, without straps; and it is said a Company has been organized for their more extended manufacture.   *Sad Irons* are a leading article at a factory in West Philadelphia; and *Hoes* are made at several establishments, and particularly at Prince's extensive Hoe factory, on the Pennepack.

In the manufacture of the miscellaneous articles which are included in the term Hardware, it will be perceived, from what we have stated, that there are numerous establishments, and the aggregate production is very considerable; but, at the same time, there are no very large factories.   The most extensive manufactory of General Hardware is that of E. HALL OGDEN, on Ninth and Jefferson streets.   He employs about seventy-five hands; manufactures, as per catalogue, nearly one hundred and fifty distinct articles, including Malleable and fine Gray Iron Castings. His works consist of a main building, fifty by one hundred feet, with two wings, one forty by one hundred, and the other fifty-six by one hundred, filled with all requisite machinery, driven by an engine of twenty-horse power. His manufactures are carefully made, and his Saddlery Hardware, in particular, is noted for its excellence.

The manufacture of Hardware is a branch deserving the attention of capitalists; for, with an abundance of skilled workmen, cheap raw materials, and a good market, a few large establishments could hardly fail to prove remunerative.

Iron is converted in Philadelphia into a variety of forms besides those enumerated; and if we were to venture into the departments of the Miscellaneous Manufactures of Iron and Steel, our article would be indefinitely extended. In the city of Franklin, Lightning Rods are, of course, a prominent article of manufacture; and one maker, Mr. THOMAS ARMITAGE, states, with a laudable pride, "I have put up sixty thousand Rods, and have shipped a great number to various parts of the world, especially to the Southern States, and have never heard of the loss of a single life, or the destruction of a dollar's worth of property, by lightning, in any building to which one of my Rods has been attached. I have received information of three to five thousand instances in which Rods that I have put up have been struck by lightning, in all of which it has been carried safely to the earth, without the slightest injury to person or property." With such a record of facts in their favor, he may well have confidence in the superiority of his Rods.

But we must take leave of the subject of "Iron and its Manufactures;" and do so by giving the following statistical aggregates, as recently compiled by the present Secretary of the Board of Trade, as aforestated. Our own investigation, as far as it extended, gives a greater product for certain items, particularly Saws, Shovels, Guns, &c.; but the general aggregate, we believe, is approximately accurate.

	Men employed.	Total production.
Forges and Rolling Mills, - - - - -	710	$1,801,150
Foundries of Stoves, Hollow Ware, &c. - - -	2,440	2,500,000
Locomotives and Steam Engines, with Foundries, -	3,008	3,428,000
Machinist and Foundry establishments, - -	1,417	1,912,000
Platform Scales and Foundry work not before included,	98	95,000
Wrought Iron Bolts and Nuts, - - - -	421	411,000
Malleable Iron, &c., - - - - - -	356	330,000
Safes, &c., - - - - - - - -	90	120,000
Rail-road Cars, &c., - - - - - -	220	300,000
Manufactures of Steel, mainly		
Saws, - - - - - - -	310	345,000
Coach and Car Springs, - - - -	170	238,000
Steel Hay Forks, - - - - -	80	78,000
Shovels, &c. - - - - - -	195	397,000
Edge Tools and Hammers, - - - -	125	127,000
Cutlery, Skates, and Instruments, - -	150	150,000
Rifles and Guns, - - - - - -	120	120,000
Other classes, - - - - - - -	500	500,000
Total, - - - - - -	10,410	$12,852,150

If to these totals we add the share of Iron making and Rolling, within forty miles of the city, which truly belong to its business—being established by its capital, and obtaining supplies as well as finding its market here—we are sure that two thousand workmen, and a production of $2,500,000, would be within bounds for that share. Ten thousand workmen, finding homes constantly within the city, are engaged in the commercial manufacture and working of Iron, and two thousand of the same class alternate between the city and the country in its vicinity. The finished work made in the city, and its immediate suburbs, has a commercial value of twelve millions of dollars; and the share belonging to the city, though located a few miles from it, of three millions more.

XXII.

Jewelry, Silver-ware, &c.

Precious Metals are first mentioned in history as a means of facilitating the transfer of property: And Abraham weighed to Ephron the silver, "four hundred shekels of silver, current money with the merchant." The adoption of gold and silver for personal adornment was subsequent to its use as money, and even to this day the idea of value, in the popular mind, is associated with these metals principally in the form of coin. It will, therefore, seem surprising to many, that the value of the gold and silver plate in the world has been carefully estimated to be *two thousand millions of dollars*, which is at least *one-fourth more* than all the coin in the world. In the United States, precious metals, of the value of at least thirty millions of dollars, are annually converted into plate or worked up into ornamental forms.

Philadelphia has long been the chief seat in America for the conversion of the precious metals into coin. The United States Mint was established in this city, in 1793, and up to the close of the year 1856, the entire coinage amounted to $391,730,571.86.

At the present time, Philadelphia is also one of the principal points for the manufacture of Gold and Silver Plate, and works have been produced in both these metals that would do no discredit to the master goldsmiths of Europe. Mr. WALLIS, in the Report on the Industry of the United States, thus testifies:—

"The manufacture of Gold and Silver Plate is more or less carried on in nearly all the larger cities, especially New York, Boston, and Philadelphia. In the last-named it partakes of the character of a settled trade, there being some twelve or fourteen establishments in which a considerable number of persons are employed, and the productions of which are of a varied, but for the most part of a useful, as well as an ornamental character. Table services, and all the articles of utility comprised in suites of plate for domestic use, form the staple articles; and these are manufactured in large quantities.

"The workmanship is usually sound; but it often happens that, on close examination, a deficiency in that nicety of finish, especially in the chasing, which characterizes the best English work, is observable. Still it is rarely found that the equally, or perhaps more objectionable practice of over-chasing, to the destruction of the artistic effect of the details, is committed. The fault is evidently that of timidity in handling; but there is a wisdom in leaving off at the right time, which the elaborate chasings of English works rarely display.

"In most of the manufactories a few European workmen are to be found; but the Americans engaged in this department of industry are usually of a superior class, and it is remarkable how soon they get into the system of those amongst whom they are thrown as mere learners. In this, as in other branches of industry, their minds being thoroughly prepared by education, they seem to seize upon and master even very difficult points in manipulation and construction, as it were by mere instinct."

Many of the magnificent Services of Gold Plate, Silver Trumpets, Horns, &c., which, at different times and in different parts of the United States, have been presented by citizens to those whom they delighted to honor, were executed at workshops in Philadelphia. But, besides Gold and Silver Plate, the manufacture of *Jewelry* is largely and successfully carried on, particularly the finer and more costly kinds, as Diamond and Pearl Jewelry. The taste displayed in setting diamonds and pearls, and in Cameo, Enameled, and Filagree work, and the weight and purity of the solid gold work, would astonish those who are familiar only with the work of this description executed in New England.

The London Commissioners refer, in particular, to the jewelry made by Messrs. Bailey & Co., of this city, and highly commend their original designs and workmanship.

There are fifteen lapidaries in the city, constantly occupied in cutting and preparing the various stones—Rubies, Sapphires, Agates, Emeralds—besides the large quantities that are imported from abroad.   The value of the gold jewelry annually made in Philadelphia, including gold chains, amounts to $1,275,000.

The articles embraced within the term Jewelry are exceedingly numerous; and we can only allude to those which may be considered special branches, particularly Gold Chains and Pencil and Pen Cases.

Gold Chains are made by at least four houses—DREER & SEARS, STACY B. OBDYKE, E. C. BONSALL, and NEWLIN, BISHOP & Co. The first-named firm has been engaged for many years in the manufacture of gold chains of the finest quality and workmanship.   During the past few years they have extended the range of their manufactures, and now produce chains of every merchantable degree of fineness.   The present firm is a continuation of the old house of Dreer & Hayes, favorably known to the trade since 1833.   Their manufactory, in Goldsmiths' Hall, is one of the most complete and best arranged in the country, lofty, well-lighted and ventilated, and supplied with the most scientific tools and machinery, driven by steam power : they have a capital invested in the business of $150,000, employ one hundred and twenty-five hands, to whom they pay annually $70,000, and produce chains of the value of $200,000 annually.

Messrs. Dreer & Sears carry on, in addition, an extensive business in assaying and refining the precious metals, and sell bullion annually to the amount of $300,000.

Gold and *Silver Pencil and Pen Cases* are made largely in Philadelphia, but, as we are informed, only by one firm—Messrs. GEORGE W. SIMONS & BROTHER, Sansom Street Hall. This firm claim to have been first to use steam advantageously in this branch; and their present machinery, which is new, costing upward of $25,000, is of the first class, and embodies all the latest improvements. Messrs. Simons & Brother had the misfortune to lose their entire stock of tools in the great fire which consumed the Artisan Buildings in Ranstead Place, on April 1st, 1856, not a single appliance of manufacture remaining available; but in six weeks from that time, as we are informed—with a steam engine, much heavy shafting, and other machinery, and all the peculiar tools of the art to be constructed anew—their factory was again in good running order in its present location; and ever since they have been occupied in perfecting and simplifying its details, till now it is confessedly a model in its appointments. In addition to Gold and Silver Pencils and Pen Cases, Messrs. Simons & Brother manufacture Gold, Silver, and Steel-top Thimbles, Finger Shields, Tooth and Ear Picks, Watch Keys, Gold Pens, Cane Heads, Bracelets, Breastpins, Ear Rings, Finger Rings, Sleeve Buttons, Studs, Guard Slides, Charms, Seals, Badges, etc. In the manufacture of Thimbles they have effected many improvements, particularly in the application of steam; and it is said that as many Gold Thimbles are now sold annually as there were Silver ones a few years since, while the consumption of the latter has increased in the same ratio. They employ sixty hands.

Next to Gold Plate and Jewelry, the largest consumption of this precious metal is in—

Watch Cases and *Dials*. This branch of manufactures has rapidly augmented in the United States within a few years, and now consumes a large amount of Gold. English and Swiss Silver Watches are almost invariably imported complete and ready for sale, but Gold Watches are usually cased here. William Warner,

23

of this city, was the first American manufacturer of Watch Cases, having established the business previous to 1812 ; and Philadelphia continues to be now one of the chief seats of this manufacture. In purity of Gold, in excellence of workmanship, and in elaborateness and beauty of ornamentation, it may safely be said, the cases made in Philadelphia are not surpassed by any.

The price of Cases chiefly depends on the weight and fineness of the Gold ; but in some instances much labor is expended upon them. The engraving alone costs from $5 to $60. A Case of common Gold can be sold at $20, or less ; while others, like the magic cases, are worth from $75 to $125—the labor expended upon them costing more than half that sum. The business requires a large capital, as the material and labor must be paid for in cash, while the cases are usually sold on a credit of four months.

There are at present sixteen Watch-Case manufacturers in the city, who have a capital invested of $375,000, employ two hundred and ninety-four men, and make Cases annually of the value of $942,000.*

* A reference to two or three of the principal establishments in this branch may not be inappropriate :—

E. TRACY & Co., Goldsmiths' Hall, Library street, are one of the most prominent firms making both Gold and Silver Cases, and possess facilities for completing every part of all styles in their own factory. They are also Refiners and Assayers, and prepare all the Gold and Silver they use by dissolving them, which is said to be the only sure process to obtain them entirely pure. They usually employ about fifty workmen ; and for quality of gold, as well as beauty and excellence of workmanship, their Cases are second to none. They make a very large number of Cases for the *American Watch Company*, of which Mr. Tracy is one of the three proprietors—a Company that has succeeded in competing with foreigners in a department of manufactures that was before untried in this country.

T. ESMONDE HARPER, formerly Harper and McLean, S. E. corner of Walnut and Dock streets, is another house that may be referred to with satisfaction, as illustrating the prominent and excellent Philadelphia houses in this branch. Mr. Harper acquired his knowledge of the art from Mr. Wm. Warner, who, as previously stated, was the first manufacturer of Watch Cases in this country. He usually employs about fifty hands, and manufactures largely for Philadelphia, New York, Boston, and Baltimore, both of the finer and most costly Cases, as Magic, Miniature, Hunting ; and of the lower-priced, as Detached Lever and Cylinder Cases. In Engraving the Cases, he employs several of the rarest and most ingenious machines that have ever been con-

In the manufacture of *Gold Leaf*, there are now nine firms, who employ one hundred and twenty-five hands, and produce an annual value of $175,000. The malleability of Gold is such that it may be beaten into leaves one two-hundred-and-eighty-thousandth of an inch in thickness; in other words, a pile of 280,000 leaves will measure but one inch in thickness. Gold Leaf is made into books containing 500 Leaves, the leaves being 3⅜ inches square; so that each book contains *five thousand six hundred and ninety-five* square inches of Gold Leaf, sufficient to *carpet* a small bed-room, and yet the weight of Gold *is less than four pennyweights.* In the process of hammering or beating, membrances of parchment, vellum, and gold beaters' skin (a peculiar substance prepared from the outer membrane of the large intestine of the ox) are interposed between the hammer and the Gold.

The largest Gold-beating establishment in the city, and, we believe, the largest in the United States, is that of HASTINGS & Co., on Fifth street. They employ forty-five hands, and have facilities for beating out 200 packages, or 4,000 books per week. The quality of leaf produced is very superior.

Of *Gold Foil* the annual production in Philadelphia is $150,000. The principal firm in this branch—which has become to some extent a distinct business—is CHARLES ABBEY & SONS, the oldest established and most extensive manufacturers of Gold Foil for dentists' use in the United States. Mr. Charles Abbey, the founder of the firm, is a veteran in the business, having been uninterruptedly engaged in manufacturing Gold Leaf and Gold Foil—for the last twenty-three years, Gold Foil only—since 1816. His three sons, who with him now compose the firm, are also skillful and practical manufacturers, rendering the partners a host

structed. They are the invention of a monk in Switzerland, but introduced into this country some years ago, and now made exclusively in this city.

Another very extensive house is that of JACOT & BROTHER, 109 S. Second street, who supply with Cases some of the largest Watch Importers and Jobbers in all the principal cities of the Union. They commenced business about twenty years ago, and turn out about six thousand Cases per annum. To obtain a sufficient number of skilled workmen Mr. J. frequently visits Switzerland, and brings back a number of workmen and their families, advancing the money for the expenses of their journey, to be refunded out of their subsequent earnings.

within themselves.   Mr. Abbey's connection with the manufacture of Gold Foil may be said to cover the entire period of its use for purposes of dentistry.   When he commenced, dentistry was scarcely known as a distinct profession.   This firm prepare all the Gold they use, not purchasing from refiners, as is customary ; and that they make a most superior article is proved by the fact that " Charles Abbey & Sons' Foil" is quoted in the trade's circulars at a higher price than that of any other manufacturer.   They supply all the prominent dentists in the United States, and receive orders from England, France, Germany, Sweden, and from other parts of Europe.

Gold Spectacle Frames are a leading article of manufacture with a number of firms—N. E. MORGAN, BUTLER & McCARTY, McALLISTER & BROTHER, W. BARBER, and I. SCHNAITMANN. The quality of those made here can be very highly commended.

Gold Pens are an exclusive branch of manufacture by two or three firms, and are also made by several others.   *Hair Jewelry* is made by a half dozen, and some of the specimens are exquisite.

2. SILVER AND PLATED WARES.

The *Silver Ware* made in Philadelphia, it is claimed and generally acknowledged, is at least equal, in workmanship and design, to the very best made in this country ; while many connoisseurs, who have visited the most celebrated silversmith shops in the old world, ascribe to none of them precedence over those of Philadelphia.   The articles are generally made of a fixed standard, several degrees purer than coin, and, consequently, possess great intrinsic value, aside from mere workmanship.*   The manufacture of Spoons

* Messrs. Bailey & Co., the leading Jewelers and Silversmiths of Philadelphia, claim the distinction of having first introduced the use of Silver of the full British standard, say from 925-1000 to 930; the American standard being but 900.   They now work no other, a test being made monthly by J. C. Booth, Esq., Chief Assayer of the Mint.   One advantage of thus raising the standard is, that it successfully secures the trade from importations of Silver from England, for purchasers are assured, by a full gurantee, of receiving Silver as pure as that stamped by the English government.   This improvement in the quality of Silver also renders the manufactured articles more beautifully white, susceptible of higher polish, and less liable to oxidation and consequent discoloration.

The house of BAILEY & Co. has been in existence over twenty years, and

and Forks, by machinery, is largely carried on, the shops being provided with "rolls," and all other improved machinery that has as yet been introduced. A great deal of Silver-ware made in Philadelphia is retailed in New York as Parisian.

Within the last few years, since the discovery of the process of Electro-plating, the wares produced in an inferior metal, but covered over with a film of silver, have become quite popular. In England the principal improvements effected in this class of goods are identified with the names of Elkinton, Mason & Co. ; in the United States, the most successful experimenter was John O. Mead, who, with his two sons, constitute the firm of JOHN O. MEAD & SONS, leading Silver-platers of Philadelphia. The following recital of facts, extracted from a memorandum now before us, tends to establish the point.

" Previous to 1836 Mr. John O. Mead was executing all the silver-plating and gilding by the old process of quicksilver and acids, for the N. P. Ames Manufacturing Company, which then, as now, was employed principally in making swords, cannon, and military equipments for Government. In 1836, Mr. N. P. Ames was tendered an appointment as one of a committee to visit England and Germany for the purpose of acquiring such knowledge as would be necessary for establishing a Government Manufactory of these articles, but, though declining the honor, he followed the committee in the next vessel. While in England he was invited to attend certain lectures instituted by Government, where the subject of depositing silver by electricity was discussed, and its feasibility theoretically but not practically demonstrated. On his return to the United States, in 1837, he brought with him one of Smee's bat-

its reputation at the present time, throughout the Union, is unsurpassed by any other similar establishment. Though the South has been their principal customer, their wares are well-known throughout all parts of the Union, and everywhere favorably. A Communion Service, finished by this firm to order for one of the wealthy churches in Charlestown, Mass., adjacent to Boston, attracted by its beauty a wide-spread attention. All the processes —the designing and drawing of the patterns, the melting and refining of the metal, to the last finishing touch of the graver, are executed upon their own premises, and under their personal inspection. The Silver Ware made in this single establishment amounts to $100,000 annually. This firm is now about completing a marble store on Chestnut street, which will be one of the finest and most attractive of the many attractive buildings on that street.

28*

teries, and such investigations relating to the subject as had then been made, which, however, had not resulted in the discovery of any process by which silver could be deposited on any *base* metal, as copper, German silver, &c. After about a year of close study by day and night, Mr. Mead, to whom the matter had been submitted, aided by scientific suggestions from Professor Silliman, discovered that prussiate of potash was the alkali that would hold up silver and not oxidize base metals, and considered the point gained. This was in 1839; but the difficulty yet to be surmounted was a means of depositing any given weight of silver that might be desired. The discovery of the cyanide solution in 1840 solved this difficulty, and enabled Mr. Mead to deposit any required amount of silver on base metals, and subsequently on any metal direct without the intervention of any other metal—a result which even yet few houses can or do accomplish. When Mr. Mead had perfected his experiments, he instructed others in the process, and was the means of putting into successful operation a number of concerns now flourishing in New England."

It is one of the advantages of electro-plating, that all ornaments, however elaborate, or designs however complicated, that can be produced in silver, are equally obtainable by this process. Messrs. Mead & Sons are now producing articles of every kind and variety, from the most elaborate Epergne to the plainest article of Tea or Dinner Service, in the greatest perfection. Their manufactory is a very extensive one, over *two hundred* hands having been employed in it at a time; and, in their warerooms, near the Girard House, may be seen all the latest and most beautiful patterns, rivaling, in style and finish, those of solid silver. They make about *fifty* different patterns of tea-sets, and their plated-ware exceeds, in durability and variety, as well as in richness of design, that of any of the New England concerns. Services of plate are constantly being furnished by them to private families, hotel proprietors, steamboat and ship builders; and wares of their manufacture have been shipped to England, Turkey, Persia, and China.

MEYER & WARNE, another prominent firm, have manufactured, for some years, a new article of Plated-ware, of which the judges at the Franklin Institute Exhibition of 1856, spoke as follows:—

"It appears that the manufacturers have substituted a new method in making, by which the expense of chasing is dispensed with. Although the mode of manufacture is not fully known to the judges, yet sufficient is known to enable them to say they consider it a decided improvement; because any pattern can be exactly reproduced, and at a cost which will enable the maker to sell a handsome article at a moderate price. The judges would, therefore, desire to speak in the highest terms of this improvement in manufacturing."

Since that period the firm have perfected the process, and are now producing a metal that is truly remarkable for its strength, whiteness, and cheapness, while it has a ring somewhat resembling silver.

Mr. HARVEY FILLEY has long been identified with the manufacture of Plated Wares in Philadelphia, and has established an enviable reputation. We notice that he announces an article which he terms Nickel Silver.

Britannia metal and Britannia metal goods are made by four firms. This metal is composed chiefly of tin, antimony and copper, which are melted and mixed, and then cast into bars. Many improvements have been effected in the manufacture of Britannia-ware, the most important of which is the art of *spinning* the hollow ware into form. All the fine work is now *spun*, but the process is by no means peculiar to this city.

The statistics of Works in the Precious Metals and their imitations, for this city, are approximately as follows:—

Persons employed, - - - - - - -	1,700
Product, viz.:—Gold Jewelry, Pens, Spectacles, &c., - - -	$1,275,000
" Watch Cases, - - - - - -	942,000
" Leaf and Foil, - - - - -	325,000
Silver Ware, - - - - - - -	450,000
Plated and Britannia Ware, - - - -	380,000
Total, - - - - - - -	$3,372,000

This statement, it will be apparent, does not include the additional value produced by assaying and refining, includes nothing for the Mint, which employs 125 persons, nothing for the silver-plating of Door-plates, Knobs, Bell-pulls, Cutlery, &c., but simply the product of the manufacturing establishments in Gold and Silver. The amount thus only partially given far exceeds the annual product in Providence, R. I., which, it has been heretofore supposed, was the chief seat of the manufactures of Gold and Silver in the United States.

XXIII.

Lamps, Chandeliers, and Gas Fixtures.

In nearly every Exhibition of American Manufactures, which has been held in the last quarter of a century—we presume it will be conceded by every one—the most attractive, artistic and brilliant feature of the display was the Chandeliers, Candelabras, Girandoles, &c., made and deposited by Philadelphia houses.

Previous to 1830 the whole trade in Chandeliers was in the hands of foreign importers; now the American market is entirely supplied by home manufacturers; and so great has been the progress made, and so high the perfection arrived at by those of Philadelphia, judging solely from the frank and unequivocal testimonials of intelligent foreigners, respecting their excellence in design, workmanship, and finish, that we anticipate a period not remote when Philadelphia Lamps and Chandeliers will compete successfully with those of Europe in European markets. We probably cannot adopt any more effectual method of conveying to persons who may be ignorant, if any such there be, of the degree of perfection attained in this important manufacture by the two principal houses, viz.:—CORNELIUS & BAKER, and ARCHER, WARNER, MISKEY & Co., than to submit the following extract from Whitworth & Wallis' official Report on the Industry of the United States.

" The manufacture of ornamental Brasswork, as applied to the purposes of lighting, forms a progressive and important branch of industry in several large cities, but more especially in Philadelphia.

" The chief portion of the work is cast, little or no ornamental stamping being attempted. It is scarcely possible, however, to conceive better work than the generality of these ornamental Brass Castings. At Philadelphia especially, the greatest attention has been paid to this point; and a peculiar advantage is derived here from the fact that the sand obtained in the vicinity of that place is of so fine a character as to require no sifting for use, and the finest castings are easily made, so far at least as material goes. The pattern is simply modeled in wax, and from this a brass pattern is cast direct, no white metal being used. The brass pattern is carefully and thoroughly chased, and from this all future work is produced. Thus, the shrinkage and variation of size between the white metal pattern and the brass casting, often found to exist in castings made from the former, is avoided, and the register of the two sides of a branch, or other portion of a Chandelier or Gas bracket requiring to be fitted together, is more perfect than it otherwise would be. The brass pattern, too, takes a sharper and more decisive chasing than white metal; and as the castings are never chased, as, from the fineness of the sand, they are sufficiently sharp and effective without it, the accuracy of the pattern is of the first importance; and all that is required to be done, after the castings leave the foundry, is to file off the very small amount of superfluous metal retained in the casting, and fit the parts together.

"The bodies of Chandeliers, whether vases or dishes, are invariably spun up from the flat metal plate, instead of being stamped, as is usually the case in England. This is the old method of producing these portions of Lamps and similar articles, and appears to have been introduced into practice in America by German workmen. It is not confined to small bodies, but is used for the production of larger sizes than are usually considered practicable. Very large bodies, however, are generally hammered up.

Discs of plate metal, for the purpose of spinning up, are cut by a machine with two wheels, having the sharp edges working against each other, after the manner of a pair of shears. Those circular cutters work with great ease and rapidity, giving great facilities to the workman, and presenting an elegant method of doing laborious work with the greatest possible ease and certainty.

In annealing the spun work, after the first process of raising from the flat plate, it was formerly found that the metal cracked, more particularly in the bottom angle. As the first form from the plane is a simple truncated cone, the second process of spinning, after annealing, gives the requisite curves to the sides. To prevent this cracking during the annealing process, it has been found that the simple bending or squeezing in of the sides of the cone, until the circle becomes a somewhat elongated ellipse, and then placing a quantity within each other, has the desired effect, and cracking rarely, if ever, takes place. This is stated on the authority of Mr. Cornelius, of the firm of Messrs. Cornelius, Baker, and Company, by whom it has been successfully adopted in practice. If any additional proof were required as to the value of accurate scientific knowledge, as applied to a manufacture in which mechanical invention and great skill in chemistry and metallurgy are so essential to complete success, it will be found in the fact that this gentleman received an education specially adapted to the requirements of the business which his father, the late Mr. Cornelius, was about to establish in Philadelphia; and that his studies in abstract science and the various discoveries and expedients, both mechanical and chemical, as resulting therefrom, have given the house, of which he is now the head, an immense advantage over both foreign and domestic competitors. The system, order, and accuracy which prevail throughout this establishment, is full evidence of the influence of a mind reaching as far beyond the ordinary traditions of the workshop and foundry, in the scientific sense, as in the practical result it goes beyond the mere dilettantism of speculative science *sans* application.

In this establishment, consisting, as it does, of three distinct factories, one for casting and soldering, another for machine-work, and a third for filing and fitting, seven hundred workmen are employed. Commenced about thirty years ago, it has gone on increasing its sphere of action to the present time, a constant attention to the scientific principles of metallurgy and mechanism having tended to its prosperity from the beginning ; when, in the establishment of a business, at that period so novel in its character, and so doubtful as an enterprise, it was more than probable that every thing, except traditionary modes of action, would be rather repudiated than encouraged. In point of economy, however, science has been found to be the cheapest as well as the best assistant—as indeed it ever will be when combined with a thorough knowledge of the work to be done by its aid.

" In the dipping process, as pursued in these works, great modifications are made in the character and strength of the acids used. It was found that, from the variation of temperature at Philadelphia, ranging, as it does, from below zero in the winter to 96° and 98° in the shade in the summer, nitric acid became unmanageable during the hot season, as its fumes were given off so rapidly as to injure the health of the workmen. The accurate scientific knowledge, however, brought to bear upon this point—one, too, involving the very existence of the trade, except at a frightful destruction to human health and life—has obviated every difficulty, adapted the acids to the temperature, and the dipping department is comparatively free from noxious fumes, even under the highest of the above temperatures.

" The result is equally satisfactory as regards the color of the work when dipped, some novel effects being produced, and a singular purity of color obtained.

In lacquering, considerable improvements have also been made. It was found that the lacquers made after the English formula lost color very quickly, from the extremes of temperature already noted ; and that during the months of July and August, when the dew-point of the barometer is reached in Philadelphia, the red-lacquered work always streaked in the direction of the marks of the spinning tool on the broad surface of metal. After a series of experiments, carried through several months, Mr. Cornelius succeeded in making a lacquer, which he states to be quite permanent under any variation of temperature.

" The usual methods of decorating burnished surfaces by varnish pencilings and dippings, for colored effects, are adopted here, as in Europe. The lacquering is all done by men, no females being employed.

" The columns of Table Lamps are made of sheet metal formed into tubing, and fluted upon a mandril by means of a wheel acting in the direction of the axis. This is said to have been adopted by Messrs. Cornelius, Baker, and Company before it was attempted in this country; and they were enabled by these means to go into the market with a great advantage in price over the cast column lamps imported from Europe, as the cost of that portion of each article was two-thirds less in Philadelphia than in Birmingham.

" The manufacture of Lamps for the consumption of lard still forms a considerable item of trade, especially for the western markets; but these are not so much in demand as formerly, owing to the more extensive use of gas. At least one hundred and fifty patents have been taken out, from time to time, in the United States, for contrivances for effecting the consumption of lard for the purposes of lighting. The lard Lamp manufactured by this firm, however, appears to have been one of the most successful; the principle of the candle being kept in view, and the heat applied in the direction of the point of illumination.

" Great attention is paid both by Messrs. Cornelius, Baker, and Company, as also Messrs. Archer and Warner, to the perfect accuracy of all their Gas Fittings. The Gas-Works of the city of Philadelphia are celebrated for the perfection to which the manufacture of gas is carried, and the thorough scientific principles upon which every detail of the establishment is carried out, not the least important of which is the uniform gauge of all Fittings; so that any part becoming defective, is at once repaired without trouble, and, of course, at a less expense than when a constant variation in the gauge of the Fittings is permitted.

" At Messrs. Cornelius, Baker, and Company's manufactory, the sawing of the slit of the Gas Burner is executed with the greatest accuracy; and although not done with such rapidity as by the process usually adopted in England, the greatest exactitude is obtained as to the quantity of gas the Burner is capacitated to consume; and whenever a new saw is substituted for an old one, the slit is carefully tested as to its capacity for consumption by means of a Gas-meter placed by the side of the workman. In short, guessing is avoided in every thing connected with this establishment.

"In the fitting of the pipes, the same accuracy and care are manifested. The screw is turned in a lathe to prevent the possibility of splitting the pipe, which is more or less invariably done by the ordinary screw-plate; but the crack, being very minute, is not discovered until after having been some time in use, when the gas begins to escape, and continues until a permanent leakage is established.

"In order to secure the joints completely, a composition of wax, resin, and Venetian red is applied to the tap, which is sufficiently hot to melt it; the pipe is then screwed in, and the joint is at once sounder and cleaner than when cemented by the application of white lead, the usual material employed for this purpose in England.

"Among the workmen employed by Messrs. Cornelius, Baker, and Company, are eight or nine Englishmen, the rest being Germans, French, and Americans; but the majority are undoubtedly native workmen.

"Messrs. Archer and Warner employ about two hundred and twenty-five workmen, and are engaged in precisely the same trade as Messrs. Cornelius, Baker, and Company. The remarks as to the character of the work produced by the last-named firm, especially Gas Fittings, and the perfect division of labor, which is not so general a feature in American as in European manufactories, applies with equal force to both establishments; and though that of Messrs. Archer and Warner is not so extensive, its operations are carried on in a systematic and efficient manner, the results being shown in the articles produced, which are excellent of their class." (For a description of this establishment see APPENDIX.)

The two principal firms above alluded to employ about one thousand hands, and do a business annually of one million of dollars. Besides these, M. B. Dyott and Hiedrick & Horning make Gas Fixtures to some extent, in connection with the manufacture of Brass and Composition Lamps; and numerous parties make Gas Burners and various minor articles appertaining to the use and consumption of Gas. The manufacture of Carriage, Engine, and Locomotive Lamps is carried on at a half dozen establishments. *Lamp Shades* are made by V. Quarre, and others; and the entire business done in Chandeliers, Gaseliers, and Lamps we estimate at $1,300,000, employing twelve hundred and fifty hands.

XXIV.

Leather and its Manufactures.

The manufacture of various kinds of Leather, particularly Sole Leather and heavy Upper Leather, has long been a leading pursuit in Pennsylvania. The Tanneries, which reveal themselves here and there in ravines along the highways and by-ways of the State, some traveler has remarked, are almost as plentiful as the old-fashioned water-propelled grist mills, or the country taverns. In 1840 the capital invested in Tanneries in Pennsylvania amounted to $4,255,055, by which, at that date, 5,226 operatives were employed. The abundance of Oak, particularly the White Oak and Chestnut Oak, has facilitated and rendered profitable the business of tanning ; and the excellence and cheapness of oak bark, probably more than any other circumstance, explain the immense production of Leather which finds its principal depot in Philadelphia. Quercitron Bark, which is simply a product of the ordinary Black Oak, is now largely exported to Europe, where it commands high prices.

The two principal processes for the manufacture of Leather, it is perhaps needless to remark, are denominated *Tanning* and *Currying.* The latter is mainly a mechanical process, and the former a chemical one, though requiring more or less manipulation in order to facilitate the chemical action. In Philadelphia the principal branch carried on is Currying ; much of the leather tanned in the interior of the State being brought to the city in its rough state, and requiring the art of the currier to smooth and adapt it, in pliability and softness, to its various uses. The firms engaged in currying Leather within the limits of Philadelphia number at least thirty-five, some very extensive, while there are but ten tanneries, four of these being employed in making Sole Leather exclusively, two in Calf Skins, one in Belting Leather, while the others make both Sole, and Calf, and Sheep to some extent.

It would be interesting, did our space permit, to note and trace the effects of the various improvements that have been made by mechanical means in the manufacture of Leather. The steam engine has been generally introduced into the factories of leather-dressers and tanners, and is now used for grinding bark, for softening foreign hides, and in giving motion to many machines for

29

washing, glazing, and finishing Leather. Important results have also arisen from the invention of ingenious machinery for splitting hides and skins. This is effected by means of a long sharp knife, kept in rapid motion about the sixteenth of an inch from the edge of a smooth bar of iron, over which the skin is drawn by a revolving cylinder. By another machine, the skin is pressed between the revolving rollers, and presented, as it emerges, to the edge of a long straight knife, nicely adjusted between the upper and under surfaces of the skin, and kept in motion backward and forward, to facilitate the operation of splitting. But the most remarkable of modern improvements in this connection are what has been termed "time-shortening inventions." The feat has actually been performed of butchering a kid, dressing the meat, and tanning the hide, all in the self-same single hour ; and it has been repeatedly and unequivocally demonstrated that *good* Leather can be made in a comparatively brief period of time, by the aid of machinery, bringing the skins into rapidly repeated contact with the tanning liquor by means of a revolving cylinder, which catches them up and dashes them down alternately.

The distinction which Philadelphia is justly entitled to claim in this branch of manufacture, is in the production of the finer kinds of leather. *Calf Skins* are made of a most superior quality, unequaled elsewhere, it is believed, in this country, and not excelled by the celebrated French. One of our manufacturers, it will be remembered, entered into competition with the French and all others, at the World's Exhibition at London, in 1851, and carried off the Prize Medal. All parts of the West, as well as less remote States, are chiefly supplied with this leather from this city.

Deer Skins are very largely manufactured into Leather, which is used for Gloves, Suspenders, Drawers, &c. About 60,000 deer skins are annually converted into leather in Philadelphia alone. The hair is considered to be the best material for stuffing saddles.

Sheep and *Lamb Skins* are tanned in all the various modes ; in bark, in alum or salt for white leather,* and are also largely ap-

---

* *Alum* or *White Leather* is made to a considerable extent, and used for Saddlers' facings, lining Shoes ; for Masons' aprons, for covering necks of bottles and spreading plasters, and for various other purposes. The skin is softened in lime-water, washed several times in pure water, and after-

propriated for Parchment and Chamois Leather. Two of the manufacturers have hydraulic presses for expressing the grease from the skins, thus facilitating the operation of dyeing in brilliant colors, which cannot well be done while any grease remains in the pores of the skins. The raw material is principally furnished from the flocks of our own country, though Sheep skins are imported to some extent from the Cape of Good Hope.

But the branch of the Leather manufacture in which Philadelphia may fairly claim a decided pre-eminence is that of *Morocco*. At least *one and a half millions* (1,500,000) of Goat skins are annually converted into leather in Philadelphia; and the excellence of quality is no less remarkable than the quantity. The Goat skins are chiefly obtained from the East Indies, and three-fourths of the whole amount imported into the United States are brought to Philadelphia. The East Indian skins are small and have short hair, and are peculiarly suited for ladies' and children's shoes. The Goat skins from Tampico are highly esteemed, being large and heavy; while those from Curaçoa, though smaller, are very superior, and used chiefly for making kid for gloves and gaiter uppers. Those from the East Indies comprise perhaps four-fifths of the whole importation.

The skins are principally imported into Boston, brought to Philadelphia to be made into Morocco, and many of them again returned to Boston to be converted into shoes. Boston and New York are both largely supplied with Morocco from Philadelphia, and also the principal cities in the West, from Pittsburg to St. Louis. The climate and peculiarities of water in Philadelphia seem admirably adapted for this manufacture; and with the aid of the highest skill, attracted hither as the chief seat of the manufacture, contribute to produce results that are apparently not attainable anywhere else in this country.

There are now twenty-five Morocco manufactories in Philadelphia, located principally on Margaretta, Willow, and St. John Streets, employing six hundred and thirty males and seventy-five females, who produced last year 125,000 dozen of Morocco and

ward in fermented bran liquor. Yolks of eggs, flour, alum, and salt, are used. In France and England 6,000,000 eggs are used annually in preparing Leather for Gloves.

Kid skins, averaging $9 25 per dozen, which would amount to $1,156,250. Several of the firms have very complete establishments for making Fancy Leather for Shoemakers, Hatters, Book binders, Coachmakers, Saddlers, &c.

The entire product of Leather in 1857 we state as follows :—

Sole Leather, Calf Skins, Upper Leather, Skirting, &c., Tanned and Curried, - - - - - - - - - - -		$1,175,000
Morocco and other manufactures of Goat Skins, 125,000 doz. a $9.25,		1,156,250
Sheep Skins, - - - - - - 50,000 doz. a $6,		300,000
Deer Skins, White Leather, Parchment, Vellum, &c., - - -		135,000
Total, - - - - - - - - - -		$2,766,250

The principal manufactures of Leather are considered under their appropriate captions, viz.:—BOOTS and SHOES, SADDLES, HARNESS and TRUNKS, &c. The miscellaneous manufactures consist principally of Gloves and Clothing, Belting, Hose, &c. The Gloves and Buckskin goods made in Philadelphia have a deserved and wide-spread reputation. Their qualities and merits are familiar to all dealers. The interests of this branch were for a time depressed and injured by the large quantities of inferior articles thrown upon the market, through auction houses, by the Gloversville manufacturers; but, by adhering to the principle of making only articles of the first quality, which in this class of goods are alone of any value to the consumer, the Philadelphia manufacturers have maintained their reputation and increased their business. The annual production of Buckskin Gloves, Mittens, Drawers, Suspenders, &c., in Philadelphia, including Kid Gloves, which are made of excellent quality, will exceed $150,000. The product of Belting, Leather Hose, &c., may be reckoned at $175,000, and consequently, the entire product of Leather, Buckskin Gloves, &c., will be stated approximately at $3,091,250.

XXV.

Marble, Stone, Slate, Soapstone, &c.

Marble, as a building material, is used more extensively in Philadelphia than in any other American city; and the preparation of it for this purpose alone would constitute a prominent, and perhaps a flourishing pursuit. One cause that has contributed more than any other to bring this material into such extensive use in Philadelphia, aside from its beauty, is its cheapness and the facility of obtaining it, there being several very old and excellent quarries within fifteen miles of the centre of the city. The *Hitner Quarry*, which is nearest to the city, has produced some very fine White Marble; but on account of the extreme depth of the quarry, reaching in some places to two hundred and forty feet, the working is very expensive. Contiguous to this is the *Lentz Quarry*, which produces a marble not very desirable on account of the dark-blue spots. The *Fritz Quarry*, recently purchased by the "Pennsylvania Land and Marble Company," is an old quarry, and has produced a very fine white and blue variegated Marble, known as the "Pennsylvania Clouded," formerly much used for mantels and chimney-pieces in old Pennsylvania houses of the better kind. The *Dager Quarry*, in the same vicinity, produces a similar Marble, but not quite equal to it in beauty. On the west side of the Schuylkill are the Henderson Quarries, extensive, deep, and expensive to work, but which produce very good qualities of White and Blue Marble, found in alternate layers in the same quarry, the bed of blue being first, and the white underneath it. Numerous important buildings in the city exhibit specimens of Marble from these quarries; as, for instance, the Mint, Exchange, and steps and ashlar of the Girard College; and, among the new buildings, Henry Korn's store in Third above Market.* A Black Marble, formerly much used for hearths, wall-plates, and shelves of mantels, when *Black and Gold* was the fashion, is also found in these quarries in boulders or detached masses, but not equal in quality to the Irish Black. One mile distant from these is the Brooks' Quarry, from which the Blue Marble composing the front of Levick, Raisin & Co.'s store, on Market street, was obtained.

* When the building which formerly occupied the site of the present University of Pennsylvania was torn down, a corner stone of Henderson's Blue Marble, two feet eight and a half inches by one foot eleven inches, was discovered, with the following inscription cut on the face:

" *This Corner Stone* of the house to accommodate the PRESIDENT OF THE UNITED STATES, was laid May 10th, 1782, when *Pennsylvania was happily out of Debt:*

THOMAS MIFFLIN then Governor of the State."

The products of the Pennsylvania quarries, however, constitute but a small proportion of the Marble consumed by the Marble-workers in Philadelphia. Large importations of different varieties of Marble, but principally veined Italian, are annually made from Leghorn, and sold on arrival at public auction, at prices varying from $2 to $4 per cubic foot. One establishment, that of Mr. JOHN BAIRD, consumes annually over 15,000 cubic feet of Italian Marble. The quarries of J. K. and M. FREEDLEY, at West Stockbridge, Mass., supply a good quality of ordinary building Marble, which is extensively used in Philadelphia; and the Vermont quarries, particularly those at Rutland, from which the finest varieties of American Marble are obtained; and those at Manchester, owned by FREEDLEY, MACDONALD & Co., supply this market with large quantities sawed to sizes for gravestone and other purposes, and from hence reshipped to the South and West. It is not an unusual circumstance for quarry operators in New England to consign a cargo of Marble to this city on a venture; and as ventures do not always arrive exactly at the time of demand, the jobbers and dealers in Philadelphia can frequently purchase on terms so favorable, that they can in turn supply customers in the South and West with Marble in slabs cheaper than either could purchase it in block at the quarries. The wholesale dealers in this city, however, are generally owners of quarries—Mr. S. F. PRINCE is the only jobber who has no interest in any quarry.

The trade in Marble, as an important pursuit, is of comparatively recent origin; but probably in no other has the adoption of improved facilities been more rapid and general. Less than twenty-five years ago, all Marble was sawed by the friction of a saw without teeth, aided by sharp sand, pushed backward and forward by manual force. Now, Marble is sawed, rubbed, and polished by steam power; and a block of Italian Marble has been converted into four hundred superficial feet of slabs in twelve hours. Holes of any required size are now drilled by machinery, and perfect joints are made by the aid of lap-wheels. The rapidity with which a rough block of Marble can be converted into highly-finished products, is only less astonishing than the time-shortening tanning process referred to in the preceding article.

There are now six steam mills in Philadelphia for sawing and preparing Marble; and some of them are the most extensive, complete, and best-arranged mills of the kind in the entire Union. The proprietors of these mills are EDWIN GREBLE, JOHN BAIRD, LEWIS THOMPSON & Co., S. F. JACOBY & Co., J. & E. B. SCHELL, and ELI HESS.

Greble's Works are located on Chestnut above Seventeenth street, and consist of a four-story brick mill 88 by 40 feet, with a two-story addition in the form of an L, 26 by 115 feet, of which the lower part is occupied for offices and as the stonecutters' department; and the entire second story, a large and handsome room, is appropriated as a mantel wareroom, in which may at all times be seen about one hundred different patterns of mantels. In connection with the Marble-Works on Chestnut street, he has in another location a very extensive yard devoted to the preparation of Brown Stone building-work. Some of the most elegant fronts Philadelphia can boast were executed in his establishment. In the two concerns, about one hundred and twenty men are usually employed.

Mr. Edwin Greble is one of the oldest established and consequently best known Marble-workers in Philadelphia. He commenced the business in 1829, and for more than a quarter of a century has been uninterruptedly engaged in the Marble trade, prosecuting a constantly-increasing business. He was among the first in Philadelphia to use Italian Marble in monumental work; and, since the discovery of very excellent varieties of American Marble, and its durability has been well established, he has been among the most zealous in recommending that to favorable consideration for like purposes.

Mr. Greble's Works were formerly located on Willow street, above Broad, where he owned the second mill for sawing Marble by steam-power built in Philadelphia.

BAIRD'S MARBLE WORKS are so well known throughout the Union, that a detailed description would be superfluous. The enterprising proprietor has provided first-rate equipments for working Marble, and then has shown a master's hand in advising the public of the important fact. It was the fortune and pleasure of the Author to refer to these Works some two years ago, and a brief extract from his remarks on that occasion must suffice for a present description :

" In the workshops a large force of skillful workmen may be seen engaged in executing various designs, and converting rough blocks or pieces of marble into various beautiful forms by the mallet and chisel, which machinery, with all its triumphs, has not as yet superseded. In designing and carving Mantels, Monuments, and elaborate works of Art, Mr. Baird has been unusually successful, through the agency of accomplished workmen; and some of the most exquisite specimens of the Phidian art, of which this country can boast, are the products of his workshops. It has been his practice not only to secure the best native and foreign artists in carving and designing, and to stimulate their ambition by rewards and liberal remuneration, but to encourage the study and practice of both these arts by establishing schools for the benefit of his apprentices. The fruits of his enterprise in this respect may be seen in his Mantel warerooms, and in the Monuments and Tombs which adorn our Cemeteries. His warerooms contain upward of 130 different patterns of Marble Mantels, made from all varieties of marble, common and rare, from the clouded Pennsylvania to the Carrara statuary, and ranging in price from $10 each to $1,000 and upward per pair. The designs in most instances are original, and the carving on the most costly renders them worthy of a place among the chef-d'œuvres of the art. The flowers and fruits appear to want nothing but color to start into life, and the heads and scrolls are worthy specimens of the sculptor's skill. In Monumental Art, the triumphs of the proprietor of those Works are written on the Cemeteries of Philadelphia, the Mausoleums of the South, and the resting-places of the dead throughout the Union. Whether it be a people's testimonial of gratitude to heroes who sacrificed their lives in battling with the pestilence, or a contribution from patriotic mechanics to the pile erecting in honor of Washington, or the more numerous and diversified mementos of affection to departed relatives or friends, the same wealth of resources, the same masterly execution, are visible in all. In the workshop devoted exclusively to this branch may at all times be seen a greater variety of finished Monumental and Tomb work of Italian and American Marble than in any other establishment that we know of in this country."

LEWIS THOMPSON & Co.'s Works are employed principally, and we believe exclusively, for sawing Marble into furniture tops. These works in their entirety are among the most remarkable in Philadelphia. We shall therefore reserve a description for the APPENDIX.

The Marble-Works of S. F. JACOBY & Co., J. & E. B. SCHELL, and ELI HESS, are all provided with good facilities, and turn out large quantities of sawed Marble, which are sold to the South and West in slab or converted into finished products. The trade with the South and West is rapidly extending, as the facilities of Philadelphia for supplying Marble become more widely known. The demand also for new forms of Marble-work, as Tiles, Mosaics, etc., is likewise increasing.

But the marked characteristic of this trade in Philadelphia, entitling it to a high rank and position as a pursuit, is not in the Mills, well-equipped as they are, but in the artistic ability and taste which have been displayed in designing and executing ornamental and monumental Marble-work. Long before the Marble-workers in New York and other cities were seemingly aware that uniformity in design was not a merit, those of this city employed special designers; and the genius of at least one, who for twelve years was solely occupied in making monumental designs for one firm, has afforded copyists abundant and profitable occupation. The sums expended by Struthers, or Baird, or Greble for original designs, probably exceed the expenditure, for a similar purpose, of the whole trade in other cities. Of the *execution*, too, in Monumental work, it would be impossible to speak in exaggerated terms. The sculptors of Philadelphia might, with confidence in a verdict favorable to themselves, submit the question to any jury, even of intelligent competent foreigners, who would take the time necessary to form a correct judgment, whether the workmanship ordinarily displayed by them in carved Marble-work (except in statues, in which they have had but little experience) is not superior to that ordinarily executed in Italy, the home of Sculpture.* Certain it is, that the tombs imported from abroad, though perhaps the most costly, are not the most noteworthy and finely-chiseled Art-objects in our Cemeteries. Nearly all the Sarcophagi in which repose the ashes of the greatest of American heroes, jurists, and others, were executed by Philadelphia sculp-

---

* Even in Statues no very great inferiority can be admitted. Mr. THOMAS HARGRAVE has recently executed two Statues—one of Moses Delivering the Law, and another of Christ the Mediator, that, we are told, will bear favorable comparison with the best works of the kind executed in Italy.

tors; for instance, the Sarcophagus of Washington, of Chief-Justice Marshall, of Chief-Justice Tilghman and Bushrod Washington, of the Rev. Mr. Whitefield, and others. At this moment the chisels are busy in shaping Sarcophagi that will, if possible, invest with increased interest the burying-places of HENRY CLAY and JOHN M. CLAYTON.* But the theme is susceptible of unlimited elaboration.

There are now about sixty marble yards in Philadelphia, employ-

---

* Both of these works, we are happy to learn, are being executed of *American* Marble, and by Mr. WILLIAM STRUTHERS, who is well-known as one of the most successful and eminent of the workers in Marble in Philadelphia. He is successor and representative of the house of John Struthers, and J. STRUTHERS & SON, established more than a half century ago. The Marble work of nearly all the elegant and costly public buildings for which Philadelphia is distinguished was executed by this firm—the U. S. Bank, now Custom House, U. S. Naval Asylum, U. S. Mint, Chestnut-street Theatre, Philadelphia and Western Banks, Philadelphia Exchange, Mechanics' Bank, Philadelphia Saving Fund; Girard Buildings, Chestnut above Eleventh; Farmers' and Mechanics' Bank, Bailey & Co.'s new Marble store, and many others. Their skill in this branch, however, has not been monopolized by Philadelphia, but may be seen in many of the Marble buildings of the United States: the State Capitols of North Carolina and Ohio; the Commercial Bank, Natchez, Miss.; the United States (Branch) Bank at Pittsburg, Penn. Mr. Struthers is also largely engaged in executing building work in Sandstone; and we believe was the first to introduce the stone of this description, now so popular, from the Albert and Pictou Quarries, British Provinces. Many of the elegant stores and mansions which enhance the architectural beauty of the city, as for instance, Morris L. Hallowell's, J. Stone & Sons, L. J. Levy & Co.'s stores; Pennsylvania Rail-road Company's building on Third street and Willing's Alley; John Grigg's mansion, in Walnut street; Wm. Welsh's mansion, in Spruce street; J. Hare Powell's mansion, in Walnut, &c. But the branch of his general and extensive business which entitles Mr. Struthers to special distinction, because excellence in it is more rare, is Marble Monumental work. To enumerate all the important Monuments which have been executed in the yard of J. STRUTHERS & SON, would require far more space than our limits can afford. Art-objects, of the highest character in point of taste and workmanship, have been sent by this firm not only to all parts of the United States but to England, the West Indies, China, and Syria.

As an exception to the characteristic American rule of frequent change, we may state that the spot, 360 Market st., now occupied by Mr. Wm. Struthers, as a Marble Yard, has been used as such uninterruptedly since 1798.

ing, on an average, eight hundred and forty hands, and executing work to the amount of $860,000 annually. Nearly one-half of the amount is done by four firms.

Two or three of the marble-workers are also extensively engaged in executing Building work in Sand-stone; and the statistics of the entire business, including workers in Brown-stone, Granite, Flag-stones, &c., would amount, on an average, to 1,150 hands, and a product of $1,160,000.*

3. SLATE, SOAP-STONE, ETC.

In our introductory remarks we adverted to the Slate Quarries, in Lehigh County, and stated that the very best qualities of Slate are obtained in Pennsylvania. In other places, the preparation of Slate for Roofing purposes constitutes the principal item of its manufactures, but in Philadelphia, School Slates are also made extensively. A new method of framing School Slates

---

* Nearly all varieties of Stone used in this country, for building material, have their representatives in buildings erected in Philadelphia. The following are of

Stone, from the Albert Quarries, British Provinces.—Cowperthwaite's Building, Chestnut st., above 6th; Howell & Brothers, S. W. corner Chestnut and 6th; Pennsylvania Rail-road Company's, 3d and Willing's Alley; J. Stone & Sons', Chestnut, above 8th; Western Saving Fund, Walnut and 10th; Morris L. Hallowell's, Market, below 4th, Gans, Leberman & Co., 3d above Market; Geo. W. Ball's mansion, Chestnut street; R. W. D. Truitt's, Chestnut st.; W. J. Duane's, Locust st.; Thos. Beaver's, Logan Square.

Stone, from the Pictou Quarry.—Joseph Harrison, Jr.'s, mansion on 18th, and houses on Locust st.; Simes' Building, 12th and Chestnut; William Welsh's mansion, Spruce, below 12th; Thomas Thompson's, Spring Garden, below 12th; Womrath's stores, Arch st., above 4th.

Connecticut Stone, Middlesex Quarry, Portland, Conn.—Schuylkill Navigation Company's building, Walnut, above 4th; L. J. Levy & Co.'s, new store, Chestnut, above 8th; two large Warehouses, north side Chestnut, below 3d; Farnham, Kirkham & Co.s', Chestnut, below 3d; John Grigg's mansion, Walnut street, opposite Rittenhouse Square; J. Hare Powell's, Walnut, opposite Rittenhouse Square; Peabody's and others, on Walnut, opposite Rittenhouse Square; Farquhar Building, Walnut, below 3d; and many others.

Miscellaneous Stone.—Ingersoll's house, Walnut street, Paterson Quarries, N. J.; Athenæum, from Little Falls, N. Y.; Dr. Jackson's store, Arch st., Caen stone; Dr. Jayne's stores, &c., Quincy Granite.

was invented a few years ago, by Mr. EDWIN YOUNG, and is rapidly growing into favor. The Slate is nearly oval, and is framed with a single strip of hard wood, fastened together by a secret metal clasp, not seen from the outside, constituting the strongest fastening known. About 3,000 cases of these Slates, averaging ten dozen each, are made annually, of the value of $27,000, at the factory in Philadelphia; but when introduced into the Public Schools, as they probably will be, the product will be much increased. Slate has also been converted into *Billiard Tables*, for which it is said to be the best material.

Enameled Slate Mantels are now made by Arnold & Wilson, manufacturers of Hot-air Furnaces, Parlor Grates, &c., who have recently erected ovens for baking on the enamel. The advantages of these are a high degree of beauty, combined with strength and cheapness.

Of *Soap-stone* the manufactures are quite limited, though the material obtained at a quarry now within the corporate limits of Philadelphia, about two miles above Manayunk, is of the best quality. This quarry is one of the oldest in the country; was opened before the Revolution; but until it came into the possession of its present enterprising owner—SAMUEL F. PRINCE—its value was scarcely appreciated. Its products now amount to about 6,000 tons annually, and are disposed of principally to Iron manufacturers along the Schuylkill, in Trenton, &c.; and five hundred tons were shipped last year to Pittsburg, and small quantities on order to England.

This stone, hitherto little used for economic purposes, is adapted for many; particularly for fire-stone, kitchen sinks, wash-tubs, bath-tubs, and especially for baths, and sizing rollers used in cotton mills. For the last purpose it possesses the advantage of not being affected by the acids ordinarily used in sizing, and of not warping, contracting, or expanding by changes of temperature and moisture.

Both Slate and Steatite, or Soap-stone, have not attained their maximum of appreciation, and offer excellent opportunities for the enterprising to establish new manufactures.

XXVI.

Oils.

The following revised Report gives a brief but comprehensive description of the Oil manufacture in Philadelphia:

Sir:—The manufacture of Oils in Philadelphia comprises Linseed, Lard and Tallow, Red, and Rosin Oils—Linseed constituting much the largest business.

In the manufacture of *Linseed Oil* there are five mills employed, generally possessing very improved machinery, and two of them make each 1,200 gallons of Oil and ten tons of Oil Cake per day. The material used for making this Oil is flaxseed, or linseed, imported from Calcutta, that being the only point from which the best article can be obtained. The process of the manufacture of this Oil is as follows: The hard seeds are passed first between cast-iron rollers, in order to crack the shells. The rollers are sometimes of different sizes, so that different velocities may be given to their surfaces; this enables them to draw the seeds in, and to perform their work more quickly. Above the rollers is a hopper containing the seed, from which the rollers are fed. In some places the rollers are not used, but the seed is at once subjected to two vertical mill-stones, or runners, revolving on a horizontal bed. When the seed is sufficiently bruised by either or both of these means, it is placed upon heated tables, and then into wool bags, and afterward either in what is called a hydraulic press, or *wedge press*, and pressed until the seeds come out of the bags in the form of flat cakes. The Oil thus obtained is of the best quality, and is kept distinct from that obtained by the after processes.

The residue, which is known as Oil Cake, is largely exported to Europe, where it is highly valued as food for cattle, this compensating partially for the importation of the raw material. Its value, however, in this country as food for stock, has not been fully appreciated, but is growing more into favor of late years.

In addition to the regular establishments making Linseed Oil, the manufacturers of Zinc Paints, Colors, White Lead, etc., frequently make sufficient Oil to supply their own necessities.

There are eight concerns in the manufacture of *Lard, Tallow, and Red Oils*, of which the annual product amounts to nearly as much as that of Linseed Oil. The Lard or Tallow is placed into bags, and then under a press heated to a temperature from forty to sixty degrees, remaining as long as the Oil will drip from it; thus the Oil is pressed out, the extract being known as Lard and Tallow Oils, and the balance as *Stearin*. The Stearin is then submitted to a second process,

30

being placed under a powerful steam press, which is also heated by steam, and the extract is termed Red Oil, largely used in the manufacture of soaps.

Tallow and lard are not the only materials used in the production of Red Oils, but Palm Oil is largely employed, which is obtained from Guinea, Africa, in a state thick like lard somewhat, undergoing precisely the same process as the former.

An Oil is also made from rosin, or gum of the pine-tree, and known as Rosin Oil. This gum contains several ingredients which, when submitted to the process of distillation, are separated and are each useful, but are of entirely different natures, being Oil, Acid, Naphtha, Pitch, and Tar. The Oil is, by various manipulations known to the manufacturers, converted into Lubricating, Tanners', and Painters' Oil. The Naphtha, after being properly refined, will burn with a clear, brilliant light, nearly equal to gas, but much cheaper, and well adapted to be used where gas has not yet been introduced. There are four establishments within the city proper, who have facilities for manufacturing from the rosin of commerce, about two thousand gallons per day. Another establishment erected at " Chester," about eighteen miles south of this city, is of much larger capacity than either of the preceding, and from which some twelve hundred gallons could be produced daily. These Oils range in price from twenty to fifty cents per gallon. The capital invested is over $150,000.

The bleaching and pressing of Sperm and Whale Oils is another considerable item connected with the Oil business of Philadelphia. One of the firms, of which there are four, is successor to the parties who claim to have been the first to introduce the process of the chemical bleaching of Oils in this country.

The refining of these Oils from the crude state greatly improves their burning and lubricating qualities, giving to them a white, clear appearance, by taking out that residue known as foots' and spermaceti.

Oils pressed and bleached in the coldest months of winter, or by means of a freezing temperature by ice, are the only Oils used in cold weather that remain always entirely fluid.

The manufacture of Rail-road and Cart Greases, made of Rosin and other Oils, is carried on to a considerable extent. These Greases, which are now extensively used for the oiling of machinery, vehicles, etc., have grown much into favor of late. They are considered much more economical than Oils, and, as a lubricator, vastly superior.

There are but two establishments in this business, R. S. HUBBARD & SON and TAWS & BEERS, the former of which is the largest in the

United States. They both produce daily about 3,200 lbs., at an average price of eight cents per pound.   O. W. KIBBIE.

The following is a summary of the aggregate production :

	Establishments.	Capital.	Hands.	Annual Production.
Linseed Oil and Oil Cake,	5	$450,000	125	$687,500
Lard and Tallow,	8	200,000	60	276,730
Rosin Oil,	5	150,000	15	600,000
Sperm and Whale, bleaching and refining,	4	200,000	30	187,000
				300,000
R. R. & Cart Greases,	2	40,000	15	80,000
Total,	24	$1,040,000	245	$2,131,230

XXVII.

Paper Hangings.

Decorative Paper-Hangings came into use about 200 years ago, and are said to have been copied from the Chinese. The manufacture of them in this country, however, only dates about thirty years ago ; and Philadelphia claims the credit of having first established it. The progress made, however, in design and elaboration of workmanship, has been so rapid, that now the importation of foreign Papers is an unimportant item—said to be not more than five per cent. of the whole amount consumed, and confined to French goods of the first quality. For more than twenty years, Philadelphia has supplied all the principal American markets, including New York and Boston, with the best American made Papers ; and though since the establishment of the business in this city, a large number of factories have been started at various points in New England and in New York, she continues to produce a large share of the superior qualities—generally quite equal to the best French manufacture. The cheap and low grades, which constitute the bulk of the production in other places, our manufacturers do not make to any extent. The medium qualities and the finest Velvet, Velvet and Gold, and Satin-surfaced Papers are made, but none of the "one cent per square yard" goods. The printing usually is what is termed *Block Printing*, though some first-class machines are in use in the principal establishments.

The processes of manufacturing Paper Hangings are briefly as follows :—The paper comes from the mill in rolls about 1,200 yards long, and from twenty to thirty-five inches wide ; costing from nine to fourteen cents per pound, the average price being about eleven cents. The stock generally used here is said to be heavier, though costing less than that employed in the best French papers, and therefore free from the absorption of moisture, which almost invariably disfigures the surface of those made in New York and Boston. The pattern having been first carefully drawn, is then pricked, and the outlines of the various tints are pounced each on a separate wood block made of pear-tree, mounted in pine. These blocks are pressed on the sieves of color and then applied to the paper, each block following the other on the guide-marks left by the previous impression. An idea may be formed of the enterprise and labor required to produce some decorative Paper Hangings, when we state that on a single one of them, representing a chase in a forest, including the animals, birds, and attributes of the chase, exhibited at the World's Fair, 12,000 blocks were employed.

In making *Flock Paper*, the pattern is first printed in size, and then with a preparation of varnish or Japan gold size. When this is partly dry, colored flock, prepared from wools, is sifted on the varnish pattern, to which it adheres. When gilding is introduced, the leaf-metal is laid on the varnish pattern ; or, if worked in bronze-powder, it is brushed over with a hare's foot.

The designs are principally original, and are largely supplied by the Female School of Design established in this city, and which has already made important contributions toward elevating the standard of correct taste.

During the last year there were six Paper Hanging factories in Philadelphia, besides various small establishments where a few hands are employed. The proprietors of the principal factories were HOWELL & BROTHERS, BLANCHARD & ROCK, LOUIS BELROSE, BURTON & LANING, HART, MONTGOMERY & Co., and J. E. VAN METER. Misfortune has sorely visited one or two of the establishments, but it is probable that no important changes will be made. The factory of Howell & Brothers, situated at Nineteenth and Spruce streets, is a four-story brick structure, three hundred and ninety-six feet by eighty—undoubtedly the largest in the United States, and probably larger than any similar European manufactory.

The following are the statistics of the business, as made up for us by a leading manufacturer :—

Blank Paper consumed, 1,250 tons at 11 cents per lb., - -	$275,000
French and American White Clay and Whiting, average 364 tons, at $10 per ton, - - - - - - - - -	3,640
Colors, - - - - - - - - - - -	44,040
Flocks, (Shearings of Broad Cloth,) - - - - - -	1,860
Oil, - - - - - - - - - -	4,600
Gold Size, - - - - - - - - -	3,150
Gold Leaf, - - - - - - - - -	20,100
Glue and Sizing, - - - - - - - -	31,500
Coal, 1,875 tons, at $4, - - - - - - -	7,500
Hands employed, 456; one third males—wages, - - - -	123,240
Cost of Printing Blocks and Designing, - - - -	11,000
Total, - - - - - - - - - -	$525,630

Annual product, $800,000.

XXVIII.

Rope, Cordage, Twines, &c.

The term Cordage usually comprehends all the various sizes of Rope, Cords, Twines, Lines, &c. In this city, there were, as early as 1810, no less than fifteen Rope-walks ; and at the present time there are about that number of Cordage manufacturers ; but now one single establishment turns out annually a greater product than all then made. The materials used are Manilla, Russian, Italian, and American Hemp ; and, for Fishing Cords and Twines, Cotton, Flax, and the best qualities of Linen Thread. Manilla Hemp is the fibrous inner bark of a species of Plantain, growing in the Phillipine Islands, whence it is imported into this country. The American Hemp used is grown chiefly in Missouri and Kentucky. A considerable amount of Russian Hemp is also used ; and *Jute* is now employed to a considerable extent, in the manufacture of Cords, Bagging, &c.

The present condition of the business may perhaps be best illustrated by reference to one or two of the leading establishments. The largest manufacturers in the city are Weaver, Fitler & Co., who are also among the very largest in the United States. They are the successors of one of the oldest Rope manufacturers in the city ; Mr. Weaver's father having founded the establishment in

30*

1816. The firm have now two factories in operation; one of these, for making Manilla and tarred Cordage, is located on the Germantown Road, near the first toll-gate; and the other, for the manufacture of Fine Yarn, Jute Rope, Cords, &c., on Seventh street, above Columbia Avenue. These two factories are capable of turning 4,500,000 pounds of different kinds of Rope per annum, or about seven tons daily. The product embraces every size and description of Cordage, from a bed-cord to the largest size gang of rigging. Marline, Hambroline, and Spun Yarn are also extensively made by them.

This firm were the first to introduce, in this city, the use of machinery for spinning yarns for Manilla Rope and Cordage, which has nearly superseded the former slow process of spinning by hand. The quantity turned out by their present machinery, employing about two hundred hands, would, by the former process of hand-spinning, require at least eight hundred, and perhaps one thousand men. Their machinery is all of the latest improved construction, and is said to be more complete than that in any similar establishment in the United States. The firm pride themselves upon the manufacture of a superior article of Cordage; and their reputation, in this respect, in the South and West, is well established.

Messrs. SPROAT, M'INTYRE & CO. are also large manufacturers of many descriptions of Cordage; and were the first to introduce here the manufacture of fine yarn Jute Rope, which they make from $\frac{1}{4}$ inch to 1 inch in size. This establishment commenced business in 1850, as J. & H. Sproat, under which name it was conducted until 1857, when Mr. John M'Intyre, who occupied their present place of business, 23 North Front street, and had a Rope-walk on Frankford Road, was admitted into the firm. The manufacture is carried on still at the latter place, and also at their factory at Lambertville, on the Delaware, where water-power is derived from the feeder of the Raritan Canal. These two factories use fifty bales of Jute Hemp in a week, weighing each three hundred pounds, and each making that number of pounds of $\frac{1}{4}$ to 1 inch Rope. They employ about seventy hands, one-third of whom are females. This firm claim to make more Twines and Lines of every description, such as Wool Twine, Hemp and Mineral Water Twines, Baling Twine, Broom Twine, &c., as well as Linen Yarns, Venitian Filling, and Carpet Chain, than any other house in the city. The Mineral Water Twines made by them cannot easily be surpassed; and in the employment of fine Jute Hemp for this purpose, they claim to have preceded all other manufacturers. They have lately introduced Boon's Patent Laying Machines, and also Boon's Forming Machines ; and they now believe they possess a larger set of machinery for this branch than any other establishment in the State.

Besides these two principal manufacturers, there are at least a dozen others, who make nearly every description of smaller Cordage, and who generally aim to produce articles of superior quality, so common a feature in Philadelphia manufactures. We have seen, in some of these smaller establishments, Cords, Twines, Lines, &c., for various purposes, truly remarkable for accuracy and smoothness of finish.

The whole capital invested in the business is $450,000. The other statistics of the business for 1857, are furnished us as follows :

Raw Material.

1,500 tons Manilla Hemp, at $175, - - - - -		$262,500
1,250 " Western Hemp, at $150, - - - - -		187,500
		450,000
300 hands employed, averaging $300 a year each, - -		90,000
		$540,000

Product.

1,500 tons of Manilla Cordage,		
1,000 " of Hemp,		
2,500 " average price 10 cents per lb., - - -		$560,000
Besides the above, Twines and Cords of various kinds were made, amounting to at least - - - - - -		250,000
Total, - - - - - - - - -		$810,000

The prices of Cordage, the present year, are much less than those during last year.

XXIX.

Saddles, Harness, Whips, Trunks, &c.

The manufacture of Saddlery in this country is distinguished from that in any other part of the world by the immense variety of styles and qualities which are produced. We are informed by a leading manufacturer, that of *Saddles* there are probably not less than five hundred various styles and qualities, with a proportionate quantity of Bridles, Bridle Mountings, Martingales, Girths, Circingles, Stirrup Leathers, Saddle Bags, Medical Bags, &c. Of Harness, for Coach, Gig, Dearborn, Sulky, Stage, and Omnibus, there are perhaps three hundred styles and qualities; while, in coarse Harness, for Carts, Drays, Wagons, and Plows, there is also great diversity.

It is a fact well-known to persons who are familiar with the history of Industry during the past few years, that the Saddle and Harness-makers of Philadelphia have invariably carried off the "palm" at local Exhibitions and Fairs; and the fact that the Prize Medal was awarded to a Philadelphia firm at the World's Fair, in London, cannot be unknown to any observant person, who has traversed Seventh street, North of Chestnut. The special causes conducive to superiority in the Harness manufacture are manifold; all the raw material consumed, especially the Leather and the Hardware, are made here of the very best quality; the workmen have permanent employment, and the manufacturers have an established reputation for faithful work, which they are determined to maintain. The solvency and character of the trade in Philadelphia, enable them to buy at the very lowest rates; and the system of manufacturing involves much less ostentation, and, consequently, less expense than in many other cities where the sales-house and factory are distinct and separate establishments, even if owned by the same parties. In this city, the goods are generally manufactured and offered for sale under the same roof. The ingenuity of the manufacturers too has been repeatedly and successfully called into exercise, and the very best of the new styles of Saddles made in the North were first originated and introduced by one of our large houses; while improvements upon the old English styles render those made in Philadelphia in several respects superior to the foreign. In the new styles, of the Spanish and Mexican order generally, the utmost care is taken to guard against injury to the horse, and also to produce (which they have, beyond all other places,) the most comfortable and pleasant Saddle, for both horse and rider. Hog-skin continues to be the principal Leather consumed in the best Saddles, on account of its softness and capacity for exposure to the sun and rain; though Buckskin is also frequently used for the seat, and for the horns of Ladies' Saddles particularly.

For the manufacture of SADDLE-TREES, there are two establishments—the proprietors being CONDIT PRUDDEN, and AARON SCHELLENGER. The last-named has been in the business for twenty-eight years. His Trees are all cut out of the solid material by the axe and the "shave," no bent work being made in this shop. He employs eight hands.

Mr. CONDIT PRUDDEN has had an experience of over thirty years in the manufacture of Saddle-Trees, and about one-third of that time in Philadelphia. He conducts the business on a more extensive scale, it is said, than any other Saddle-Tree maker in the Union. His manufactory is a four-story building; the machinery is propelled by steam-power; and the capacity of his works is sufficient to turn out 1,200 finished "Trees" per week. About thirty hands are at present employed. A list of "Trees" made in this establishment would include all the ordinary descriptions, and some patented ones. Saddle-Trees are shipped from Philadelphia to New York and New England, and largely to the West and South.

The Saddle and Harness manufacture employs a capital of three-quarters of a million of dollars, nine hundred and sixty hands, and yields a product of $1,500,000. Two of the oldest houses, W. S. HANSELL & SONS, established for forty-five years, and M. MAGEE & Co., have branch establishments in New Orleans. Orders for fine work are occasionally executed for Europe; while the coarser qualities are shipped to various parts of the South and West, and also sent in considerable quantities to Mexico and Cuba. The manufacture consumes annually over *one hundred thousand sides* of Leather.

2. WHIPS.

The manufacture of Whips is a business entirely distinct from that of Saddles and Harness; but the relations existing between them are so intimate, that they may properly be considered in the same article.

The Whip manufactories of Philadelphia are said to be the first established, and among the most extensive in the Union. The principal factories are those of PEARSON & SALLADA, and CHARLES P. CALDWELL; but, outside of these, large quantities of Raw-hide and common Whips are made by individuals, for saddlers' use. The factory of Messrs. Pearson & Sallada is said to be the largest in this country ; but we have no particular information with regard to it. The factory of Mr. CHARLES P. CALDWELL is located at Mantua, in West Philadelphia, and there all kinds of Whips are made, from those which sell at $1 25 per dozen, to those which sell at $600 per dozen. The materials used for the stock are Whalebone, Rattan, Fancy Woods, Leather, Gut, Gum, Pitch, Glue ; for the lashes, Leather, Gut, and Thread ; for the handles, Wood, Ivory, and Bone ; and for the mountings, Gold, Silver, Ivory, Pearl, &c. Machines are used for plaiting or weaving the gut covering. This machine is a circular frame, around which is a series of bevel cogs, driven by a crank-handle in the hand of the operator ; the whip stands in the centre, and receives its gut from numerous spools which surround it ; the machine at the same time plaiting the gut over the stock. They are of different capacities : one plaiting sixteen threads, another twenty-four.

The great difference in the cost of Whips, some selling as high as $50 each, is mainly in the character of the mountings. Mr. Caldwell uses, in all his Whips, the very best material ; and the reputation of his manufactures is unsurpassed by any in the United States. He employs about thirty-five hands. Fully one third of his products are sent to New York and Boston.

Canes are also made in both these factories, and the materials and mountings are often exceedingly rare and costly.

The Whip and Cane manufacture suffers severely from the enormous expansion of ladies' skirts, and the consequent demand for Whalebone hoops. The price of Whalebone has recently quadrupled, and that which formerly could be purchased for thirty-five cents per pound now costs $1 20 per pound in its rough state, and $1 75 ready cut. It appears that neither Whips nor Canes could prevent the advance ! The annual product of the Whip and Cane manufacture in Philadelphia, estimating for that made outside of the factories above-mentioned, is at least $175,000.

3. TRUNKS AND PORTMANTEAUS.

In this branch of Leather manufactures, a capital of $100,000 is invested, two hundred and fifty hands are employed, and an aggregate value of $313,000 annually produced. The same care in the selection of materials, and attention to finish, that we remarked in the Saddlery and Harness manufacture, are noticeable in this; and not merely in the finer qualities, but neatness and taste, as well as strength, characterize the cheaper varieties.

Philadelphia Portmanteaus are deservedly famous for their combination of strength and capaciousness, with lightness. At the World's Fair, in London, it will be remembered by many, that a Portmanteau, made in this city, costing $500, received the First Premium, notwithstanding competition from all countries. New York procures the ordinary qualities of Trunks and Portmanteaus from Newark, and other places, but the very finest are obtained exclusively, we are informed, from this city.

The London Commissioners remark: "The workmanship and finish of the best class of goods are unexceptionable; and even in the cheaper and lower qualities the style in appearance is a matter of much consideration, and displays a decided advance, in point of taste, upon the unsightly character of the cheaper kind of traveling conveyances of England."

XXXI.

Ship and Boat Building.

It is no idle nor foundationless boast to say that the Ship-Builders of Philadelphia have contributed materially to the present commercial prosperity and supremacy of the United States. The history of commerce will establish the fact incontestibly, that the rapid rise of the Commercial Marine of the United States is due mainly to the superior swiftness of American vessels; and it must moreover be conceded that the Ships constructed in Philadelphia were, for a long period of time, the "crack sailers" of the ocean. The *pivot board*, so essential to the speed of Sailing Vessels, was originated and brought to perfection in this city; and cotton duck and horizontal canvas, which are esteemed the best materials for Sails, were invented by a citizen of Philadelphia. The speed of Philadelphia-built Ships is demonstrated by the records of short passages, and their staunchness is established by the low average rates of insurance.

During the last ten years the attention of the private Ship-

Builders of Philadelphia has been largely directed to the construction of Steamers. Within that time, the firm of Birely & Son has built one hundred and seven Steamers, having an aggregate tonnage of 21,018 tons. A list of the Vessels that have been constructed at this port would include a number of important ones, as the Steamship "Pennsylvania," "City of Richmond," the Steam Propeller "S. S. Lewis," 1039 tons burden; the Steamer "Star of the South," built for R. F. Loper, of Philadelphia, for Boston; the "General Knox," famous for her good qualities and speed; the "Carolina," which her builders, C. & N. Cramp, claim is the fastest Propeller afloat; the Clipper Ships "Manitou," 1,500 tons, the "Bridgewater," 1525 tons, and the Propeller "Phineas Sprague," 1,000 tons, built for the Philadelphia and Boston line by the young firm of Birely & Linn, who are now building a 1,200 ton Propeller for the same line. The various Propeller and Steamship Engines and Iron Boats that have been constructed here, were adverted to in the article on "Iron and its Manufactures."

But the supremacy which Philadelphia claims over all other cities in Marine Architecture, is in her Government Navy Yard work. The Navy Yard in this city is alike remarkable for its success in constructing vessels that are acknowledged to be the equals of any in the world, and for the neglect and positive opposition of the Government in providing equipments. The area is eighteen acres,* enclosed on three sides by a substantial brick wall, the other fronting upon the river. The space, it will be perceived, is ample for the construction of the largest Vessels, though many of their essentials, as engines, chains, anchors, etc., are made outside of the walls. The moulding lofts for modeling Ships-of-War are the most spacious in the country. There are two ship-houses, one of which, 270 feet long, 103 feet high, and 84 feet wide, is the largest in the United States. The other is

---

* At our request, the Commandant's courteous Secretary, HENRY S. CRABBE, had the whole area of the Navy Yard measured, and ascertained that it contained eighteen acres within the present walls; but he informs us, that an assurance is given, by parties interested, that the two large lots lying contiguous to, on the south side of the Yard, will be purchased by the United States this season. and added to it.

210 feet in length, 80 feet in height, and 74 feet in width. The *Sectional Floating Dry Dock* is unsurpassed by any other. There are nine sections, equal in dimensions, 100 feet long, 30 feet wide, and 11 feet deep. The water is pumped out in about two hours by four high-pressure engines, of 12-inch cylinders, and 24-inch stroke, and 8-feet beams. They are of simple construction, but answer their purpose admirably. The conveniences for docking Vessels are excellent. Two or even four Vessels can be docked at the same time, and placed on the Rail-ways in the yard at the head of the Dock by the Hydraulic Engine, which has already proved its power by hauling up the "City of Pittsburg," a first-class Steamer of the Commercial Marine, and placing her far above the reach of the water, convenient for the operations of the mechanics. The Basin is constructed chiefly of Wood, and does not retain dampness—a serious objection to those of stone. The entire cost of this noble Dock was $813,742.*

The number of workmen employed at the Navy Yard ranges from five hundred to thirteen hundred, averaging eight hundred.

At the present time there are about twelve hundred men at work, principally on the new Sloop "Lancaster." Their occupations are divided as follows: Shipwrights, Sawyers, Borers and Carpenters' Laborers, Smiths, Joiners, Gun-carriage Makers, Caulkers, Reamers, Spinners (oakum), Pickers (oakum), Sailmakers, Mast and Spar-makers, Riggers, Painters, Boat-builders, Plumbers, Block-makers, Engineers, Masons, House Carpenters, and ordinary Laborers.

The list of Vessels that have been built at this Yard includes not one that failed from bad construction or inferior material. The history of the Yard is a succession of successes. We have not the necessary data to enumerate them all, but herewith submit a

---

* At Christian-street Wharf, Messrs. J. Simpson & Neill have a Sectional Floating Dry Dock, constructed on the same principle as the above, and capable of docking the largest merchant vessel. The water, we are informed, can be pumped out in one hour, and an eighteen hundred ton merchantman docked in the same time.

31

List of Vessels of War of the U. S. Navy, built in Philadelphia, and now in service, as per Official Register, for 1858.

Name.	Guns.	Tonnage.	When Built.
SHIPS OF THE LINE.			
Pennsylvania..	120	3241	1837
North Carolina....................................	84	2635	1820
FRIGATES.			
United States	50	1607	1797
Raritan	50	1726	1843
SLOOPS OF WAR.			
Germantown	22	939	1846
Vandalia..	20	783	1828
Dale	16	566	1839
SCREW STEAMERS.			
Wabash......(1st class)...........................	40	3200	1855
Lancaster ...(2d class).....	18	2360	1858
SIDE-WHEEL STEAMERS.			
Mississippi......(1st class)	10	1692	1841
Susquehanna.........."	15	2450	1850
STORE VESSEL.			
Relief..		468	1836

Besides these we might mention the " Arctic," memorable for her connection with the Kane expedition; the " Shubric," used on the Coast Survey ; and the " Princeton," celebrated in her day as the swiftest Vessel afloat, and also for her connection with naval operations performed at Vera Cruz. She was rebuilt at the Gosport Navy Yard, and is now degraded to be a Receiving-Ship at this station.

It is a very significant fact, established by the history of the Naval Marine, that, while various attempts have been made in other American cities at constructing Naval *Steamers*, Philadelphia is as yet the only port where it can be said the work has been done successfully. This, however, is not surprising, inasmuch as all the elements conducive to success and economy in Marine construction are concentrated here in an unusual degree. Cheap coal and iron, the best mechanical ability, experience, and facilities, combine to render Philadelphia practically the most available point in the entire Union for the construction of Vessels of every grade and description.

Besides the Ship-Builders, there are six concerns, all located in the old district of Kensington, employed chiefly in Boat-Building, which includes Yawl Boats, Whale Boats, Life Boats, etc. In building Yachts they have been very successful. The Yacht " Decoy," designed and built here, we are informed, has challenged the world year after year without a response. Holmes' Patent Life Boats, very highly spoken of by competent authority, are built at this port.

The minor branches connected with Ship-Building are all carried on here. Sails are made unsurpassed in quality and unequaled in reputation. The sails of the Yacht "Maria," of the New York Squadron, and of the Yacht "America," famous for her triumphs over the English in a contest for the supremacy of the seas, were made by Mr. Maull of this city. The entire business of Ship and Boat-Building, including Masts and Spars, Sails, Blocks, and Pumps, judging from the best data we can obtain for an average of five years, has amounted to $1,760,000 annually.

During the last year, and even for two years past, it is well known that the shipping interests have been seriously depressed, and the construction of new Vessels at all ports was exceedingly limited. At Philadelphia, during the last year, one hundred and forty-seven new Vessels, having an aggregate tonnage of 17,917 tons, were admeasured by the United States officers.

XXXI.

Soap and Candles.

"The quantity of Soap consumed by a nation," says the celebrated Liebig, in his Familiar Letters on Chemistry, "would be no inaccurate measure whereby to estimate its wealth and civilization. Political economists, indeed, will not give it this rank; but, whether we regard it as joke or earnest, it is not the less true that, of two countries equal in population, we may declare with positive certainty that the wealthiest and most highly civilized is that which consumes the greatest weight of Soap." It is not, however, merely by the quantity consumed of this important article, that the distinguished chemist would establish its claims to represent the civilization of a people. The vast train of chemical, manufacturing, and commercial operations called into existence for its economical production, and the cheaper, more extended, and altogether new arts and processes incidentally growing out of these, would, even with political economists, entitle it to this rank.

The materials used in making Soap are alkalies, and fatty substances, or oils, both of animal and vegetable origin. Of the former, Potash, Soda, and a small proportion of Lime, are employed. The artificial production and cheap supply of Soda, the alkali chiefly used, from common salt, introduced about the beginning of the present century, has since that time completely revolutionized the business in Europe and this country, and probably within the last twenty years quadrupled its amount. Of fats and oils, Tallow, Lard and Fish Oils, Palm, Olive, Linseed, Cocoanut, Sassafras, and other Oils, and Rosin, are the principal. Their chief agency is to serve as a vehicle for the alkali, upon which the detergent properties of Soap mainly depend : while the combination of the latter with the fatty acids, generated in the process of saponification, subdues its caustic qualities, and preserves the skin, and the texture and colors of fabrics. Rosin enters into the composition of common Yellow Soap, and, in due proportion, improves, while an excess deteriorates the quality of Soap—an adulteration largely practiced, because of the cheapness of common Rosin. Lime, a portion of which is necessary, with the commercial Soda Ash, which contains only about fifty-four per cent. of pure alkali, injures, by giving undue causticity, if too freely employed. The "Concentrated Lye" made by the Pennsylvania Salt Manufacturing Company, which is represented in Philadelphia by LEWIS, JAMES & Co., is warranted to make Soap without Lime, and with little or no trouble.

In this city, there are about thirty-five establishments engaged in the manufacture of Soap ; and few branches of our manufactures have grown more rapidly with the prosperity of the city. We have been assured that there is more Soap now made here in one month, than there was ten years ago in a whole year. At that time, we were greatly dependent upon New York, New England, and Western Soap makers; and Colgate's Soap of New York crowded every store ; but now our own manufacturers supply our market, to the exclusion of nearly all competitors, and have besides large supplies for exportation. A few of them manufacture almost entirely for exportation to the West Indies, South America, &c. They make all the varieties in common use, and some make Soap of superior quality. Palm Oil is extensively employed for making Soap and Stearin Candles ; and Olive Soap of remarkable power, soluble in strong brine, and therefore well adapted for marine use, is an important article made here.*

---

* An article of this kind, known as Chemical Olive Soap, manufactured by Mr. WM. CONWAY, 316 South Second street, is deserving of particular notice, inasmuch as it has become a staple article in nearly every grocery

Numerous experiments have been made toward producing Castile Soap in this city, but apparently without success.

The manufacture of CANDLES is so very generally associated with that of Soap, that the branches may be considered inseparable. The advances that have recently been made in chemical science have wonderfully influenced the manufacture of both articles; and by the separation of constituents, purification, distillation, pressure, and other arts and appliances, known to the initiated, it is possible to attain very remarkable results from very unpromising materials. The most impure fats, as well as Palm and other oils, may be made to yield, by the skillful Candle-maker, a product from their solid portions but little inferior to those made from wax, which is too expensive for ordinary use. *Dip Candles* are nearly obsolete; but the manufacture of Moulded Tallow Candles is still an important part of the business of nearly all the Soap-makers. The cheapness and brilliancy of gas have, however, superseded their use in most of the principal cities, while the introduction of

store in the city, on account of its superior detersive qualities, and its adaptation alike to the use of hard and soft water. This Soap is the result of a series of experiments commenced about three years ago, in consequence of the popularity of an Eastern Soap of similar name, extensively sold here at that time, which it has almost wholly superseded.

With the self-reliance which characterizes all successful enterprises, and a thorough practical acquaintance with his business, to which he was brought up, assisted by a competent knowledge of the chemical principles involved, Mr. Conway resolved that what had been done elsewhere could be done by himself. His first attempts were followed by successive improvement, until the efficient and economical Soap now made by him leaves little to be desired in the way of amendment, and after much pecuniary loss in establishing its name and merits, he is being rewarded with substantial success. It is claimed for it that it is superior to any common Soap for washing in any water, hard or soft; for the reason, among others, that it contains more alkali, which at the same time is so completely neutralized by the other ingredients, that the fabric is not in the least injured by it. Mr. Conway is also a large manufacturer of Candles, and of Palm, Variegated, White, Yellow, Pale Brown, and other common Soaps, as well as of Fancy and Perfumed Soaps. Of his improved Chemical Olive Soap, his sales to the city alone average one hundred boxes per day, besides the quantities sent elsewhere.

31*

Camphine, and the various illuminating oils, has also tended to limit the manufacture of Candles. *Wax Candles* are made by only one person in this city.

At least two establishments in this city are engaged in the manufacture of *Adamantine Candles* and *Olein Oils*, with all the modern appliances for doing a large business. DAVID THAIN & Co., and C. II. GRANT & Co., have a capital invested in this business of $400,000, employ 116 hands, and produce annually, of Candles and Oils, $570,000. The entire capital employed in the Soap and Candle manufacture is $950,000; and the aggregate product is $2,057,600.

XXXII.

Sugar Refining.

SIR: Your request for accurate information with regard to the Sugar Refining business in Philadelphia has received due attention, and I take pleasure in transmitting the result of my researches respecting the subject.

There are five large Steam Refineries, besides two extensive establishments which extract Sugar from Molasses. Those engaged in Refining Sugar are—J. S. LOVERING & Co., T. A. NEWHALL & Co., BUTE & SMITH, EASTWICK BROTHERS, and J. R. ROUDET; in extracting Sugar from Molasses, G. L. BROOM & Co., and FELTUS & ZIMMERLING. The buildings used by these firms are very extensive, and the combined steam power amounts to over 500-horse power. The number of men employed in the different works is about 700, and the amount of raw Sugar imported from the West and East Indies, St. Domingo, &c., and used, will reach 1,000 hogsheads per week, from which nearly 5,000 barrels of Loaf and the different grades of Clarified Sugar are produced, the greater portion of which is for the Philadelphia market. Each barrel of Sugar weighs about 240 pounds.

By the introduction of machinery and steam the process of purifying and refining Sugar underwent an entire revolution; and this improvement, with the substitution of aluminous finings in place of bullock's blood, which supplied a fertile source of deterioration, has wonderfully increased the quantity of production and raised the standard of quality. The Raw Sugar, from the West Indies, is imported in cases and hogsheads; from the English Islands in hogsheads; from South America chiefly in bags, as also from Manilla and the Mauritius. These latter bags are double, and made from the leaves of reeds, plaited or woven into suitable material.

The first operation of the Refiner, after removing the Sugar from the hogshead, boxes, &c., is dissolving the Sugar in a pan by means of steam passing through a perforated pipe in the bottom of the pan. The color is then extracted from the solution by means of chemical and mechanical means, when it is passed to what is known as the vacuum pans, heated by steam, for the purpose of being boiled. By this means the liquor is so concentrated that the Sugar is only held in solution by the high temperature, so that on cooling a rapid crystallization takes place, which produces that uniform fine grain, such as is required in Loaf Sugar. The syrup, after boiling sufficiently, is poured into the moulds, which are of the funnel or sugar-loaf form, for the purpose of assisting the separation of the mother-liquor. The syrup or liquor which runs from the mould is again boiled, from which the lower grades of Sugar is produced. The syrup coming from this second process is sold for molasses. The production of molasses is about one fifth from each hogshead.

The value of Refined Sugar manufactured in this city in one year, taking for the basis of calculation the data given above, with the prices which ruled in 1857, and a working period of ten months, would be from $5,500,000 to $6,000,000; and the business of the year, including Molasses, amounted to $6,500,000. It will be remembered, however, that the high price of Sugar during the early part of last year diminished the demand; and at present prices, which are much lower than in 1857, the Refineries here are of sufficient capacity to produce $10,000,000 annually, if constantly in operation.

The art of Refining, it is believed, has attained a higher standard in this country than in any part of Europe, and the excellence of this manufacture is not approached by any imported article. Within a recent period, our own city has advanced greatly in both the quantity and quality produced. A few years since but a single Refiner had a name here, and a well-deserved one;* now several others are approaching

---

* The firm alluded to, it is perhaps needless to remark, is J. S. LOVERING & Co.—a name well-known in the principal markets of the world, and we may say in the scientific world. Their Refinery is one of the very largest in this country; but we have made this reference more particularly to describe their Barrel-making establishment at Bridesburg. The grounds enclosed for the Works, a memorandum before us states, contain about nine acres. The Maple Logs are kept in large quantities in a pen leading from the Frankford Creek, and which is immediately back of the Saw mill where the timber is cut into planks the thickness of the width of the stave. The

the high standard with rapid strides. The PENNSYLVANIA REFINERY, corner Race and Crown streets, with very slight additions to its apparatus, has nearly trebled its capacity of production, and has for several years fully reached the highest standard, as regards quality, attained either in New York or Philadelphia.   M.

XXXIII.

Tobacco Manufactures—Cigars, Snuff, etc.

SIR: The manufactures of Tobacco in Philadelphia are limited to Cigars, Snuff, and Smoking Tobacco. Chewing Tobacco is made in the neighborhood of the plantation; and the reason that so much of inferior quality is made, is that the demand exceeds the supply. All the first quality grown is required for wrapping the frost-bitten, unripe, and otherwise injured leaves, which are deposited in the centre of the plug.

The manufacture of Cigars is a great business in Philadelphia. I am practically acquainted with the trade, and since your queries were received have made it a subject of careful investigation. My opinion is there are about 1,000 Cigar manufacturers in Philadelphia; thirty of whom employ from ten to sixty-five hands; the others from one to five

pieces are then taken to another room and placed in steam-tight boxes, into which the engines exhaust their steam, and thus in a short space of time the timber is sufficiently steamed to be cut into staves. The blocks are then taken into the stave-cutting room, where one man with a huge knife, worked by steam, cuts eighty staves the proper shape every minute. This knife has a curved blade, which gives the staves the proper curve. This knife is over five feet in length, and acts perpendicularly; and the feeder stands at a table and presents the piece of timber to the knife, a gauge stopping the slack, so that the knife cuts the stave the proper thickness. The staves are then put in iron cars and run into drying kilns, two hundred feet long, and heated by hot air. They are, after being dried, placed upon a revolving table, which joints them so that they fit together in such a manner as to give the proper bilge. After passing through another machine, which cuts the chime and prepares it for the head, the staves are passed to the setters-up. About thirty-five men and boys are employed, and with this force from three hundred to five hundred Barrels are made daily. The maple logs are brought from the head-waters of the Delaware; 1,500,000 feet, costing from $8 to $10 per thousand, are used annually. The hoops are of ash, and brought from the shores of the St. Lawrence. The machinery is moved by three steam engines; one of sixty-horse power, and two of thirty-horse power each.

or six. The whole number of employees, journeymen and girls, engaged in making Cigars, is fully 4,000. If each hand makes 1,500 Cigars per week, a minimum amount, the weekly production is 6,000,000, or 312,000,000 Cigars per year. A fast hand will make five hundred Cigars per day. The average labor expended upon each thousand Cigars costs about $3 50: the weekly production of 6,000,000 would cost $21,000; and the yearly cost for labor on 312,000,000, would be $1,092,000.

The average cost of each thousand Cigars is $8 00; that of 312,000,000 is $2,496,000. A profit of twenty per cent, makes the annual production about $3,000,000. The largest factory employs sixty-five hands, and manufactures 4,000,000 Cigars, of which the average cost is $16 per thousand. This firm has invested in the Cigar business alone $80,000.

About one-third of the leaf tobacco, for making Cigars, is obtained from Cuba; the rest is of American growth. The Cuban is of course used for the superior qualities. The best Cigars are known by their pure color, and the white solidity of the ashes. The best Cigars made in Philadelphia need only the foreign brands, and Custom-house marks, to sell as real Havanas. They are shipped to all parts of the West and South, and the best qualities are sold largely in New York as imported.

The Cigar branch alone employs a capital of $1,800,000.

Machines have not as yet been found to work well. A Liverpool house is said to have a patented machine in operation, which will make 5,000 Cigars per day; and in Prussia machines are extensively used, which is one reason why German Cigars are so cheap, and so badly made that few will smoke.

In the manufacture of *Snuffs* there are four mills, that employ fifty hands, and have a capital invested of $80,000. Garrett's Mill has been established probably a century. *Smoking Tobacco* is cut in the Snuff mills, and also by mills devoted exclusively to the purpose. The product will average 5,000 pounds per day, worth ten cents per pound, which amounts to $156,500.

The Cigar, Snuff, and Leaf Tobacco trade undoubtedly employs a capital of $3,000,000, turned twice a year, which produces a business of $6,000,000 per year.

XXXIV.

The Umbrella and Parasol Manufacture.

Archæologists have demonstrated that those portable protections from the sun and rain, called Umbrellas and Parasols, probably commenced with the latter of these inventions, and in a region where the intensity of the light and heat rendered a shade almost indispensable. In the contrivance of such a shelter, the pole and top of a tent seem to have originally suggested the well-known form, which, in its general features of a dome or canopy, still remains unaltered. The materials used in the early Parasols were exceedingly heavy, and one or more attendants were required to carry them over their possessors; hence, the ownership of a Parasol was at one time indicative of high rank. But we are compelled to pass by much curious information, with respect to the early history of these articles, and their gradual introduction into common use, and proceed to consider the present state of the manufacture, particularly in Philadelphia.

The Umbrella and Parasol manufactories in Philadelphia, it is supposed, are more extensive than any others of the kind in the United States; and their products have proverbially a better reputation for quality than any others.   It is probable there are more than a hundred places in Philadelphia where Umbrellas and Parasols are made to some extent, but the very extensive establishments are limited to four or five.   The causes that have contributed to the supremacy of Philadelphia in this manufacture, are principally those which have led to a like result in other branches; but there are also special and particular reasons for the superiority.   The *sticks* and *metal mountings* made in Frankford, a populous suburb of the city, are unsurpassed for excellence and efficiency. The *stretchers*, made from the best Pennsylvania iron—the wire, drawn at Easton, and formed, forked, and japanned at the House of Refuge, under the superintendence of a firm in this city—are tougher, and less disposed to rust or oxidize, than any in the world. The mechanical genius of the manufacturers has also been active, and a number of very important improvements, which facilitate the manufacture, have originated here.   Of this description we might instance the "Sorting Machine," invented by Wright Brothers & Co., for adjusting the strength of the ribs, or *setts*, worked by balance-weights, and by determining the strength of each rib, insures the perfect and regular shape of their goods.

The firms most extensively engaged in this manufacture are WRIGHT, BROTHERS & Co., SLEEPER & FENNER, WM. A. DROWN & Co., SIMON HEITER, and WM. H. RICHARDSON.   The first two are probably the largest, and certainly among the largest Umbrella manufacturing concerns in the Union.

The quantity of material annually consumed in the manufactory of WRIGHT, BROTHERS, & Co., is enormous. The *Pennsylvania Inquirer*, referring to them, stated—

"This house produces an average of 2,200 Umbrellas and Parasols a day, or about 700,000 per annum: and consumes one million yards, equal to 570 miles, of Silks, Cottons, and Ginghams ; upward of 200,000 pounds of Rattan, and about seventy-five tons of Horn, Bone, Ivory, and other materials, for ornamental mountings. Of Whalebone, the house alluded to above consumes over 100,000 pounds, equal to about one-thirtieth of the average products of the whale fisheries of the world.

Such are the extent and variety of the mechanism used, and the perfection and nicety with which it is adapted to the purpose, that, with the help of ample steam-power, all this vast quantity of material changes its form, and 700,000 Umbrellas are manufactured in the establishment of the Messrs. Wright, with the help of only 450 hands constantly employed under one roof. All parts of the Umbrella are now arranged with mathematical accuracy by the machinery used, some of which was invented by one of the proprietors of the establishment, from whom the above information was obtained, and can be used by no other manufactory. The system to which all parts of this manufacture is reduced is now so perfect as to place the cost of production very low, and far below competition from hand labor and ordinary machinery—in addition to forming the article with a beauty and accuracy only to be obtained from the best mechanical means."

Messrs. SLEEPER & FENNER are a prominent firm, who have been identified with this manufacture for about twenty years, and now rank among the most extensive makers of Umbrellas in this country. The products of their establishment are sold largely in Boston, and other parts of New England, and are highly and deservedly appreciated, where appreciation is a compliment.

WM. A. DROWN & Co. are the successors of Erasmus J. Pierce, one of the pioneers in this manufacture. Previous to the last war, he was engaged in the business, in Baltimore ; but his residence in this city dates, we believe, from 1815. At that time, forty Umbrellas per day was a large product—fully as much as the demand would warrant. Mr. Pierce retired from active participation in the business about 1836, and at the time of his retirement was accounted among the very largest manufacturers. His successors, Messrs. Wm. A. Drown & Co., are noted for the fine styles of Umbrellas and Parasols which they produce; and their goods are well-known throughout the entire country.

The house of SIMON HEITER, though less extensively engaged in the manufacture than some of the others to whom we have referred, is well known to the trade, and takes a very respectable rank. All the styles usually made in this country are produced in his manufactory; and by means of connection with houses in Europe, he is in early and constant receipt of whatever novelties are originated in the workshops of Paris or elsewhere.

WM. H. RICHARDSON has been connected with the trade for many years, and during the period of this connection he has introduced several novelties that can be highly commended. One is the Walking-cane Umbrella, a very ingenious and neat affair, well adapted to the use of pedestrians and travelers. It consists of a convenient size of Umbrella, in a very handsome rosewood case, which, when the Umbrella is hoisted, forms the handle, and, when closed, becomes a handsome cane. Another arrangement for the comfort of travelers is an Umbrella, the handle of which may be readily converted into a rest for the head; it is intended to be used in cars as a head-support, and thus facilitates sleeping while traveling. Another novelty to which we beg leave to invite attention is an Umbrella that can readily be packed in a trunk, and also the Holland Rectangular Steel Tube Umbrella, which, with the frame covered with strong silk, is said to weigh only *nine ounces.*

The Umbrella and Parasol manufacture in Philadelphia employs directly about 1,500 persons, and indirectly, and in all its branches, 2,500. A large proportion of the employees are females, whose earnings average from $2 to $5 per week. A capital of about $700,000 is invested, and the average annual product is about $1,275,000, though in 1853 it was nearly two millions of dollars; the sales of one firm alone exceeding a half million of dollars. The value produced in Philadelphia is nearly equal to that of Paris in 1847, when the product was stated at £296,000.

The circumstances that have contributed to the development of the Umbrella manufacture in Philadelphia, we have in part already alluded to. The establishments in Frankford for the manufacture of Metal Mountings, Tips, &c., are deservedly noted, and supply not only the manufacturers of this city but of New York. The Ivory and Bone Turners, and Carvers, perform their part well in ornamenting the handles. In the establishment of HARVEY & FORD, undoubtedly the most extensive in the United States, 150 operatives are employed, and their carved Ivory-work successfully rivals the finest of England or France. The quantities of material annually consumed in this branch may be inferred from the following statistics, recently given in the *Ledger*, for two of these shops, viz. :

"Ivory, 30,000 lbs., worth about $80,000; Walrus, a large quantity of which is also used for like purposes with Ivory, 6,000 lbs., worth $3,000; Boxwood, 9 tons, worth $270; Vegetable Ivory, 30 tons, worth $3,000. One hundred and forty men and boys are employed, who receive for their labor $29,700 per annum. In the two shops, one hundred and five Turning Lathes and eleven Saws are at work. Eight thousand bushels of dust are sold every year for $5,200. The dust is used by farmers as an excellent manure."

A considerable proportion of the finished work in Ivory and Bone, it is proper to explain, is used for the handles and mountings of Whips and Canes, and various kinds of Surgical Instruments.

There is also an extensive establishment in the city for the manufacture of WHALEBONE and RATTAN, and is said to be the only factory in the country where Whalebone is prepared for all purposes to which it is adapted, viz. : Umbrellas, Parasols, Whips, Canes, Dresses, Hoops, Bonnets, Hats, Hair Pins, &c. This manufactory, of which the proprietors are George W. Carr and Samuel Warrington, trading under the firm-style of GEORGE W. CARR & Co., was established in 1842. The machinery and fixtures are principally original, and said to be unknown to other manufacturers. Steam, supplied by a twelve-horse engine, is used in all the various processes of Boiling, Dyeing, Drying, and Heating.

Previous to the great advance in Whalebone, this manufactory consumed annually from 150,000 to 200,000 lbs. ; but at present,

32

the consumption is much reduced by the introduction of substitutes of much lower cost. Rattan is now a leading article in the manufacture of Umbrellas, Parasols, Chair Seating, Skirt Hoops, &c., and this firm consume annually about 200,000 lbs. The manufacture of Skirt Hoops is largely carried on by them, and for covering the Skirts they use the Whip Braiding machines before referred to. Messrs. G. W. Carr & Co. employ fifty to sixty hands—men, boys, and girls.

XXXV.

Wagons, Carts, Drays, and Wheelbarrows.

Within comparatively a few years the demand for Wagons of a peculiar construction has elevated the business of Wagon-making into the rank of manufactures. The wheelwright and the blacksmith are no longer able to supply the combined wants of the United States Government, Express Companies, and Emigrants: and establishments are required that can purchase lumber and iron in large quantities, and which are provided with all the requisite machinery and appliances for turning out heavy vehicles with expedition and rapidity. The excellence of the timber furnished from the forests of Pennsylvania, New Jersey, and Delaware ; the reputation of the builders, established even prior to the time when Conestoga Wagons transported all heavy goods from the Eastern States to the West ; the present facilities of the Wagon-making establishments, and the immense stock of well-seasoned lumber always kept on hand, make Philadelphia, at the present time, the best and principal seat of the Wagon manufacture.

There are now forty-five establishments in Philadelphia where Wagons, Carts, Drays, &c., are made, but, we believe, only three that carry on the business on a large scale, and supply the South, the United States Government, and distant markets generally ; viz. :—WILSON, CHILDS & CO., SIMONS, COLEMAN & CO., BEGGS & ROWLAND. Two other houses, BENJAMIN FRANKLIN, and W. HOSKINS, have executed orders of some importance for remote sections.

These establishments send their products to almost every part of the South, including Texas ; and even Mexico obtains a second edition of our old Conestogas, which are there drawn by mules. The United States Government has for some time obtained its

Wagons from Philadelphia, inasmuch as those built here are found to be the most serviceable on the Western frontiers, where only those of the best material and construction can withstand the abrasion of travel. A considerable proportion of those required for the transportation of the Utah expedition are being made here, and the factories at the present time swarm with industrious artisans. We shall describe two of the principal establishments in the APPENDIX.

The statistics of the product are given as follows :—

Value of Wagons annually sent South, and for the U. S. Government,	$525,000
" " Wagons, Carts, Drays, &c., made for the city and vicinity,	290,000
Total product, - - - - - - - -	$815,000

XXXVI.

Wood-Working—Building Materials, &c.

The working of Wood in Philadelphia, and throughout the United States, is especially remarkable for the application of labor-saving machinery, by which the most important results are attained from apparently very simple means. All the implements and machines designed for the purpose, and in use in this country, from an ordinary axe to a planing machine, are far in advance of those in use in Europe. A house in Liverpool is now importing the best American Wood-working machines, and making great efforts to introduce them generally into England.

The abundant supply of Lumber in Pennsylvania, and the sources of that supply, were stated in the Introductory. It is probably our duty to describe the facilities that are in use in the various and numerous Wood-working establishments; but we are reminded that the leading branches of productive industry have already consumed more than their allotted space ; and we are satisfied that nothing like justice could be done, within narrow limits, to a subject so comprehensive. We may probably describe one or two of the leading establishments in the APPENDIX; but here it must suffice for us to say that the machinery in the various Planing Mills, Sash Factories, Turning and Scroll-sawing Establishments, &c., is truly remarkable for its efficiency ; and that those establishments occupied in preparing the various parts of Wood-work required in Buildings, can supply builders at a much cheaper rate than the latter can produce them in their own workshops, without the aid of such machinery.

The statistics of the aggregate product, prepared with a good deal of labor, will convey some idea of the extent to which the business of Wood-working, in its several branches, is carried on :—

Value of Lumber Sawed, including Mahogany and Fancy Woods,		$580,000
" " Flooring and Planed Lumber, - - - - -		370,000
" " Sashes, Blinds, Doors made in factories, - - -		250,000
" " Mouldings, Turnings, in Wood, &c., - - - -		850,000
" " Barrels, Casks, Shooks, Vats, &c., - - - -		715,000
" " Boxes, Packing, estimated, - - - - - -		500,000
" " Picture and Looking-Glass Frames, stated by a leading manufacturer at $1,500,000, estimated, - - -		750,000
" " Matches—6 Match factories, estimated, - - -		125,000
" " Coffins, ready made, - - - - - - -		219,000
" " Lasts and Boot-Trees, one maker using machines, -		36,000
" " Cedar and Wooden Ware, - - - - - -		150,000
" " Patterns, Stove and other, - - - - - -		115,000
" " Show Cases, &c., - - - - - - -		55,000
Miscellaneous manufactures in Wood, viz.: Hydrant Stocks, Ladders, Kindling Wood, Shingles, Laths, &c., estimated - -		100,000
Total, - - - - - - - - -		$4,815,000

We now take leave of the Branches of Productive Industry in Philadelphia that can be called "Leading," or those of which the aggregate product and relative commercial importance entitle them to the designation. A recapitulation of the respective values produced in these branches alone would show that Philadelphia is a very great manufacturing city ; and if we were to step beyond the city limits, and compute the industrial values of the district of which this city is the commercial centre, we would have at at least the following additional aggregates :—

Dry Goods in the vicinity of Philadelphia, as before stated, -	$6,696,000
Iron—Anthracite made near Philadelphia, - - - - -	4,569,720
" Charcoal in Eastern Pennsylvania, - - - - -	1,754,280
Products of Forges and Rolling Mills in the vicinity of Philadelphia, in 1856, - - - - - - - - -	5,000,000
Miscellaneous manufactures of Iron, estimated, - - - -	3,000,000
Leather, estimated, - - - - - - - - -	2,500,000
Paper, estimated, - - - - - - - - -	2,000,000
Wood-working, including Agricultural Implements, Barrels, Handles, Tools, &c., brought into Philadelphia, estimated, - -	1,000,000
Total, - - - - - - - - -	$26,520,000

RECAPITULATION.

Agricultural Implements, Seeds, and Fertilizers... $1,003,00(

Alcohol, Burning Fluid, Camphene, &c...... 1,022,140

Book Manufacture, and its kindred branches, as already given.............$5,593,000

Profits of Publishing Books and Periodicals—amount $4,090,000, estimated rate 20 per cent... 818,000

Daily and Weekly Newspapers, estimated by a leading Newspaper Publisher 1,370,000

            ————— 7,781,000

Boots and Shoes..4,141,000

Brass and Copper...1,230,000

Ale, Porter, Lager Beer, &c...2,300,000

Bricks, Fire-Bricks, Pottery, &c..1,459,000

Carriages, including Wagons, Carts, &c...1,715,000

Chemicals, Paints, Glue, Curled Hair, Varnishes, Medicines, &c...........................7,370,000

Clothing, Mantillas, and Corsets..11,157,500

Confectionery, Fine Cakes, &c...1,020,000

Distilling and Rectifying Liquors,..3,154,000

Dry Goods or Textile Fabrics..21,318,118

Flour and Substances as Food, including Baker's Bread, and Cured and Smoked Meats.. 14,150,000

Furniture, Upholstery, &c...3,000,000

Glass Manufactures..1,600,000

Hats, Caps, Furs, &c..1,900,000

Iron, Manufactures of..12,852,150

Jewelry, and other Manufactures of Gold, Silver, &c..3,272,000

Lamps, Chandeliers, and Gas Fixtures,...1,300,000

Leather, including Buckskin and Kid Gloves, Belting, Hose, &c...........................3,091,250

Marble, Stone, Slate, &c..1,160,000

Oils..2,131,230

Paper Hangings..800,000

Rope, Cordage, &c..810,000

Saddles, Harness, Whips, Trunks..1,988,000

Sugar Refining and Molasses...6,500,000

Ship and Boat Building..1,760,000

Soaps and Candles..2,057,600

Tobacco, (Smoking), Cigars, Snuff...3,256,500

Umbrellas and Parasols, including Umbrella Furniture, Ivory and Bone Turning, Whalebone and Rattan manufacturing...1,750,000

Works in Wood—Products of Saw Mills, Planing Mills, Sash and Door Factories, Wooden Ware, Matches, Lasts, &c..4,300,000

Total value of Leading Branches of Productive Industry in Philadel'a......$132,349,488

Add for Product of Leading Branches in the vicinity of Philadelphia............26,520,000

Total for Leading Branches in Philadelphia and Vicinity $158,869,488

32*

MISCELLANEOUS MANUFACTURES.

I.

Artificial Teeth.

The manufacture of Porcelain Teeth is modern in its origin. Originally natural or human Teeth were used; also calves' and sheep Teeth; next ivory, or Teeth carved from the tusk of the Hippopotamus. Fifty years ago there was not a Porcelain Tooth made in this country. Twenty years ago, not more than two hundred and fifty thousand were manufactured annually in the United States, and but a trifling number in Europe. Since then the demand has been continually increasing, owing in a great measure to the rapid improvements made from year to year, and the more perfect applicability to the purposes designed; and within the last eight years, it is said, the consumption has increased one hundred per cent.

Philadelphia was the first and original seat of the manufacture of Porcelain Teeth in the United States; and over one half, if not two thirds of all the Teeth now made in this country are made in Philadelphia. One firm, that of Jones, White & McCurdy, Arch street, make annually about 1,250,000 Porcelain Teeth. Their products are exported to Europe, South America, the West Indies; and in all the markets they take precedence over the European manufacture.

In the report of the last Franklin Institute Exhibition, we notice that the judges remark:

"These gentlemen claim the following as improvements in their Teeth. A close imitation of the natural organs; a great variety of shapes in conformity with nature; the thickening of the posterior edges of the canines and bicuspids, in conformity with the gradual filling out of the natural organs from the incisors to the molars; a greater capacity to withstand the extremes of temperature to which they must be exposed; the peculiar blending of the tints in imitation of nature; the shape of the bases of the gum teeth for half and entire dentures, and their more perfect adaptation to the plates with but little grinding. Also, a great improvement in the enamel surface, which is divested of that glassy, reflecting character so unnatural, and which has hitherto

(398)

been so objectionable in Artificial Teeth. Most of these improvements, especially those which relate to the shape, articulating and enamel surfaces, are of a high order of merit, and entitle them to a recall first-class premium."

It is but justice to say, that the eminent success which has attended this firm is due to their merits as manufacturers of superior Artificial Teeth ; but also in part to their standing as gentlemen, their enterprise, and to their public spirit.

The statistics of the manufacture of Porcelain Teeth, as furnished to us, are as follows :

Number of establishments, - - - - - - -	5
Capital invested, - - - - - - - - -	$175,000
Persons employed—one half females, - - - - -	125
Value of production annually, - - - - - -	$500,000

II.

Awnings, Bags, &c.

There are about eight principal establishments in this city for making Awnings, Bags, Sacking Bottoms, &c., and in addition, they manufacture Garden and Field Tents, Verandahs for windows, Wagon Covers, Flags and Banners, &c. Military Tents were, during the Mexican war, and until a recent period, made here to a very large amount for the United States Army, but are now all made at the United States Arsenal in this city. Wagon Covers and Sacking Bottoms form considerable items in the business here ; and of the former, two extensive Wagon Building establishments in this city necessarily require many for their Mexican and Southern customers.

Bags, for grain and flour, are only made to a limited extent in connection with the other branches, the recent introduction and greater cheapness of Seamless Bags having greatly abridged the demand. The covering of Hams, previous to being whitewashed for their better preservation during the hot season, also belongs to the business, and occasionally employs a number of hands. Mr. JOSEPH H. FOSTER, a principal manufacturer in all these branches, and also a practical Sailmaker, has furnished the following statistics, viz.:

Hands employed, - - - - - - - 125	
Yards of Canvas used annually, 325,000, worth, - -	$58,250
Other materials, as Rope, Twine, Rags, Hooks, Needles, &c.	3,500
Workmanship, - - - - - - - - -	30,000
Total, - - - - - - - - -	$91,750

III.

Baskets and Willow Ware.

This is a branch of business which may be regarded as yet only in its infancy in this country, but destined to become one of considerable magnitude at no distant day. Considering the numerous uses of the Willow, it is gratifying to know Philadelphia is actively leading in an enterprise which must aid in eventually rendering us independent of Europe, both for the raw material and the manufactured product. Enormous quantities of Willow are annually imported, chiefly from Havre, and then manufactured into Baskets by our German population and others; while of the finer and fancy Willow Ware, nearly the whole has hitherto been imported. For some years past, however, increased attention has been given to the cultivation of the Willow by our Basket-makers in this city, on account of the superior quality of the native growth; the foreign article being always culled of the better portions previous to shipment. Our soil and climate are found to suit the plant admirably; and there are now within the city limits not less than eighty-five acres of swamp or meadow lands under cultivation. The number of persons engaged in growing Willow, exclusively for their own use, with one or two exceptions, is about ten, whose gardens embrace from two to twenty acres each; and others are engaging in it every year. Mr. John Stinger, the largest Willow-grower and Basket-maker at present, has twenty acres, or nearly; and if persons of capital, which our pioneers have not been, were to embark in this branch of husbandry, a ready sale could be had for all that can be raised for years to come, at a remunerative price. The Willow is planted in rows four feet apart each way, or four feet by two, according to the kind. During the first year the young plants require care similar to that bestowed upon corn; but in subsequent years little attention beyond keeping rank grass from the roots. In two years it reaches a size sufficient for use; and may be cut the first year, but with diminished profit subsequently. It continues to yield for fifteen years, at an average value of $30 per acre, when it must be grubbed up and replanted.

Messrs. ROBERT & CHARLES DUNK, who are among the largest of the Willow-growers and Basket-makers, and whose family have been thirty years in the business, have experimented upon some sixty varieties of the Willow, to ascertain which are best adapted to our climate, soil, and manufacture. Few or none of those most cultivated in Europe succeed here. Those best suited to our wants, we are informed, are the *Salix Lambertiana*, *S. Cordata*, (native), *S. Pentangea*, and *S. Russelliana*. The last named is most cultivated; and as it produces all the sizes required for different kinds of work, it would supersede all others were it not liable to grow twiggy. The *S. Cordata* is best for larger and coarser Basket-work ; and other varieties are respectively used for special purposes. Of Basket-makers, there are about twelve principal ones, besides a very large number of manufacturers on a small scale. In quantity, the product in this city is supposed to exceed that of any place in this country, and the quality of the work is undoubtedly superior, arising chiefly from the greater whiteness, strength, and beauty of the native Willow.

Mr. Dunk, and some others, we believe, export nearly all they make. Much of the imported Willow is used in the covering of Carboys and other Glassware, by some of our large Glass factories.

Chairs, Settees, Cradles, Coaches, Work-Tables, Baskets, &c., are also made of *Bamboo* and *Cane*, by several persons.

IV.

Brooms.

Of Corn Brooms the product in Philadelphia is large, and the quality generally superior to the average. The Wire-fastened Broom made here, in particular, is nowhere equaled by any article of the same price. The principal supply of Broom Corn used here comes from Ohio and Indiana, and it is manufactured by between one and two hundred persons, each of whom will make from twenty-five to fifty Brooms daily. Some of our larger manufacturers employ quite a number of hands ; and we believe few, if any, establishments in this country turn out more Brooms and Whisk Brushes than Messrs. BERGER & BUTZ, of this city. Ten or twelve thousand Brooms, probably, find their way into the market weekly from the hands of Philadelphia Broom-makers, the wholesale prices of which range from $1 to $3 per dozen. To this, as to most similar branches, belongs an assortment of tools appropriate to the business, for the supply of which we are as yet dependent upon New England. Our ingenious mechanics ought to supply not only the home, but the distant demand for these articles. The same may be remarked of *Broom Handles*, which come from other quarters in very large quantities.

V.

Blacking, Ink, and Lampblack.

Blacking, it has been remarked, consists essentially of two principal constituents, viz., a black coloring matter, and certain substances which will acquire a gloss by friction. Each maker has of course proportions and methods of mixing peculiar to himself, but the chief materials used are the same in most cases. The extent to which the manufacture is carried on by some firms, is illustrated by the business of the celebrated Day & Martin, who send away on an average 150 casks, containing a quantity equal to 900 dozen pint bottles, per day. The constituents of Day & Martin's Blacking are said to be Bone-black, Sugar, Molasses, Sperm Oil, Sulphuric acid, and strong Vinegar. In Philadelphia there are five principal establishments engaged in the manufacture of Blacking—James L. Mason & Co., John Annear, J. & E. Newbert, and Charles O. Wilson, and many others who make to a limited extent. The first-named employs seventy-five persons, and turns out three and a half millions of boxes of Blacking per year.

Writing Ink is made by Messrs. Mason & Co., and by five others, viz., Bartholomew Bussier, Green & Co., Joseph E. Hoover, Apollos W. Harrison, and Samuel Schurch. The excellence of the " Columbian Writing Fluid," made by Harrison, every penman is familiar with.

Lampblack, as we previously stated, is made in connection with Printing Inks, and also at separate establishments. Matlack's Lampblack, and Martin's, are well-known, both in this country and in Europe.

The value of the Blacking, Inks, and Lampblack, made annually in Philadelphia, may safely be stated at a half million of dollars.

VI.

Boxes, Paper.

Paper Boxes are probably made more extensively in Philadelphia than in any other American city. The demand for Boxes, for all purposes, renders the variety seemingly unlimited—Boxes for Fancy Hosiery, Shoes, and Parasols ; Boxes for Shirts, Bosoms, and Collars ; Boxes for Artificial Flowers, Ruches, and other Millinery goods ; Boxes for Brushes and Combs ; for Perfumery and Fancy Soaps ; for Envelopes, Pencils and other Stationery ; Confectionery Boxes, Jewelry Boxes, Pill Boxes ; and Match Boxes, though these are generally made by the Match-makers. All these different descriptions and varieties, from the commonest and cheapest, up to the most elaborately ornamented, at $2 each, are made in the Philadelphia manufactories. A description of the largest establishment, that of George W. Plumly, must suffice for all. It consists of six floors in all, including a basement, and each room is appropriated to its own peculiar separate and distinct operations. In the basement, and under the sidewalk, there is a boiler supplying steam for heating the building, and hot water for making paste, &c.; also, a small steam-engine for pumping water, &c. Here a man and boy are engaged in covering pasteboards with white and colored paper, which is done by a machine containing two large rollers—one of which revolves in a trough filled with paste, the sheet being passed between them, covered with paste, and the paper laid on both sides. The sheets are subsequently dried and pressed. The first floor is chiefly occupied as a warehouse, counting-house, &c. In the second, the large boxes which require sewing are made and finished. In this, and in other rooms, are shears of every size and pattern; and machinery for cutting, with great rapidity, pasteboards into the lengths and widths required. The third story is devoted to another description of work—the largest that does not require sewing,—and has machinery for a variety of purposes; such as cutting boards into circular pieces for tops and bottoms of round boxes, machinery for scoring and cutting out the corners preparatory to making square boxes, &c. The fourth story is devoted to still smaller work, and the fifth to pill and other round boxes. Here the most perfect machinery is found. One machine, invented by Mr. Plumly, for cutting round pieces for ends of boxes, operates with such rapidity that it is said 5,000 can be cut in an hour. The upper stories are subdivided, and one part of each occupied by the men who cut and prepare the work; the other by the women and girls who finish the boxes. The cheapness with which boxes can be made is remarkable; some of very neat appearance can be made at about three cents per dozen; and yet each is made of several separate pieces, and each has to be many times handled, covered with colored or fancy paper, labelled, and packed. Although most of the manipulations must be done by hand, yet within the last few years a great variety of machinery has been invented for the purpose, which gives increased facilities to the operations.

The Pasteboard is principally obtained from mills at Chambersburg, Harrisburg, and Williamsport, and costs upon an average $50 per ton. The Glazed and Fancy Papers, of which the consumption is considerable, are principally imported.

Mr. Plumly makes every description of Boxes; and not only supplies in part this city, but executes orders from Boston and other parts of the East, and from the chief cities of the South and West.

The whole business employs 325 hands, and the annual product is about $175.000.

VII.

Brushes.

Few articles of manufacture admit so great a diversity of forms, sizes and qualities, or so wide a range of uses, as the production of the Brushmaker; and of none does it hold more true that the best article is the cheapest. From the delicate Pencil of the artist, to the "Whitewash," or the "Scrub," the variety in style and ornamentation is exceedingly great.

The manufacture in this city includes the usual variety of Hair, Paint, and the commoner kinds of Brushes, and employs about a dozen principal concerns, besides a large number of individuals who make to a limited extent. In this, as in other branches, our manufacturers have aimed at the production of substantial and reliable work. In the important article of a Paint Brush, particularly, some of them have successfully striven to excel; and we believe Clinton's Improved Copperbound Paint and Varnish Brush is not surpassed by any in this country, while those of other makers are generally preferred to similar Brushes made elsewhere.

Steam has not as yet been introduced to any extent in the Brush factories, and fewer Brushes are made in the Penitentiaries and Almshouses, in this State, than in some other places; but in this city, the annual production in the House of Refuge, and Blind Asylum, is increasing. The present product, therefore, including that in Public Institutions, amounts to about $225,000 annually. The Bristles are imported principally from Russia; a cold climate being indispensable, it is said, to their perfection.

Tooth Brushes, chiefly of the open-backed variety, are made here of a quality superior to the imported.

VIII.

Buttons.

Buttons of nearly every material of which these useful little articles are usually made, including Metal, Pearl, Bone, Horn, Paper, and every variety of Plain and Fancy Covered and Silk Buttons, are made in Philadelphia.

In the manufacture of Pearl Buttons, Philadelphia takes the lead in America. We have been shown in stores Pearl Buttons of Philadelphia manufacture, which, in neatness and beauty of finish, we were informed were superior to any foreign article, and certainly could not well be excelled; and Buttons made in this city, are not unfrequently sold in some of our cities as imported French or English. Some improvements in the process of finishing, unknown elsewhere in the business, have been introduced here, whereby the cost is lessened, and our manufacturers enabled to compete with the foreign.

The Pearl employed comes from the East Indian and China seas, and also from Panama and the Gulf of Mexico. The former is the finer in quality, as well as more expensive, and is chiefly used for the best qualities of Shirt Buttons and Studs. There are about eight persons engaged in this branch, employing usually about forty hands, and they make probably two thousand gross weekly of the various sizes, from small Shirt Buttons to large Coat Buttons; and with a remarkable range of prices, from forty or fifty cents to twenty dollars per gross. A difficulty exists in the want of a uniform supply of material, which becomes at times very scarce. Mr. EDWARD MARKLEW, and Mr. W. GIBBS, have facilities for manufacturing almost any amount, and we believe dealers generally will be satisfied with a trial of the Buttons made in this city.

Bone Buttons and *Bone Moulds,* of every size, color, and description in use, are made extensively by several persons, to the amount of many thousand gross yearly. Mr. J. WITZEL employs several improved machines, the invention of Mr. E. Wahl, of this city. Of these, the most important is a machine for drilling and countersinking the Button on both sides at one operation, by which one person can accomplish the labor of three. From 25,000 to 30,000 Buttons can be thus finished up by its aid in one day. One of these is now, we believe, in successful operation in Germany by a Philadelphia Button-maker. An improved Facing and Cutting-out machine, by the same inventor, securing greater ease, accuracy, and speed than the old ones, seems also an important auxiliary in producing a neat and smoothly finished article. We are not aware that *Horn Buttons* are made by more than one person in this city, his product being some eighteen gross per week. Probably the demand is not very great.

The manufacture of *Metallic Buttons* is chiefly confined to those used on military and other equipments.

Messrs. GEIERSHOFER, LŒWI & Co., who have been for some time engaged in the manufacture of Covered Buttons of every variety, have now facilities which enable them to challenge foreign competition. Silk and Fancy Buttons are manufactured by a large number of persons in this city, and have been elsewhere referred to.

Paper, or Papier-mache, Buttons are made here also to a limited extent.

33

IX.

Cedar-Ware and Wooden-Ware.

The manufacture of Cedar-ware, though not very extensive in this city, will nevertheless well sustain the reputation of our mechanics in the minor as well as the larger branches of productive industry, by the undoubted excellence both of the material and workmanship. The chief supply of Cedar, for this business, is derived from Virginia and Carolina. There are about ten principal establishments engaged in the manufacture, besides many smaller ones.

Of these, Mr. C. Dreby is the largest, employing ordinarily about fifteen hands. He is a large producer of all the important articles belonging to this class, including Bath Tubs, Wash Tubs, Staff and Barrel Churns, Buckets, Pails, Measures, Chests, &c., &c.

The manufacturers referred to, employ together about sixty hands on this work, and have invested a capital of about $60,000. The annual production is not far from $100,000.

Wooden-ware, including all the various Wooden Housekeeping articles not made of Cedar, employs quite a number of small establishments. Only one, we believe, uses steam power in the business. Among the articles of this class may be enumerated—Kitchen and Ironing Tables, Meat Safes, Scouring Boards, Step Ladders, Clothes Horses, Towel Racks, Butlers' Trays, Plain Foot Stools, Towel Rollers, Potato Mashers, Rolling Pins, Cricket Bats, Close Tools, Toy Building Blocks, Ironing Boards, Tailors' Press Boards, Wash Boards, Bungs and Spigots, Embroidery Stands, &c., all of which, with numerous other like articles of excellent quality, we have seen in the factories of two of our principal manufacturers, W. J. Walker, and J. Lewis & Son, in our tour of inquiry. Cherry Wash Boards, of a quality superior to any of the same kind from abroad, are made here, but only to a limited extent. Large quantities made of other material come here from other markets. There are no factories for making cheap Painted Buckets or Wooden Bowls, and some other articles of this class. The land of Notions and Wooden Nutmegs is still the wholesale producer of cheap articles in this branch, although each year is rendering us more independent. A larger amount of capital, we judge, might be profitably invested in this business ; and we do not know why steam factories should not be sustained by our growing trade with all parts of the country.

X.

Combs.

Combs are made in this city, of Gold and Silver, Horn, Buffalo Horn, and Shell; but the recent introduction and popularity of India Rubber Combs have materially lessened the demand and trenched upon the profits of the business.

Horn Combs constitute the leading and staple product. They are made of all descriptions and of very good quality, at more than a dozen different establishments, and by many others individually. Some of these produce Combs equal or superior to any made elsewhere, while those of all our manufacturers have the merit of opening in conformity with the sample. The facilities for obtaining a cheap and abundant supply of material are exceedingly good, and Eastern manufacturers come here for their supply. One factory in this city, we are informed, consumes not less than four thousand horns weekly, at a cost of from four to twelve dollars per hundred.

In Tortoise Shell and Buffalo Horn, usually quite distinct branches of business from the foregoing, we have three or four principal manufacturers, whose work, for quality and quantity, has secured to Philadelphia the pre-eminence in the Comb business of the United States. Large quantities of Shell, particularly, are worked up with great taste and skill; and as much of the value of a Comb is derived from the labor put upon it, our manufacturers are only able to compete with the foreign article, by a more successful adaptation of the styles to the prevailing taste of the ladies, who do not generally fancy the French. In this our Comb-makers lead the fashion, and change their styles, particularly in the finer carved varieties, every six months.

A principal part of the Shell used here comes from the West India Islands, and the best qualities from China. The cost of the former is $6 to $7 per lb., the latter somewhat more. Buffalo Horn is chiefly derived from South America; a very small part from the Rocky Mountains. Several very ingenious machines have been adapted to various parts of the process by Mr. Redheffer, of this city, who has patents for these and other improvements in the different branches. Our manufacturers export their Combs to the West Indies, Mexico, California, and all parts of the Continent. The capital employed is between $40,000 and $50,000. The number of hands altogether employed is little short of two hundred; their average wages $7½ per week, and the annual product about $150,000.

XI.

Musical Instruments.

The Musical Instruments that are made in Philadelphia comprise Organs, Melodeons, Accordeons, Concertinos, Violins, Flutes, Guitars, Drums, and Piano-Fortes. *Organs* are made by four manufacturers—the two largest probably ever constructed in this city being that in Concert Hall, with 4 manuals and pedals, 60 registers, and 3,050 pipes, and that in Calvary Presbyterian Church, with 3 manuals and pedals, 44 registers, and 1,865 pipes. In tone, workmanship, and action, the Philadelphia-made Organs possess as many excellencies as any that can be found either in this country or Europe.

Melodeons are made to some extent by two houses; and *Accordeons* are made largely by ANTHONY FAAS, whose instruments are claimed to be better in every respect than the French and German articles imported. Their range of notes doubles that of the foreign Accordeon; their construction is stronger, and by an echo attachment similar to the pedal of a piano, the tone may be sweet and delicate as that of a flute, or changed to the deep and powerful volume of the organ. Six Silver Medals have been awarded this manufacturer, a favorable indication of the growing appreciation of Philadelphia Instruments. These Accordeons are sold only by JOSEPH SERVOSS, 16 North Second street. *Patent Concertinos*, a modification of the Accordeon, are manufactured by Mr. C. M. ZIMMERMAN, who received a first premium at the World's Fair, in London. Over one hundred men and a capital of fifty thousand dollars are employed in this business.

Philadelphia is the principal city in the Union for the manufacture of *Violins*. The principal makers are JOSEPH NEFF, JOSEPH WINNER, MATHIAS KELLER, C. M. ZIMMERMAN, JOHN PFAFF, and A. M. ALBERT, most of whom enjoy the highest reputation with Musicians and Music dealers. For brilliancy of tone their Violins are famous, and are extensively used in Orchestras. Mr. Neff has received for his Violins five Silver Medals, and the Diploma of the New York Exhibition in 1853.

Flutes and *Guitars*, of the very best quality, are made by KLEMM & BROTHER, and by Mr. J. BERWIND. The latter also makes a kind of Eolian Harp, little known in this country, called the *Cithern*, with thirty-six strings.

German Silver Band Instruments are made by Klemm & Brother, who largely import Brass Instruments. A manufactory of these is about being established.

Drums make little noise, though their quality, it is said, cannot be *beaten* by any in the world. There are two manufacturers of these; one of whom has a patent contrivance for straining the head of the Drum to a uniform tightness. Mr. Zimmerman has large contracts with Government for his Military Drums. Tambourines and Banjos, as well as Musical Chairs, are made by him.

The most important branch of the Musical Instrument manufacture, however, is that of *Piano-Fortes.* The business was established by Mr. Thomas Loud, in 1820; and there are now twenty manufacturers in the city, none of whom however prosecute it on a very extensive scale compared with the manufacturers in Albany, and other places. Of novelties, perhaps, the most remarkable and valuable is a small Piano, two thirds the ordinary size, which has been invented by Messrs. Goldsmith & Co. They claim that in quality it is equal to the others, while costing one third less. It has a double sounding-board; and in a peculiar manner the strings are brought across the main bridge and attached to another section of the bridge: a method claimed to give an increased volume of sound, and a delicate vibration.

A. B. Reichenbach, 1230 Chestnut street, claims to use an action distinct from all others; having less friction, and being less liable to get out of order. The pedal and damper are peculiar in their construction; and not only in the tone and quality of the Piano, as a Musical Instrument, does he aim at perfection, but he also pays special attention to making it beautiful as parlor furniture. The serpentine leg veneered with rosewood is one of the strongest and most beautiful in use, and is made, we believe, only by Mr. Reichenbach. The Pianos of the Philadelphia trade are justly celebrated for their handsome appearance, power, and exquisite tone. The capital invested in Piano-Fortes is $150,000; men employed, 250; product annually, $315,000.

XII.

Oil Cloths.

The manufacture of Floor Oil Cloths is limited to two houses— Thomas Potter, and James Carmichael. The aggregate capital is $170,000, the employees 150; and the materials consumed annually are stated as follows: 67,000 gallons Linseed Oil, 17,500 gallons Spirits Turpentine, 329 tons of Whiting, 164 tons Yellow Ochre, 21,000 lbs. Glue, 52,250 lbs. of Lamp-black, 525,000 yards of Cotton Cloth, 50,000 yards Cotton Drill, 49,000 yards of Cotton Duck, 202,000 yards Linen Canvas, 2,650 tons of Coal; and the annual product is $289,000. The Philadelphia establishments in this branch are peculiar, inasmuch

33*

as they manufacture a variety of articles—Table Oil Cloths, Stair Crash, and Enameled Oil Cloths in imitation of Leather, used for covering Carriages, and also for covering Desks, Tables, and Cushions, as well as Floor Oil Cloths. The establishments in other places limit their production to one class of goods, that is to Oil Cloths only, or to Enameled Cloths only. Stair Crash, it is said, was first made in this city, and nearly the entire production in this country is still confined to the Philadelphia establishments.

XIII.

Perfumery and Fancy Soaps.

The manufacture and consumption of Perfumery and Soaps, are not necessarily evidences of a love for personal cleanliness. Cologne, the dirtiest city in the world, is providentially the great manufactory of Perfumery. The name of Jean Maria Farina is synonymous with Cologne bottle; and in the London Exhibition there were four J. M. F.'s, each claiming to be the original. To such an extent is speculation in the name carried in that city of seventy-four distinct smells, that children entitled to the surname of Farina are bargained for as soon as born, and christened Jean Maria; and at times this event is said to be anticipated.

In Philadelphia the manufacture of Soaps and Perfumery is an important and extensive business. The materials employed are palm oil, tallow and lard, cocoa-nut oil, caustic alkali, sal soda, soda ash, and various essences, and essential oils of oranges, lemons, &c. In quality the manufacturers claim, and we think justly, that the Fancy Soaps made in this city are unrivaled. A New York critic remarked upon the Soaps exhibited at the Crystal Palace, in 1853:

"BAZIN of Philadelphia, the successor of the well-known Roussell, makes as fine Soaps as any in the world. Very little inferior to them, if at all, are the Soaps of JULES HAUEL, of the same city. There is no doubt that these productions must eventually succeed in driving the French Soaps out of the market. In the inferior and cheaper class of Soaps, the articles made by Colgate and Hull seem to be most generally in use. The Soaps and Perfumery manufactured by the Messrs. TAYLOR, of Philadelphia, are as fine as any that can be produced. We have already alluded to a Gothic window contributed by them, the panes of which are composed of transparent, or rather translucent Soap, in a great variety of tints. The effect is almost as perfect as that of stained glass."

The manufacture of *Perfumery* is usually carried on conjointly with that of Fancy Soaps. Messrs. GLENN & Co., successors of L. W.

Glenn, are, we believe, the oldest house in this business, having been esblished upward of thirty years ago. APOLLOS W. HARRISON is a well-known exclusively wholesale manufacturer. The following statistics of some of his annual expenditures will illustrate how extensively the business is carried on by certain houses in this city, viz.: for paper boxes, $9,000; for wooden boxes, $3,000; for paper and printing, including lithography, $6,700; for glass, $19,000, &c. Mr. Harrison employs from sixty to eighty hands, has twenty-five traveling agents in various parts of the United States and the Canadas, and his sales of Perfumery and Soap, during the last year, amounted to $140,000. Messrs. R. & G. A. WRIGHT, another popular and well-known firm, have a factory one hundred feet square, and which is believed to be the largest of the kind in France, England, or America. Messrs. A. HAWLEY & Co. make, it is said, a greater variety of *Fruit Essences* for flavoring Mineral Water Syrups, Confectionery, Jellies, &c., than any other house. Their various Extracts for the Handkerchief, Pomades, Toilet Powders, Soaps, Shaving Creams, Tooth Pastes, &c., are very popular, and are noted not less for their durability of the odors than for the exquisiteness of the perfume. Their Pomades, it is claimed, will keep five years entirely unchanged. Mr. Hawley is a Chemist, and devotes great attention to the Chemical processes involved in the manufacture of Perfumery and Toilet Soaps.

The value of these articles annually manufactured in this city, estimating for what is done by other than the regular houses, is about $850,000.

XIV.

Roofing.

The importance of a good Roof cannot well be over-estimated; and in a great city the selection of a material that is *Fire Proof*, as well as Water Proof, seems to be a duty which a builder owes to the public. Shingles, of course, from their combustible nature, if for no other reason, cannot be recommended. Of Metallic Roofs there are a great variety presenting claims to public attention and public confidence. *Slate* is used to some considerable extent, and *Tin* still more extensively, as a material for Roofing. Zinc has not been found well adapted to the climate; but as a protective coating for Sheet Iron, it is, as we stated, extensively used. One manufactory in this city is occupied in coating Iron for Roofing with a preparation, of which the principal ingredient is said to be Indian Rubber. The manufacture of Corrugated Iron for Roofing we have already alluded to. Within a very few years

Composition Roofs have become very popular, and the manufacture of them constitutes an important business. These roofs combine many advantages, and it is to be hoped that experience may ultimately justify the large expectations that have been formed of them.

In 1852 a Composition Roofing, which had been in use for many years in the West, was introduced to this city by Messrs. H. M. WARREN & Co., 228 Walnut street. It is known as Warren's Improved Fire and Water Proof Roofing. This article seems, in a remarkable degree, to have united the suffrages of builders and consumers of every class in its favor; and if we may judge of its merits by the degree of popularity it has rapidly attained, it must combine many points of excellence. The buildings, whether private residences, stores, warehouses, factories, depots, or public buildings, including a part of the United States Mint, which have been covered with it in this and neighboring cities, within a few years, are among the largest and best known; and the names of many of the leading business men are appended to testimonials in its favor. It has been received with equal favor in other States, and in Canada. The materials used in its construction are *Felt*, *Composition*, and *Gravel*. The two former are said to be made of such ingredients as possess elasticity and tenacity, and are combined with the latter so as to form a Roof not only durable, with no liability to crack or decay, but one which is impervious to both fire and water—a combination never before obtained in a Composition Roof. Its fire-proof qualities have been repeatedly subjected to severe *tests*, specially instituted for the purpose, and it seems to have passed the " ordeal by fire" with perfect impunity. The advantages upon which the manufacturers base its claims as an improvement upon all others, are, that, in addition to being *Fire* and *Water proof*, it is also cheap and durable, and requires a less pitch, and consequently less area to be covered, and less masonry upon the walls, while furnishing more facilities for light and ventilation ; besides being more accessible on ordinary and extraordinary occasions than any other. They also claim that it will not expand and contract by heat like Metal Roofs, and will bear more than double as much heat without danger to the boarding beneath ; that it requires only an inclination of one inch to the foot, and may be walked upon or used for drying purposes without injury ; that it is a great advantage to firemen when adjoining buildings are on fire; that it is not injuriously affected by changes of temperature, or by the jarring of machinery ; that it is adapted to every climate, and is easily and quickly repaired ; that Gutters of the same material may be formed on the roof; and finally, the cost of it is only about one half that of tin, and less than that of any other Fire Proof

Roof now in use. If it possess these qualities, of which there seems to be no doubt, Messrs. Warren & Co. well deserve the success which has attended its introduction.

XV.

Straw and Millinery Goods.

Sir: The manufacture of *Straw Goods*, in Philadelphia, was an important and an increasing one, prior to the late commercial revulsion. The product of the factories of White, Custer, Wilcocks, Rogers & Fraley, and other firms, had obtained a deserved celebrity in the South and West, and large orders were filled here from parties who had previously made their purchases further East. The business embraces not only Bonnets for ladies and Flats for girls, but Hats and Caps for men and boys; and includes also the Stiffening, Pressing, and Shaping of Panama and other imported Hats. You inform me you have already noticed the manufacture of Straw Hats, which is the larger part of the business.

The Straw Goods manufacture requires much room for Bleaching, and can be more advantageously carried on in a country village: consequently one of our well-known manufacturers removed to the village of Bridgeport, Montgomery County, where he erected the most commodious and complete factory; and when the business was prosperous employed about two hundred persons, mostly females. The manufacturers in the city are also engaged in making Silk Bonnets, Bonnet Frames, &c., and conduct a Jobbing or Retail business, or both. The Braid for Bonnets is chiefly imported, and known as English, Florence, Italian, Neapolitan, &c. The Bonnets and Hats made here, especially those for children, exhibit excellent, in fact remarkable taste. Some establishments are devoted largely to Bleaching and Pressing Hats and Bonnets, and conforming them to the prevailing styles. The largest of these is, perhaps, that of James Telford, who also makes Silk Bonnets and Bonnet Frames. The annual production of Straw Goods was over $600,000—being $350,000 for Hats, and $250,000 for Bonnets.

Millinery Goods.—The manufacture of *Ruches* alone employs two hundred hands, most of whom are females. Joel Thomas, who has been engaged for fifteen years in making Ruches, was the first to make it an exclusive business; and for many years he chiefly supplied New York and Boston, as well as Philadelphia. By introducing machinery, which is unequaled elsewhere, he has been enabled greatly to increase the production. So perfect is this machinery, that one man can goffer from six hundred to seven hundred dozen in a day; and the establishment can turn out one thousand dozen of finished Ruches per day, besides other Millinery goods. Favorable arrangements for the importation of the raw material enables him to compete successfully with rival makers in other cities. Footings and Edgings are also joined to a large extent, which are known to the trade as Joined Blonds. Ruches are also made by hand by a number of others: and the total annual production exceeds $150,000.

Of Artificial Flowers, the manufacture has declined in importance within the last few years. The preference for French Flowers, in the importation of which a considerable business is done in Philadelphia, has retarded the native production and compelled a number of establishments to abandon the business. There are now not over seven houses in the city that can be called manufacturers of Artificial Flowers. There are others where something of the kind is done, but not to any extent. The largest establishment employs fifty girls, and another forty-five. The number of persons employed is about two hundred, and the annual product is about $85,000.

Artificial Flowers, scarcely distinguishable from the natural flower, of variously tinted paper, for mantels and other ornamental purposes and festive occasions, and also Wax Flowers and Fruits, &c., are made by Mrs. A. M. HOLLINGSWORTH, who keeps a very great variety of *Materials for Flowers* of every kind—paper muslin, silver and waxed leaves, stamens and pips for Flowers, cups for roses, tissue, carmine, blue, glazed, and mottled papers, cut Flowers, sheet wax for Wax Flowers, and materials for Fancy Leather Work and Potchiomania, &c., &c.

Bonnet Wire, which is said to be superior to that imported, or made elsewhere in this country, is made by M. Bird; also by Joseph Moore, and others.

Bonnet Frames are made to an extent that causes us to wonder where the fair heads are which will wear them when covered and trimmed. One firm alone makes 16,000 dozen yearly; and the whole quantity made for sale out of the city is at least 100,000 dozen, requiring of course 1,200,000 female heads to fasten them to. Not only the West and South, but some parts of the East, are supplied with Frames made in Philadelphia.

Silk Bonnets, and Bonnet Frames, are generally made by the same parties, the largest of whom are LINCOLN, WOOD & NICHOLS, J. HILLBORN JONES, I. S. CUSTER, JAMES TELFORD, AARON E. CARPENTER, and T. MORGAN; one of whom made, during the last year, Bonnets and Frames of the value of $50,000; another is making the present year over $40,000. This manufacture is carried on in connection with the jobbing and retail trade, and by some in connection with the manufacture of Straw Goods. Besides the larger manufacturers, our most celebrated Milliners make up Bonnets during the dull seasons, which they supply both to the Philadelphia Jobbers and directly to Western and Southern merchants. The well-known taste and skill of Philadelphia Milliners have obtained for the Bonnets made here a high reputation. It is a fact, not generally known, that many families in the South and West have the Clothing and Boots for the gentlemen, and the Dresses, Bonnets, and Shoes for the ladies, made in this city, which has a deserved reputation for superior articles.

The annual production is as follows:—

Straw Goods, $600,000—deduct Straw Hats already enumerated, $350,000—Bonnets alone, - - - - - -	$250,000
Ruches, and other Millinery Goods, - - - - -	150,000
100,000 dozen Bonnet Frames, - - - - -	110,000
Silk Bonnets, - - - - - - - -	100,000
Artificial Flowers, - - - - - - -	85,000
Total, - - - - - - - - -	$695,000

This is simply the product in the manufactories in this branch. The articles of Millinery made up in this city by individuals is, of course, very large.

XVI.

Surgical and Dental Instruments—And Appliances, etc.

The manufacture of Surgical and Dental Instruments, Trusses, Splints, Bandages, and the various Appliances which constitute the *armamenta* of the Surgeon, demand on the part of the manufacturer who would rise to eminence, other elements of success than mere mechanical skill. The Surgeon has to deal with living tissues of great sensibility; and whether he be called upon to remove diseased or injured parts, to restrain unhealthy or irregular development—to aid the curative power of nature, or to compensate for lost members by artificial contrivances—the judgment, tact, inventive power, and manual dexterity of the Instrument maker, and even some knowledge of the anatomical relations and functions of the parts, are important aids to the Surgeon in carrying out his ideas. The perfection, finish, strength, and reliability of the workmanship, are of the highest importance both to Surgeon and patient, often involving the success of the one, and the comfort, if not the very life of the other. Hence he is compelled to be something more than a mere artisan, to keep pace with the advance of surgical knowledge, ever aiming at the simplicity which characterizes the Surgery of the present day. At the same time, the delicacy of construction of many Instruments, upon which often depends the success of the operator, as in Ophthalmic Surgery for example, requires a nice mechanical hand, and the best of material. In view then of the qualifications required, it is no small compliment to say, as we can say with truth, that the best Cutlers are and always have been Philadelphia houses.

Philadelphia has become noted for its manufacture of Surgical and Dental Instruments, partly by reason of the number of its eminent Colleges of Medicine, which have made it the chief seat of medical learning in this country, and partly by the superior skill of the Instrument maker; most of the improvements in these Instruments, originating in this country, having been made in this city. Nearly the whole of the West and South is supplied by this city, on account of the cheapness and superiority of the manufactures; and greater facilities for cheap and rapid communication with Southern ports, is alone needed to secure a still larger share of the Southern trade. It is everywhere the custom of Druggists to keep more or less of Surgical Instruments on hand, which they procure from the manufacturer at a discount, and sell at his card prices. Some of the Philadelphia houses are thus engaged in supplying an extensive wholesale trade; while others confine themselves more to a retail business, or make to order. To the latter class also, more especially, it belongs in part to prepare for the many Professors, Hospital and practicing Surgeons of the city and vicinity, the various *Splints, Bandages, Trusses* and appliances for surgical injuries and deformities, with their required modifications for special cases. The manufacture of Trusses, Bandages, Spinal, and other apparatus, is conducted as a separate business by a number in this city. There are two who devote themselves exclusively to the manufacture of *Artificial Limbs.* Mr. John F. Ord has been long and favorably known as a maker of these; while B. Frank Palmer's Artificial Limbs have achieved a world-wide reputation. So successful is the imitation of the natural motions of the joints, and so light and elegant the construction, that we have heard eminent Surgeons declare that they would prefer this excellent substitute to some natural legs they endeavor to save. About five hundred Limbs are fitted annually at the manufactory in this city, and at the two branch offices. This ingenious inventor informs us that he has originated a Steel-trap Leg, which a man may go on by day and catch rats with at night; and recommends, as a new branch of Productive Industry, the manufacture of Artificial Back-bones for the benefit of weak politicians and clergymen.

The number of establishments engaged in the manufacture of Surgical and Dental Instruments is at present eight. Mr. HORATIO G. KERN is now, we believe, the most extensive maker of both these classes of instruments, not only in the city but in the United States. He has an Instrument for every tooth in the head, and can furnish Dental Cases at prices varying from $50 to $500. His house has been established for upward of twenty years.

Mr. J. H. GEMRIG is also a celebrated maker of Instruments for Surgical and Dental Professors. The greater part of the Instruments used in the Jefferson College, the University of Pennsylvania, and the Pennsylvania Hospital, were made by him; and the esteem in which his workmanship is held by the Professors in those institutions, and others of distinction, is evidenced by the frequent complimentary allusions to it by many of them in their lectures to their students. His Instruments for operations upon the eye, in particular, are preferred to any of European make, by one whose success in some branches of Ophthalmic practice, is equal to that of any living Surgeon. His Pocket Surgical Cases, are also specially characterized by neatness, compactness, and excellence of finish. All the new Instruments, as they are introduced from Europe, are reproduced by him for the use of our Surgeons, at much less cost than the imported.

The other principal makers are GEORGE SNOWDEN, PUGH MADEIRA, M. KUE-MERLE, D. W. KOLBE, and LOUIS V. HELMOLD. Besides these, Mr. F. LEY-POLDT, 508 E. North street, makes the manufacture of *Scarificators* and *Spring Lancets* a specialty. Both of these Instruments he makes of a variety of patterns, sizes, and materials; but his Patent Scarificator he believes combines several qualities which render it superior to all others. It is one third smaller, yet with the same number of ordinary-sized Lancets, which are also protected by a plate from the effects of verdigris. It is more readily kept clean, and is so simple and durable as to be taken apart and put together again with facility, and consequently is easily repaired. His prices are also very moderate.

The amount of capital employed in the Surgical and Dental Instrument business is $200,000. The number of workmen engaged is two hundred, and the total annual production is set down at $350,000.

A branch of this business, comparatively new but of growing importance, is the manufacture of Gold and Silver Instruments, most of which were until recently imported. There are now three persons, G. P. PILLING, J. S. WARNER, and JOSEPH A. FITZGERALD, who give their chief attention to this, and make such articles of Silver as the Instrument makers in this and the other cities of the Union may require for their cases. One of these, George P. Pilling, employs seven hands principally on this work.

In addition to those already mentioned, there are seven or eight persons who more exclusively make Trusses, Surgical Bandages, Supporters, Splints, and other Instruments for surgical maladies, resulting from natural causes, disease or accident, and requiring mechanical appliances for their relief or cure.

Trusses, are articles especially in demand, and have undergone endless modification testing the ingenuity of many makers. Messrs. HORN & ELLIS are the largest wholesale manufacturers of these. Another branch of the business we refer to with pleasure, inasmuch as the articles, which were formerly wholly imported, are now made in this city of a quality even superior to the foreign. We allude to the Elastic Stockings, Belts. Knee-Caps, Anklets, Armlets, and Suspensory articles for the treatment of varicose enlargement of the veins, dropsical swellings, rheumatism, and the support of weak parts. These are made of Vulcanized Indian Rubber Thread, which first receives a covering of cotton or silk, and is then woven into a porous and elastic fabric, either with silk or cotton of the required size and shape; and when applied to the parts for the purpose named, or as a retaining apparatus over spirits, &c., they exert, without lacing, a gentle and equable pressure, and form a neat and convenient application pervious to air and to the perspiration, and are very durable.

Mr. B. C. EVERETT, of the Philadelphia Surgical Bandage Institute, 14 North Ninth street, has the sale of these for the manufacturer, and keeps a great variety always on hand, as well as of Trusses, Shoulder Braces, Deformity Instruments, and every thing in this line, in which he has had long experience.

Dr. M. McCLENACHAN, 50 North Seventh, in addition to Trusses, and other articles of this class, manufactures *Improved Spinal Apparatus*, and Abdominal Supporters, &c.

J. LEANDER BISHOP, M.D.

XVII.
Tin, Zinc, and Sheet-Iron Ware.

The manufacture of articles from Tin, Zinc, and Sheet Iron, is sufficiently extensive in the aggregate to be called a leading branch, but the subject calls for no particular remark. The latest Business Directory furnishes a list of about one hundred and fifty Tin-workers in Philadelphia; but it is probable there are two hundred places in the city where Tin-ware is made. The oldest establishment in the business is that of ISAAC S. WILLIAMS, on Market street, founded by Samuel Williams and Thomas Passmore, in 1796. This house is exceedingly well provided with facilities for executing heavy orders expeditiously; and has furnished with Culinary Utensils some of the first-class hotels in New York, and the largest steamboats on the Western waters. Mr. Williams is said to be the most extensive manufacturer in this city of Planished-ware, of a superior quality. This ware is made by repeated hammering of the ordinary tin-plate upon highly-polished steel anvils by hammers, also highly polished. This condenses the fibre or grain of the tin, and renders it capable of a high

34

polish, and at the same time improves its quality. Planished-ware is also made by another process, more analogous to rolling or burnishing, which it is said, renders it nearly equal in appearance to the former, and somewhat cheaper. It is however scarcely so durable.

Within a few years a great revolution has been effected in the manufacture of Culinary and Miscellaneous Tin-ware, by the introduction of machinery. By the aid of Dies, Presses, Lathes, and other contrivances, the separate parts, or the whole, according to the degree of complexity of an article, are at once struck up into the required shape, plain or with devices, as may be desired; and the work of the tinman is reduced to the simple act of soldering or uniting the several parts.

There are establishments in this city where the tin-worker may thus purchase, or order, in any desired quantity, in sets, the component parts of nearly every article in his line, ready shaped to his hand. While the use of machinery has thus taken away many of the former characteristics of the trade, it has increased its amount, and greatly extended the uses to which Tin-ware is being adapted. Every day nearly introduces some new article into the market, or some novel form in which the article can be appropriated, through these new processes of manufacture. Tin Toys, of great variety and neatness, are now largely manufactured here in this way.

Tin is also employed as a material for *Roofing*, and when laid in Paint on both sides it is regarded by many as a superior article for the purpose.

The consumption of Tin for Blacking-boxes is a considerable item. One manufacturer of Blacking consumes 1,170,000 sheets of Tin per year. The manufacture of Essence of Coffee, which is already a large and growing business here, calls for a large amount of small Cans for packing. Druggists, Grocers, Spice and Mustard Packers, also require Tin Cans, Canisters, &c., to a considerable amount. "Self-sealing Cans and Jars" are made very extensively by the well-known firm of ARTHUR, BURNHAM & GILROY, who have introduced to the public a number of patented articles of great utility. The popularity of their "Old Dominion Coffee Pot," which is said to be superior to all others for making Coffee, adds to the importance of Tin-working as a pursuit. But the various uses of Tin are too familiar to all to need enumeration.

Japanned Ware is made extensively at two Tinware establishments in the city; and there are also a few persons who conduct this branch separately, some of whom are not excelled for the beauty and excellence of their work.

Zinc is used principally for the lining of Refrigerators, Filters, Bath

Tubs, Cisterns, &c., and for Coating Iron by the Galvanic process to which we have referred.

The working of Sheet Iron into Stoves, Stove Pipe, Coal Scuttles, &c., occupies many persons; like Tin, this material is being constantly put to new uses, and many articles formerly made of Wrought Iron are now made of Sheet Iron. Hoes, and other Garden Tools, and Hinges of various kinds, are of this class.

We have received statements from a number of persons in the business as to the aggregate product, and they range from one million to one million five hundred thousand dollars. We state it at $1,200,000.

For the manufacture of *Playing Cards*, Messrs. SAMUEL HART & Co. have an extensive factory that consumes annually 200,000 lbs. of printing paper, which, if extended in one continuous sheet, it is said would reach 1,060 miles; 1,550 lbs. of colors, 600 gallons of boiled linseed, and other oils; 1,800 lbs. glue; 1,450 lbs. soap; and 200 bbls. of flour. Machinery is used, which performs the work of 300 persons; consequently this firm is able to conduct their large business with but fifty hands, whose aggregate yearly wages are $10,500. They produce 15,000 to 20,000 packs of finished Cards every week.

Messrs. GEORGE J. BURKHARDT & Co. have an extensive factory at Broad and Buttonwood streets, for building *Vats* for Brewers, Distillers, Tanners, Sugar Refiners, &c.; Tanks for Water Stations on Railroads; and Reservoirs for supplying Bath-houses, &c., for Hotels and Public buildings; in fact every description of similar vessels, whether the capacity required be 200 or 20,000 gallons. The manufactory is equipped with all the requisite machinery for turning out such work expeditiously, and steam is used in nearly every department. The material employed by Mr. Burkhardt is *White Cedar*, which he says "experience has taught to be the most durable, and in comparison with any other material is as four to one; or in other words, Cedar vessels will last from thirty to forty years, while Pine, Hemlock, Poplar, or Spruce, will decay in from six to ten years."

Manufacturers at a distance, who cannot otherwise conveniently procure White Cedar, can, by sending the size and number of vessels required, obtain from this firm the material dressed to shape, and ready for setting up.

Besides the many factories already noticed, or alluded to in the various branches of manufactures, there are many others which want of space compels us to pass by with scarcely an allusion—the manufactories of Hubs, Spokes and Felloes, Tatham & Brothers' great Lead Pipe

Factory, Spark's Shot Factory, the Starch Factories, Spain's Churn Factory, and the North American Paper Bag Company's Manufactory; Kochersperger's Steam Laundry, in which $5,000 were expended for Pipes alone to convey heat to the drying room; Daguerreotype Case Manufactories; the manufacturers of Mathematical and Optical Instruments, and numerous others, employed in making the various articles enumerated in the INDEX. A volume would hardly contain all that might be written upon the Miscellaneous Manufactures of Philadelphia. In by-ways and rooms concealed from the public gaze, there is at all times an army of industrious artisans busily engaged in transforming rude materials into objects of utility, or productions of taste and skill—"Inventions for delight, and sight and sound"—and aiming by superior dexterity in their handicraft operations to compensate for the lack of machinery and business facilities.

We now proceed to recapitulate, with some detail, the results of our investigations, with respect to the value of the articles annually manufactured in Philadelphia. They are given as our own conclusions, after laborious and careful examination, based partly on information furnished by manufacturers as to their own business; partly from a mean of estimates of those having some knowledge as to the business of individual manufacturers in the several branches; and partly upon calculations founded upon a knowledge of the number of hands employed in an establishment, and the average production per hand. Errors doubtless there are, but the aggregate generally will be found approximately accurate; certainly far more correct than any Census that ever has been, or probably ever will be taken.

Aggregate Value of Articles produced in Philadelphia, for the year ending June 30th, 1857.

Agricultural Implements, Seeds, &c., (estimated)...............................$500,000
Alcohol, Burning Fluid, and Camphene 1,022,140
Ale, Porter, and Brown Stout 1,020,000
Artificial Flowers 85,000
Awnings, Bags, &c. 91,750
Assaying and Refining Precious Metals, including actual expenses of U. S. Mint, $430,000.......................850,000
Barrels, Casks, Shooks, and Vats......715,000
Beer, Lager and Small..................1,280,000
Blacking, Ink, and Lampblack, (estimated)......................................500,000
Bolts, Nuts, Screws, &c....................411,000
Book and Periodical Publishing, *exclusive of Paper, Printing, Binding*, &c...818,000
Book Binding, Blank Books, and Marble Paper..............................1,230,000
Boots and Shoes............................4,141,000
Boxes, Packing, (estimated)..............500,000
Brass Articles 830,000

Bread, Bakers, (including Crackers,) Ship Bread, &c.........................$5,600,000
Bricks, Common and Pressed............812,000
Britannia and Plated Wares.............380,000
Brooms, Corn and other...................104,000
Brushes..225,000
Candles, Adamantine & Oleine Oils....570,000
Caps...400,000
Cards, Playing 118,000
Carpeting, Ingrain.........................2,592,000
Carpeting, Rag.................................504,000
Carriages and Coaches......................900,000
Cars and Car Wheels........................550,000
Chemicals, Dye-Stuffs, Chrome Colors, and Extracts............ 3,335,000
Clothing......................................9,640,000
Coffins, Ready-made.......................219,000
Combs...150,000
Confectionery, &c..........................1,020,000
Copper Work...................................400,000
Cordials, Bay Water, &c..................200,000
Cotton and Woolen Goods, exclusive of Hosiery, Carpetings, &c.........14,813,968

Cordage, Twines, &c....................$810,000
Cutlery, Skates, &c.........................150,000
Daguerreotypes, Cases, and Materials, (estimated).............................600,000
Edge Tools, Hammers, &c...............127,000
Earthenware, Fire-Bricks, &c 647,000
Engines, Locomotive, Stationary and Fire....................................3,428,000
Engraving and Lithography..............570,000
Envelopes and Fancy Stationery.......150,000
Flooring and Planed Lumber............370,000
Flour....................................3,200,000
Fertilizers............................503,000
Fringes, Tassels, and Narrow Textile Fabrics...............................1,288,000
Furniture, (estimated).................2,500,000
Furs......................................350,000
Gloves, Buckskin and Kid................150,000
Glue, Curled Hair, &c...................775,000
Gold Leaf and Foil......................325,000
Glassware...............................1,600,000
Hardware, and Iron Manufactures not otherwise enumerated............1,169,000
Hats, Silk and Soft.....................800,000
Hose, Belting, &c.......................175,000
Hosiery.................................1,808,150
Hollow-ware, exclus'e of Stoves,&c..1,250,000
Iron, Bar, Sheet, and Forged.........1,517,650
Jewelry, and Manufactures of Gold..1,275,000
Lamps, Chandeliers, and Gas Fixtures..................................1,300,000
Lasts and Boot Trees.....................36,000
Lead Pipe, Sheet Lead, Shot, &c.......235,000
Leather, exclusive of Morocco.........1,610,000
Machinery..............................1,912,000
Machine Tools...........................350,000
Mahogany and Sawed Lumber..........580,000
Maps and Charts.........................400,000
Marble Work.............................860,000
Mantillas and Corsets...................330,000
Matches, Friction.......................125,000
Medicines, Patent and Prepared Remedies................................1,300,000
Millinery Goods, including Bonnet Frames, Wire, &c., but excluding Straw Goods & Artificial Flowers...360,000
Mouldings, &c...........................300,000
Morocco and Fancy Leather............1,156,250
Musical Instruments.....................485,000
Mineral Waters..........................350,000
Newspapers, Daily and Weekly, (estimated,..............................1,370,000
Oil Cloths..............................289,000
Oils, Linseed, Lard and Tallow, Rosin, and R. R. Greases..............2,131,230
Paints, Zinc, and Products of Paint Mills..................................770,000
Paper...................................1,250,000
Paper Hangings..........................800,000
Paper Boxes.............................175,000
Patterns, Stove and Machinery..........115,000
Perfumery and Fancy Soaps............850,000
Picture and Looking-Glass Frames, (estimated)...........................750,000
Preserved Fruits, &c., (estimated).....350,000
Printing, Book and Job................1,183,000
Printing Inks...........................160,000

Provisions — Cured Meats, Packed Beef, &c..........................$4,000,000
Rifles and Pistols......................120,000
Saddles, Harness, &c..................1,500,000
Safes...................................150,000
Sails...................................135,000
Sash, Blinds, Doors, &c.................250,000
Saws....................................510,000
Scales and Balances.....................145,000
Shirts, Collars, Bosoms, and Gentlemen's Furnishing Goods.............1,187,500
Shovels, Spades, Hoes, &c...............397,000
Show Cases...............................55,000
Sewing Silks............................312,000
Silver-ware.............................450,000
Soap and Candles, exclusive of Adamantine Candles.....................1,487,600
Springs, Rail-road and Coach............238,000
Spices, Condiments, Essence of Coffee, &c., &c..........................350,000
Starch..................................155,000
Steel, Spring and Cast..................283,500
Stoves and Grates.....................1,250,000
Sand-stone, Granite, Slate, &c..........300,000
Straw Goods, including Hats.............600,000
Surgical and Dental Instruments, Trusses, and Artificial Limbs.........350,000
Sugar, Refined, and Molasses........6,500,000
Teeth, Porcelain........................500,000
Tin, Zinc, and Sheet-Iron Ware......1,200,000
Tobacco Manufactures, Cigars, Snuff, &c.....................................3,256,500
Trunks and Portmanteaus.................313,000
Turnings in Wood........................550,000
Type and Stereotype.....................650,000
Umbrellas and Parasols, including Umbrella Furniture, Ivory & Bone Turning, Whalebone Cutting........1,750,000
Upholstery, (estimated).................500,000
Varnishes...............................230,000
Vessels, Masts and Spars, Blocks and Pumps, &c............................1,760,000
Vinegar and Cider.......................300,000
Wagons, Carts, and Drays................815,000
Watch Cases.............................942,000
Whips...................................175,000
Whisky, Distilled.......................630,000
"   Rectified.....................2,524,500
White Lead..............................960,000
Willow-ware, Baskets, &c., (estm'd)..120,000
Wire-work, (estimated,).................250,000
Wooden and Cedar-ware..................150,000
Works in Wood not otherwise enumerated.............................100,000
Miscellaneous Articles, not otherwise enumerated. For particulars see INDEX, (estimated)...............3,000,000

Total Annual Product of Manufacturing Industry in Philadelphia.............................145,348,738
Add for Leading Branches in the vicinity of Philadelphia, as before given....................26,500,000

Total for Philadelphia and vicinity...............................$171,848,738

According to the Census of 1850, the average productive power of each person employed in Manufactures in Philadelphia, was about $1,100 per annum, a rate confirmed by our own investigations; and

34*

the capital invested was about one half the aggregate of production Assuming that these relative proportions were correct, though the aggregate amounts were manifestly erroneous, and assuming they are applicable now, the respective items would stand as follows : *Capital invested in Manufactures in Philadelphia*, **$72,500,000**; *Hands employed*, **132,000**; *Product*, **$145,348,738.**

In view of this result—a result as unexpected by the Author as it probably will be surprising to the reader—a result perhaps understated but not overstated, and of which the constituents are given with sufficient particularity to enable any one of ordinary intelligence to test its accuracy by personal investigation, with the aid of a complete Business Directory: in view of this result then, we ask, do not the facts demonstrate the original proposition and assertion, that *Philadelphia is already a great manufacturing city—most probably the greatest in the Union?* The value of the *mechanical and manufacturing* industry of the entire State of Massachusetts, in 1855, including Gas, an important item, but which we have not reckoned, was about two hundred and forty millions of dollars; we therefore may confidently say, that no other single city—not even Boston, including Lowell, and one half the State of Massachusetts,—sums up so large a production of indispensable goods as are annually produced in the city of Philadelphia. We may say, moreover, that we are convinced, as the result of extended inquiry and many opportunities for comparative examination, that the goods made in Philadelphia are generally superior to the average quality of American fabrics. One reason for this superiority is, that the operations are mostly conducted in small factories, under the direct personal supervision of the owner, or in shops often illy provided with machinery for rapid production; and consequently, the fabricator must give close attention to the selection of material and character of the workmanship, and master competition by the durability and intrinsic excellence of his fabrics. We hold it to be eminently safe for any consumer or merchant to infer that Philadelphia-made goods, at the same price, are invariably the cheapest. Many other considerations are suggested by the facts which we have collected, and partially submitted, and to which we would gladly invite attention, did space and circumstances admit. We would especially entreat the merchants of this city to co-operate with the manufacturers, as they have already commenced to do by the reorganization of the Board of Trade; and by the aid of their patronage and influence, give hope, and sustenance, and vigor to the individual producers, who are doing so much to advance the industrial reputation and development of Philadelphia, though with comparatively little profit to themselves—and who, if suitably encouraged, would render Philadelphia, industrially, the Paris of America. Secondly, we would appeal to the common sense of Southern and Western merchants, whether the cheapest market is not necessarily that which possesses and combines extensive production of its own, widely-extended commercial relations with other manufacturing centres, and superior facilities for cheap transportation. Also, we would be glad to assure the foreign artisan, trained to produce habitually "Olympian-like miracles of Art," of a warm welcome on the part of the intelligent citizens of this metropolis; and especially would we desire to invite the ingenious men of New England to turn their attention hitherward, where the raw materials are cheap and abundant, and where the opportunities for achieving grand results by the introduction of improved facilities, and superior, spirited, energetic, business tactics, are illimitable. Moreover, we would appeal to the Legislators of Pennsylvania to regard the inhabitants of this metropolis no longer as drones, and ulcers upon the body politic; but as coworkers with the agriculturalists in subjugating material forces for useful purposes—to look upon the city as an ornament to the State, and to remove all disabilities now in the way of any form of business organization which experience has demonstrated tends to encourage capital to co-operate with industry; and especially to consider the expediency of sanctioning and facilitating corporate investment in industrial enterprises with limited individual liability. We would assure capitalists, that there are opportunities for the safe and profitable investment of many millions of dollars in productive industry; and that the erection of a few model mammoth manufactories, and a liberal expenditure in order to invite the attention of the world to the manufacturing advantages of the city, would repay richly by accelerating the development of enterprise, and promoting trade and commerce. We would entreat manufacturers in all places throughout this country to banish jealousies, and co-operate with each other, remembering that the demand for manufactured commodities of an immense and daily increasing population, in the Western as well as the Atlantic States, and the opening markets in the Canadas, in the South American Republics and elsewhere, cannot fail to be greater than the industry of all can supply. We would earnestly invoke the citizens of Pennsylvania, and of Philadelphia in particular, to send Representatives to their State and National Legislatures who will truly represent their material interests and intellectual progress; but all these considerations, invitations, supplications, and invocations, we must leave to the Press, the guardians of the city interests, and others far abler than ourselves. We conclude by adopting the graphic language of the Secretary (now President) of the Corn Association, whose prophetic vision saw what we hope we have demonstrated:—" Our Steam Engines are plying their iron arms in every street. In every by-way is heard the sound of the shuttle and the clink of the hammer, as the artisan contributes his mite to the vast sum of toil; whilst many a stately edifice, with its hundreds of employees and clanging machinery, sends forth a stirring music to quicken the pulse of our city life. Why then shall we not spread beyond our borders the knowledge, that in this busy hive is being made almost every article that can contribute to the wants or luxury of man? *This is the great Mart of American Manufactures, unequaled on this Continent in the extent and variety of its products. As such, let it be proclaimed!"*

INTERIOR OF FOUNDRY.—A. JENKS & SON'S MACHINE WORKS.

APPENDIX.

REMARKABLE MANUFACTURING ESTABLISHMENTS IN PHILADELPHIA.

I.

Alfred Jenks & Son's Machine Works, Bridesburg.

THESE works, of which we have already given the incidents connected with their establishment and early history, as well as a cut of their exterior, are located at Bridesburg, a flourishing town now constituting a part of the city of Philadelphia. They are built in the form of a hollow square, cover an area of 160,000 square feet, and consist of a Foundry 130 by 50 feet (for interior view see opposite page), in which about thirty men are constantly employed; a Blacksmith Shop, 120 by 50 feet, having eighteen forges and four trip-hammers, for making, in addition to other things, Bolts—of which seven hundred are made and used in the machinery work daily; a building, 190 by 32 feet, containing an apartment used as a Brass Foundry, and also for "cleaning" the castings after they have been subjected to the process of "pickling," and a well-adapted room for storing patterns when not in use. The Machine Shop is a building 225 by 38 feet, upon the first floor of which Cards are built, and Mill Gearing, Shafting, &c., constructed; upon the second floor, Spinning and Drawing Frames, and Speeders are made; and the third floor is devoted to making Looms. A capacious Elevator is employed for raising and lowering castings and other objects between the different stories of the machine shop; and a Railway connects this and the foundry. The Carpenter Shop is a building 168 by 30 feet, three stories high; and each of the various rooms and departments is supplied with tools and machinery of the most perfect construction, peculiarly adapted to the purposes for which they are designed. In the Wood-working Room are two of Daniel's Planing Machines, one of Woodworth's, and Moulding and Sawing Machines capable of facilitating and making more perfect the wood-work required for the Carding Engines, Looms, &c. All the wood used is kept for the space of two years before being shaped by the machinery, so as to properly season it; and after

(427)

it has been thus seasoned and brought to the form desired, it is placed in a commodious Drying-house, which has been recently erected, entirely fire-proof, and always kept by the heat of steam at a temperature of seventy-five degrees, for the purpose of being still more thoroughly seasoned. The Tools and Machinery for performing the work in the several shops are mostly made by their own workmen ; among which may be classed several Drills of new and improved construction, Boring Mills, and other self-acting machines, of the most beautiful design and perfect workmanship.

Since the decease of Mr. Alfred Jenks, and, in fact, for twelve years previous to his decease, the entire business has been conducted by Mr. BARTON H. JENKS, who was thoroughly educated and fitted by his father for the important and responsible trust he now fills, for which he has peculiar qualifications by his genius for invention, his skill in mechanism, and an administrative capacity evidenced in all departments of the establishment. The family has been distinguished in the fabrication of iron for nearly two centuries, Mr. Jenks being a lineal descendant of the Hon. Joseph Jenks, Governor of Rhode Island, who, a forgeman by profession, built, in the seventeenth century, a forge which was destroyed during King Philip's war. Since early boyhood, Mr. Barton H. Jenks has been engaged in experimenting, with a view to improve various portions of machinery employed in Woolen and Cotton manufacture, on which the skill and genius of his father had been impressed ; and there is now scarcely a portion of machinery in the manifold variety of parts employed in the respective stages of Cotton and Woolen manufacture, from the machines which first operate on the cotton and wool through the series which they have to successively pass in their progress to completion for the market, that has not been improved by the members of this firm, both in the superior finish and more substantial and convenient construction of the respective parts, and in the novel arrangement and combination of parts, or additions involving valuable inventions, which render the operations more perfect, and, in many cases, the production of heretofore unaccomplished, new, and beneficial results. The *Drawing Frames* and *Ring Spinning Frames* or *Throstles* of the Messrs. Jenks, for performing the respective operations in cotton manufacture which their names indicate, are peculiarly of the class spoken of ; for, while the workmanship and mechanical skill are of the highest degree of excellence, they embrace important features of invention, which enable them to produce better work than those ordinarily used. The latter, particularly, which is known favorably among cotton manufacturers as Alfred Jenks' Ring Spindle or Ring Frame, has, in a great number of cases, taken the place of the live and dead spindles, on account of its

CARD ROOM.—ALFRED JENKS & SON'S MACHINE WORKS.

superiority in the quality and quantity of yarn it produces. The spindle of this improved frame has no fly, and has a small steel ring, called a traveler, about a quarter of an inch in diameter, with a slit for the insertion of the thread, which is wound by the ring traveling around the bobbin, being held in its horizontal plane, during its circuit, by an iron ring loosely embraced by its lower end and fastened upon the traversing rail; being sufficiently large to allow the head of the bobbin, as well as the traveler, to pass through without touching. This plan of spindle may be driven 8,000 revolutions per minute with perfect security when spinning coarse yarn ; and when producing the finer numbers, 10,000 revolutions per minute is not an extraordinary speed for it to attain ; the yarn produced in either case being superior in strength and character to the yarn produced by the other throstles at a greatly reduced speed. In the construction and arrangement of the several parts of the power-loom, whether in the simplest form for weaving plain goods, or the enlarged and complex state produced by this firm, for weaving the most beautiful and elaborate patterns of fancy cassimeres, the characteristics above mentioned are manifest in a remarkable degree ; and hence the vast number of the Messrs. Jenks' Looms in operation, and the continued demand for them over all parts of the country, where the benefits derived from such advantages as the improved shuttle-box movement for changing the picks of weft, shuttle-stopper, parallel pick-motion, and a vast number of improved attachments embodied in the Looms of their make, are experienced and appreciated by manufacturers. In the *Keystone Loom*, made by this firm, by using a different shuttle from the ordinary one, silk goods may be woven with as much facility as cotton or wool ; and a Jacquard motion, if desired, may be easily attached. It is forty inches wide, and has four shuttle-drop boxes at one end of the lay, and an improved pattern-wheel for controling the boxes, which will run 1,200 picks before it ends, and can be extended to a greater capacity. Within the last few years, Mr. Jenks has brought out a number of very important machines, to two or three of which we invite particular attention.

1. THE SELF-STRIPPING COTTON CARD.

This is an important addition to the usual cotton-carding machines. It is applied just below the "licker-in," in contact with the main cylinder, and driven by a stripper-head, at the end of the cards, at a variable speed, so as to take both the dirt and the uncarded cotton off the cylinder, and then to deliver the cotton back again to the cylinder, whilst the dirt falls into a box which is placed back of the "licker-in" and feed-rollers for the purpose ; consequently, it keeps both the cylinder and the

35

cotton perfectly clean, (which are certainly two very desirable results) and but seldom requires hand-stripping. An arrangement of rollers is placed at the doffer to deliver the cotton, instead of the old mode of a comb, thereby gaining a greater speed in the card, and not injuring the carding as the comb does. This Card will produce 150 pounds per day.

2. THE IMPROVED COTTON GIN.

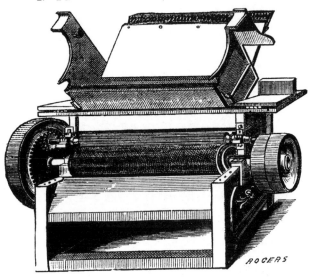

This ingeniously improved and valuable machine for ginning cotton is what is called a "*Cylinder* Gin." The cylinder is constructed in a very peculiar manner, and operates in combination with a stationary straight-edge and a spirally-grooved roller called "the agitator." The straight edge is fixed in a position parallel with, and tangential to the cylinder, with its thinner edge almost in contact with the upper side of the same, and the agitator so as to rotate rapidly at a short distance above and parallel with the straight-edge. The periphery of the cylinder consists of numerous steel-wire teeth imbedded in Babbitt metal, in positions inclined in the direction of the cylinder's motion, so that after the cylinder is "ground" or finished, each tooth presents a separate, sharp, and smooth point, tangential to the cylinder surface. When in operation, the cotton and seeds are together carried by the cylinder against the straight-edge, where they are rolled over and over by the agitator, until the teeth of the cylinder have stripped off the fibre, when the seeds immediately drop down, through a grating, into a receiving

JENKS' COTTON SPREADER.

box. The fibre is at the same time being continually removed from the cylinder in the usual manner, by a rotating brush behind the straight-edge. The teeth of the cylinder are made of the *finest steel* needle- wire, rolled into a double razor-edge section, and secured obliquely around the cylinder, with their sharper edges in directions transverse to the axis of the same; consequently, after the cylinder is " ground off "' in finishing, it presents a serrulated surface, or a surface studded over with innumerable sharp and smooth tangential teeth, admirably adapted both for entering and leaving the fibres.

It will gin any cotton, *however trashy it may be*, and take nothing through but the lint; it neither cuts nor naps the fibres in the least, leaving them nearly as long as when separated by hand; whilst it will clean as great a quantity in the same time as any other gin occupying the same extent of space, and run as easy. It will also last as long, if not longer, than the saw-gin, and cost no more for repairs.

3. JENKS' COTTON SPREADER.

This machine, a cut of which is given on the opposite page, is well worthy of attentive examination by those who are interested in its mer-its, for it undoubtedly possesses superior excellence. It is made entirely of metal, thus insuring greater steadiness and durability. The beater, shafts, blades, and feed-rollers, are made of cast-steel; the shafts which drive the feed are braced together in such a manner that the teeth in the diagonal shaft cannot break; and, by an ingenious application of the elastic principle of air, the machine is constructed to make the lap of uniform thickness, and of such compactness that any portion of it will sustain its own weight; and, moreover, the whole machine works without producing any dust in the room. (SEE CUT.)

To describe all the novel and important machines that have been constructed at these Works would require a volume; and even a recital of the improvements made by this firm in Cotton and Woolen Machinery and patented in this and other countries, would transcend our limits. Such a recital would include machines for making spools, for rifling musket barrels, for cleansing, laying, and preparing wool, and for an infinite variety of purposes. No pains or expense is spared by Mr. Jenks to bring to perfection, and develope into practical use, whatever is likely to be useful and beneficial to his patrons; and we have been repeatedly assured, by manufacturers of cotton and woolen goods, that no firm in the entire Union is more zealous in regarding the interests of their customers, more liberal in making experiments, or more entitled to general confidence and patronage.

Messrs. Alfred Jenks & Son employ in their works, at Bridesburg, when in full operation, about four hundred men, many of whom have been with them over thirty years—a fact which speaks volumes for both employer and employee. They have prepared a Catalogue of their machinery; and also separate drawings, of a size convenient for enclosing in letters, which they forward to persons desirous of dealing with them. Each machine is numbered, and accompanied by a full description—a convenience of much advantage to manufacturers, and duly appreciated by them.

II.

The Port Richmond Iron Works.—I. P. Morris & Co., Proprietors.

In the article on "Iron and its Manufactures," we demonstrated the ability of Philadelphia to do the heaviest Engineering work, by reference to some of the extraordinary machines which, from time to time, have been constructed at the various leading establishments in the city; among others, at the Port Richmond Iron Works. These Works, which rank among the largest and best equipped of the kind in the entire Union, were founded in 1828, by Levi Morris & Co., at the corner of Market and Schuylkill Seventh streets. The first engine constructed was a Vertical Lever-beam Engine, 10 inch cylinder, $2\frac{1}{2}$ feet stroke, built for John Barclay, Vine-street Wharf, Delaware, for a flour mill, and is still in existence at Wainwright's saw-mill, Kensington. At that time, there was not in the Works a single slide-lathe or power drill-press, and planing-machines were not known; the only representative of this tool, it is believed, was to be found at the Allaire Works in New York, built for fluting rollers. The original of the present planing-machine was imported from England, and purchased by the West Point Foundry Association, for their Works in West street, New York. Patterns for a similar machine were made after this model, and several sets of castings made. One of these, purchased for the Richmond Works, was fitted up here, and started about 1836. In the Foundry department the operations were also conducted with very imperfect and inefficient machinery compared with that now in use. Anthracite coal, which was introduced here about 1820, was by no means exclusively used for melting Iron. The Blowing Machinery was of a very primitive character; with unwieldy wooden bellows and open tuyeres. The best product was not more than 2,000 to 3,000 pounds of Iron in an hour, and in the course of the heat an average much below this. With the present improved Blowing Machinery, and improved furnaces, eight tons have been melted in forty-six minutes, with a consumption of coal of one pound to eight pounds of Iron melted.

The present location of these works, to which they were removed in 1846, is on the Delaware River, adjoining the Reading Rail-road Coal Wharves on the south. The buildings, which are of brick, occupy a lot having a front on the Delaware River of 145 feet, a front on Richmond street or Point Road, of 260 feet, and an entire depth or length, from the Richmond side to the end of wharf, of 1,050 feet. The remarkable feature in this establishment is the extraordinary size of the tools in use, and the perfection of the machines employed in the various shops. In the *Foundry* there are three Cupola Furnaces, the largest of which will melt twelve tons of Iron per hour ; and a large-size Air Furnace of the best description. In the *Machine Shop* there is a Planing Machine capable of planing castings 8 feet wide, 6 feet high, and 32 feet long ; a Lathe that will swing 6 feet clear, and turn a length of 34 feet ; and a Boring Mill, possessing also the qualities of a horizontal lathe, which will bore out a cylinder 16 feet in diameter and 18 feet long. This is believed to be the largest in America or Europe. In their *Boiler Shop* they have one large Riveting Machine, and facilities for making boilers or plate-iron work, of every description that may be desired. But a few years ago, Steam Boilers, made of plate-iron, were riveted exclusively with hand-hammers ; and when the City Water-works were located at Centre Square, the steam boilers were built of wood with cast-iron furnaces. At the present time, in this, as in the best shops, circular boilers are riveted in a machine, by pressure produced by a cam operating upon a sliding mandril. In their *Smithery*, they have a Nasmyth Steam Hammer, for heavy forgings ; a Tilt Hammer, for light work ; and throughout the establishment, the minor tools, consisting of Lathes, Boring Mills, Slotting and Shaping Machines, Planing Machines, Horizontal and Vertical Drills, &c., &c., are all of the best description, and combine the latest improvements.

Besides the superiority of its machinery, this establishment has been peculiarly fortunate in its mechanical engineers. This position is now filled by one of the partners, Mr. Lewis Taws, who has been connected with the establishment since 1834, and whose apprenticeship was passed with Rush & Muhlenburg, the two sons-in-law and successors of Oliver Evans, in his establishment, then located at the corner of Vine and Ninth streets. After his apprenticeship, he obtained employment at the West Point Foundry Association, in New York, where he added to his stock of information from the practice at those works, which at that time were under the able management of Adam Hall, a Scotch engineer of much eminence. By the proprietors he was sent to the West Indies, to erect sugar-mills, and remained during the grinding season ; thus obtaining a practical knowledge of this branch of the business. Subse-

35*

quently, he was selected by the same Association to erect, in North Carolina, the celebrated Gang Saw-mills, consisting of twelve to twenty-four saws, driven by direct connection with a steam-engine, running at a speed of 120 to 140 strokes per minute. Since that time, the firm with which he is now connected have built a great number of similar machines, with such improvements as have been suggested by many years experience. Flooring boards of yellow pine may almost be deemed an indigenous production of North Carolina; and these mills, which are but little known except at the South, where such timber grows, are peculiarly adapted for their production. A log of yellow pine by this arrangement can be converted into flooring boards by once passing through the mill. With the practical experience obtained by first constructing and then working the machines, Mr. Taws was eminently fitted to enter upon the more enlarged field that opened to him in taking the management, as chief mechanical director, of an establishment for the construction of machinery of every description, and for every known purpose; and with what success he has filled the position, the large amount of steam-engines and machinery of every description, constructed under his supervision, and scattered broadcast throughout this country, Cuba, and Porto Rico, and rating wherever placed as second to none, is direct, ample, and satisfactory evidence.

The firm of I. P. Morris & Co. is now composed of Isaac P. Morris, John J. Thompson, and Lewis Taws. Their list of manufactures embraces Land and Marine Steam Engines, of all sizes and descriptions; Blowing Machinery, Hoisting and Pumping Engines, Rolling Mill Work, Sugar Mills and Sugar Apparatus; in fact, all kinds of heavy machinery except Locomotives. For an enumeration of some of the machines constructed by them, see pages 289 and 326.

III.

The Southwark Foundry.—Merrick & Sons, Proprietors.

The Southwark Foundry, of which Merrick & Sons are proprietors, is another of the remarkable Iron establishments of Philadelphia, located in the Second Ward, and occupies the entire square bounded by Washington and Federal streets, and Fourth and Fifth streets.

It was started in 1836, as a Foundry (for castings) only, but was soon enlarged; and now, by various accessions and improvements, in buildings and tools, has become a first-class establishment for the manufacture of all kinds of heavy machinery.

Its buildings and yards occupy the following space:—

Iron Foundry, 115 by 107 feet, " " 95 " 53 " } area of ground floor,	17,540	sq. ft
Brass " 45 " 25 " " " "	1,125	"
Smith Shop, 165 " 40 " " " "	6,600	"
Machine " 162 " 40 " 2 stories, } " "	6,480	"
Pattern " " " " " 1 story, }		
Boiler " 150 " 64 " } area of ground floor,	12,800	"
Gasom'r " 50 " 64 " }		
Erecting " 90 " 25 " } " " "	5,550	"
" " 60 " 55 " }		
Carpen'r " 45 " 35 " } " " "	2,855	"
" " 40 " 32 " }		
Sheds, for Storage, &c., &c., " " "	10,700	"
Total area occupied by buildings, - - -	63,650	"
" " of yard room, - - - -	80,550	"
Entire space occupied by the establishment, -	144,200	"

In addition, it has a tract of land on the Delaware River, about 400 feet front and 1,100 feet deep, affording ample space for extensive iron boat yards; and on this tract there is a fine pier, 60 feet wide and 250 feet long, with a very powerful shears at the end, capable of lifting fifty tons.

A brief description of some of the objects of interest in this establishment will show that the arrangements, tools, and appliances in use, are on a scale proportionate to the capaciousness of the buildings.

The foundry has two Cranes, capable of lifting fifty tons each, and three others of thirty tons lifting power, by which any object may be transferred from one extremity to the other, or to any point on the floor. Two 50-inch Cupolas are used for melting the Iron, and are supplied by a pair of Blast Cylinders, 40 inches in diameter, and 3-feet stroke. Twenty-five tons of metal can be melted in three hours. The Ovens for drying the Cores are of immense size and capacity.

In the Smith Shop the blast is obtained by an Alden Fan. There are two Nasmyth Steam Hammers, one of 10 cwt., and one of 5 cwt. weight of ram. There are also in this shop, Bolt and Rivet Machines, for the manufacture of these articles, large numbers of which are annually used. The Brass Foundry has a Cupola and four Crucible Furnaces.

The lower Machine Shop has a Boring Mill which will bore a cylinder 11 feet in diameter, and 14 feet high; a Planing Machine, believed to be the largest in the world, capable of planing 8 feet wide, 15 feet deep, and 30 feet long—besides other Lathes and Planers, of various dimensions and power; two Slotters, Drill Presses, &c., &c. The upper Machine Shop is well stocked with smaller Lathes, Planers, Shaping and Drilling Machines, Vices, &c. The Boiler Shop is provided with a Riveting Machine capable of riveting a boiler 40 feet long, and of any diameter; with a Treble Punching Machine of immense strength; with heavy and light Shears and Punches; an Air Furnace, for heating large plates; Rolls, for bending; Cranes, &c. The largest Erecting Shed, used for putting up sugar apparatus, has a traveling crane extending its whole length. The business of making Sugar Apparatus forms a large item in the productions of this establishment; and for a list of some of the extraordinary machines that have been constructed here, see page 327. Ordinarily, from 350 to 500 hands receive constant employment at these works.

IV.

Sharps' Rifle Factory.

About six years ago, the attention of sporting and military men was invited to a new Breech-loading and Self-priming Rifle, which had been patented by a Mr. C. Sharps; and, after the most careful examination of its construction, in comparison with others, it was found to stand the tests of a first-class weapon; being safe and certain in firing, easily and rapidly loaded, simple in its construction, and constantly kept clean by its own operation. For sporting purposes, this Rifle soon became a favorite weapon; in Kansas its report was heard; the Ordnance Department at Washington expressed their admiration of the improvement; and subsequently the British Government ordered six thousand of these Rifles, for the use of their army in India. More recently, Mr. Sharps applied the principle which distinguishes his Rifles to the construction of a new Pistol or Carbine, especially designed for the use of Mounted Dragoons. The advantages claimed by the patentee for the new Pistol are numerous; among others, that it is more compact, lighter, has a more extensive range, and fires with greater accuracy than any Pistol now in use. It is single-barreled, but owing to the ease with which it can be loaded, it is capable of being fired twice as often as any revolver in a given period of time. The Pistol weighs about two and a half pounds; the barrels are six and eight inches long, and throw a half ounce ball effectively one-fourth of a mile. It primes itself for twenty rounds. There are about 1,500 Pistols now being constructed in the factory. It was recently tested, in competition with various other fire-arms, at West Point, by a board of officers appointed by the United States Ordnance Bureau, and struck a target six feet square, at a distance of six hundred yards, twenty out of thirty shots. The same Pistol was fired seventy times in seven minutes, priming it three times, every ball striking a target three feet square, at a distance of forty-five feet, with a force sufficient to penetrate eight inches of pine board. Certificates from officers in the army testify to the high estimation in which it is held by the troops that have tried it.

During the last year, Mr. Sharps, in association with Nathan 'H. Bolles and Ira B. Eddy, under the firm-style of C. SHARPS & Co., erected (for the manufacture of his fire-arms) a very extensive establishment at the West end of the Wire Bridge, near Fairmount. The building is of brick, 140 feet long by 40 feet broad, and is surmounted by a cupola, from which an admirable view of the city and surrounding country can be obtained. The machinery is of the most beautiful and accurate description; the entire cost for the buildings and machinery being about $130,000. The basement is used for the forging of the iron material of the Rifles. In the rear of the first story is placed a high-pressure stationary engine of seventy-five horse power, which forms the motive power of the establishment. The second story is used for the boring of rifle barrels, which are drilled from solid cylinders of cast-steel. The third story is the tool manufactory, where the cutting, milling, and finishing apparatus is constructed. The fourth story is the finishing shop, where the rude materials are adjusted, and from which the article issues complete. The firm are engaged at present principally in the construction of Breech-loading and Self-priming Pistols, though a large number of Rifles are rapidly approaching completion.

All the materials used in the manufacture of these arms are made in the building; even down to the screws which fasten the completed article together. The firm possess the facilities for turning out 1,000 Rifles per month. This is a most interesting and valuable addition to the manufacturing establishments of Philadelphia.

V.

Archer, Warner, Miskey & Co.'s Manufactory of Gas Fixtures.

It will be remembered by the attentive reader that the author of the Report on the Industry of the United States, whose remarks respecting the manufacture of Lamps, Chandeliers, and Gas Fixtures in Philadelphia we quoted at length, stated that the establishment mentioned in the caption of this article, was in all respects, except in extent, similar to that of Cornelius & Baker, which has been described. It is not necessary, therefore, to repeat the illustrations given in that quotation tending to show the superiority of Philadelphia manufactures in this branch. Our remarks shall be limited to a brief notice of the founding of the establishment, and of the mode of manufacturing Gas Fixtures, as conducted in it.

In 1842, Mr. Ellis S. Archer, the present senior partner in the firm of Archer, Warner, Miskey & Co., invented and patented a Lard Lamp; but could not induce any one to manufacture and introduce to the public an article, for which it was supposed there would be no demand. Undaunted by these repulses, he obtained a cellar, in one corner of which he employed a few men in making his Lamp. The value of the new invention was recognized, and the public adopted it, and those manufacturers who could see no merit in the adaptation of Lard as an illuminating material, turned their attention to it. The establishment was speedily enlarged: other branches of the business were added, and in 1848 Mr. Archer became associated in partnership with Mr. Redwood F. Warner. Determining to place their house on a level at least with the best in the United States, they saw that this could be effected by merit alone; hence, their first effort was to present in their department of Art, novelty of design combined with superiority of finish and excellence of materials. Sensibly foreseeing that the growing taste in this country required to be fed, they obtained designers and modelers of the highest talents, to whom they paid liberal salaries, and encouraged them in every way to produce graceful and effective designs, for Lamps, Chandeliers, and Gas-fittings. No amount of money was considered by them extravagant, if it secured a valuable result. The consequence of this judiciously liberal expenditure soon became manifest. From an ordinary firm, with a limited capital, doing a moderate business, they sprang to a strong position among the first houses, in their trade, in the United States. Their work is admitted by all to be equal to that of any competitor, and their manufactory, situated in Race above Fourth street, takes rank among the important ones of this city.

A systematic description, if we were to attempt one, of the mode of producing those various artistic articles for which this house is celebrated, would lead us first to the Modeling Department. In this room every idea of form and construction originates. Artists of approved merit are employed in modeling designs in wax. The form is at first rudely shaped by the hands, and then elaborated by instruments of hard wood. In the modeling room, perfect accuracy is required in the imitation of all natural forms; and for this purpose an animal menagerie is frequently assembled. A live deer, from Logan Square; an eagle, several pigeons, snakes, children, &c., were all in use as studies in modeling the Stair Railing for the private stair-cases of the Senate Chamber and House of Representatives at Washington.[*]

The model having been formed, is sent to the caster, who makes a mould of it in brass. And here may be seen the most curious method of casting yet discovered. It is, we believe, peculiar to this establishment, and the results are as beautiful as the method is singular and difficult. The whole operation of casting the wax model in brass is delicate, and requires a firm hand, and an artistic comprehension.

The model in brass is now taken to the chaser, and upon his skill much depends. The general forms are to be elaborated by his chisels; the indications of character, of fur, or fibres, or feathers, are to be expressed and defined by him. As the beauty of a portrait depends upon those delicate distinctions, analyzed only by the artist, though recognized by all, so the beauty of these ornamental castings depends upon their delicacy and correctness of detail. In chasing metal, more tools are used than in steel engraving; the tools are driven by sharp blows, with a light hammer. One of the workmen in this department, showing a hammer, informed us he had used it thirty-two years; upon a fair calculation of the daily work it performs, the number of strokes made in a day, and the length of each stroke, we believe this hammer to have traveled 36,000 miles in thirty-two years, or one-half more than the circumference of the earth. After this operation of chasing is com-

---

[*] A correspondent of the *New York Times* remarked not long since: "I mean the manufacture of *Bronzes*, which I find has assumed, in Philadelphia, more imposing proportions, and a higher artistic character, than it possesses anywhere else in the country. Everybody who has been compelled, for his sins, to furnish a house, knows, of course, that the chandeliers, argands, and general gas-fittings of Cornelius & Baker and Archer & Warner are the only American articles of the kind which can sustain a comparison with the goods imported from Paris. But in the latter of these establishments I have just seen the moulds, and the models, and some of the completed portions of a magnificent bronze balustrading designed for one of the grand stairways of the new Capitol Building at Washington, which is not merely a wonderful piece of work in itself, but is, altogether, more elaborately elegant than any thing of the kind which is to be seen in Europe. This balustrading, which is cast, not in *basso relievo*, but in full relief of arabesques and figures, is about three feet in height, and is to be carried to the length of 160 feet up the noble flight of steps leading from the new Hall of the Representatives to the corridor of the Committee Rooms. The designs, furnished by Brumidi, of Washington, are singularly bold and graceful. They represent alternate groups of infants, eagles and serpents, pursuing and pursued through wreaths of foliage and flowers, and they consequently comprise almost all those curvilinear forms and intricate traceries which lay the heaviest tax upon the skill of the draughtsman and the founder."

pleted, the model becomes a standard, from which the caster may multiply copies to the extent of the demand. Yet new patterns are continually being made; and unsatisfied with past successes, the artist is continually aiming to produce something more beautiful in the future. So far the operations of modeling, casting, and chasing, forming the preparation of the pattern, have been conducted without machinery. But when the manufacture commences, machinery is found indispensable, and we were struck with the number of labor-saving machines used by this firm, enabling them to construct the difficult parts with that degree of accuracy so essential, when numerous pieces are to be placed together to form a finished whole. Indeed, the work so finished far excels that made by the more expensive old-fashioned operations by hand.

In this description merely the general outline is described of this manufacture; even the various processes are not named; the methods of making Gas Burners, Keys, Vases, the methods of Enameling, Stamping, Screw Cutting, Japanning, &c., are very curious, and employ a large number of men.

In the manufactory of Messrs. Archer, Warner, Miskey & Co., every department is entrusted to the best workmen of their class; wherever we looked, we saw order, neatness, and industry. In every one of the numerous rooms, six fire buckets, constantly filled with water, are kept; and the penalty of using the water for any improper purposes of washing, &c., is instant dismissal from the establishment. By means of these precautions and regulations, the establishment has more than once been saved from destruction. Upon principles like these the business is conducted; and the result is, an establishment unsurpassed in the world. Every branch of the business is carried on by this firm, from connecting a service pipe with the main in the street to constructing, and suspending in its place, the most costly and magnificent Chandelier.

VI.

Remarkable Paper Mills.

THE WISSAHICKON PAPER MILLS.—CHARLES MAGARGE, PROPRIETOR.

Twenty-eight years ago a small store for the sale of paper was opened in Minor street by two brothers, Charles & W. H. Magarge. At that time Fourdrinier machines had not been made in the United States, steam had not been used as an agent for making paper, and the mills were scattered along the water courses remote from the great centres of distribution. The house established as above-noted, formed connections with the manufacturers on one side and the consumers of paper on the other, and, by a course of dealing meriting the confidence of both parties, their store became a depot to which both resorted to effect their exchanges. At the present time the house of CHARLES MAGARGE & Co. is one of the most extensive and best known Paper Commission houses in the United States; and their immense warehouse on Sixth and Carpenter streets is constantly filled with paper of every variety, kind, quality and size.

In 1845, the senior partner of this firm, Mr. Charles Magarge, in order to supply a want in the business of the house, viz.: the ability to fill orders for paper of unusual sizes at short notice, purchased a plain, substantial stone building, 175 by 50 feet, situated on the Wissahickon Creek, just below the Indian Rocks, and converted it into a Paper Mill. This mill in his hands has become the famous Wissahickon Paper Mill. A gentleman, who recently visited it, describes the modern process of paper making, by detailing the operations which are carried on at this mill:—

The visitor goes up to the second story into a room some 60 by 80 feet, in which ten girls are engaged assorting the rags. Here are numerous bales of white rags, foreign and domestic. The imported are linen; the other, cotton. In the same room these rags are cut by a machine, driven by power, which fits them for the subsequent processes. They are next sent into a rotary boiler, of two tons capacity, into which steam is admitted, and the rags boiled. Next they are cast down on a floor in the first story, where they are put into cars, on which they are conveyed to the washing engines. Two engines are employed in washing, called Rag Engines. These engines play in tubs of an oval form, of large capacity, each containing 200 lbs. of rags. The impelling power, partly steam, partly water, causes the revolution of a roller, set with knives or bars of cast steel inserted into it longitudinally. This roller is suspended on what is called a *lighter*, by which it may be raised or lowered at pleasure upon a plate, consisting of bars of steel, set up edgewise. Passing now between this roller and plate, the rags are reduced to fibre. A stream of pure water is then conveyed into the Rag Engine, and, by means of a cylinder covered with gauze wire, the dirty water is passed off. This cylinder, called a Patent Washer, is octagonal in shape, some thirty inches in length, revolving in the engine, and having buckets within it corresponding with the sides of the washer. By this process, the rags are washed perfectly clean in from three to six hours. The next is the bleaching process, performed by the insertion into this engine of a strong solution of the chlorid of lime and some acid, to cause a reaction. The pulp is then emptied into large cisterns, covered with the bleach liquor it contains, where it is allowed to remain from twelve to twenty-four hours to bleach. It is then drained, put into the Beating Engine, and reduced to a pulp, the consistency of milk, which it much resembles. This pulp is emptied into a large cistern, in a

36

vault beneath, and kept in motion by means of an agitator revolving in it. It is then raised, by a Lifting Pump, into a small cistern, from which it is drawn off by a cock, which is opened, more or less, according to the thickness of the paper intended to be made, on to a strainer, which removes the knots, sand, or hard substances that may damage the paper, and then flows upon a leathern apron, which conducts it to an endless wire-cloth, over which the web of paper is formed. This wire-cloth is kept constantly vibrating, which both facilitates the escape of water and the felting together of the fibres of the pulp. The wire-cloth, with the pulp upon it—the edges being protected by deckle straps—passes on until it comes to a couple of *wet-press* cylinders, as they are called, the lower of which is of metal, but covered with a jacket of felting or flannel; the upper one is of wood, made hollow, and covered first with mahogany and then with flannel. These cylinders give the gauze with the pulp upon it a slight pressure, which is repeated upon a second pair of wet press-rolls similar to the first. The paper pulp is then led on upon an endless felt or blanket, which travels at exactly the same rate as the wire cloth, while the latter passes under the cylinders and proceeds to take up a new supply of pulp. The endless felt conveys the paper, still in a very wet state, between cast-iron cylinders, where it undergoes a severe pressure, which rids it of much of the remaining water, and then between a second pair of press-rollers, which remove the mark of the felt from the under surface; and, finally, it is passed over the surface of cylinders heated by steam, and when it has passed over about thirty lineal feet of heated surface, it is wound upon a reel ready for cutting. All this is done in the short space of two minutes.

To manufacture paper uniformly of superior quality, four things may be said to be essential—clear pure water, superior machinery, good stock, and the requisite skill. All these are combined in the Wissahickon Paper Mill; hence the reputation of its products. As a sample of the quality of the medium-priced paper from this mill, we refer to the sheets constituting this volume on " Philadelphia and its Manufactures."

At the present time Mr. Magarge is about completing another mill contiguous to the one so well known, the construction of which is creditable alike to the enterprise of the owner and the skill of the architect and builder. Its productive capacity will be over three tons of paper per day.

2. THE RIVER SIDE PAPER MILL.—EDWIN R. COPE, PROPRIETOR.

This is a new mill, situated on the Norristown branch of the Philadelphia, Germantown and Norristown Railroad, on the banks of the Schuylkill, about two miles above Manayunk, and adjacent to the well-known Soap Stone Quarry. The mill is a substantial stone edifice, three stories in height, and consists of a main building, 50 by 80 feet; a machine house, 50 by 100 feet; an engine house, 20 by 40 feet; two boiler houses, one 24 by 42 feet; another, 27 by 32 feet.

The paper machinery was supplied from the well-known establishment of Nelson Gavit, and is supposed to be one of his master pieces. It is impelled by a vertical beam condensing and non-condensing steam engine, 150 horse power, with George W. Corlis's cut-off and regulator, attached. This engine was manufactured by the Camden Iron Manufacturing Company, and is regarded as a very superior machine. The boilers attached to this engine are classed as *dropped flue boilers* of improved construction, planned by Mr. Dialogue, Superintendant of the Camden Company. Water is supplied to the mill from a reservoir, containing ten days, or two weeks' supply, "coming in with a head of upward of forty feet." The arrangements for keeping it clear, under all circumstances, are perfect. The entire cost of the mill and machinery was $70,000. It now employs thirty hands, of whom one-half are females. The consumption of rags is 4,000 lbs. per day; and the leading production is a quality of book paper, which, in all the essentials of good paper, firmness of the sheet, purity of the pulp, color, and hardness of the sizing, can fairly compete with any of the best papers made in this country or in Europe.

The proprietor of this mill, Mr. E. R. Cope, has long been identified with the manufacture of paper in this city; and, situated as his mill is, with every facility for obtaining raw materials and coal, both by railroad and canal, with the best of water and very superior machinery, its products will take rank among the choicest and most highly appreciated.

VII.

Rogers' Carriage Manufactory.

With the exception of two or three noted establishments, our attention in our tour of observation around the city was invited to none other more frequently than to that which forms the caption of this article. Even in a brass-founder's shop we were reminded "not to forget the superiority of the light carriages constructed in Philadelphia—that Rogers builds as good vehicles as are built in the world;" and that he deserves special credit, for by the excellence of his manufactures he reflects credit upon the city. Attending Herkness' Auction Sale of Carriages, we noticed that whenever a second-hand "Rogers' Wagon" was offered, the attention of the bystanders was awakened—bidding became lively, and the price obtained was evidently satisfactory to the seller. We then recollected of having read that a light carriage, constructed by Mr. Rogers to order, for a gentleman in Switzerland, was regarded, from its extraordinary lightness and strength, as so great a curiosity that, the owner having left it for a day at a hotel a few miles from Zurich, the hostler exhibited it during his absence, at a stipulated charge for "a sight," and thus made more money in one day than his wages amounted to in six months. All these circumstances combined—the complimentary allusions to his work by competent judges, and similar allusions to his standing as a gentleman by his fellow-mechanics, excited a strong desire to know something of his manufacturing facilities, and the following is the report of a gentleman who was specially employed to describe them.

REPORT.

The entire establishment includes two buildings, a Factory and a Repository, and the combination constitutes, probably, the largest one of the kind in the country; having an actual working space of nearly 40,000 superficial feet. The factory is situated in the northern part of the city, at the intersection of three streets, Sixth, Marshall, and Master, occupying the entire square; the lot is 137 feet on Sixth street, 137 on Marshall, and 172 on Master street. The factory itself is a handsome building, of forty feet front, and the full depth of the lot. It is four stories high, well lighted, furnished with all conveniences, and with the jobbing-shops, silver-plating, and wheel-shops, and lumber sheds, forms a hollow square. Within these boundaries every part of the business is pursued; and in the factory nine distinct occupations necessary to the manufacture of a carriage are carried on. One hundred and twenty-five men are employed in these departments, including smiths, designers, body-makers, wheelwrights, carvers, painters, platers, trimmers, upholsterers, and others of occupations less distinctive. These departments we shall examine upon the plan employed in the construction of a carriage.

Entering the front door, we find ourselves in a large and handsome room, filled with Barouches, Buggies, Germantown Wagons, and light carriages of every description, completed, and ready to be sent to the Repository. Interested, at present, more in the method than the result, we are shown by the politeness of Mr. Gorgas, the intelligent foreman of the establishment, into the Body Department, in which the carriage we intend to build is commenced.

The first step in construction is the execution of a design on paper; and here the idea of a Buggy, or Barouche, is first realized upon a scale of three quarters of an inch to the foot. This done, the purchaser is at liberty to suggest any alteration in his plan; and the second step is the execution of a geometrical plan of the body upon the black board. The third step is cutting the patterns in thin wood. The skeleton in wood is now completed, and the shape and proportion are determined. In this process not merely mechanical ability of execution, but mathematical exactness of design is essential. In this room fifteen men are busily employed in wheel work, and making body patterns. Wood, however, is insufficient; its strength barely supports its own weight, and could not possibly support the strain of unequal movement, with the burden of a single person. The skeleton must be strengthened; and for this purpose it is removed from the second story to the Smith Shop, upon the ground floor. In this department the body is bound, and riveted in Iron. This shop is probably one of the finest of the kind in the country. All the iron-work of a carriage is executed here; and the bolts, iron axles, locks, hinges, tires, and springs, are made and fastened to the wood. Twelve large forges are in constant use, and thirty-five men are employed. The springs used are all of one kind, and found practically superior to any that are patented. The principle in all springs is identical, and the important difference is in the quality of manufacture more than the mere form.

The process ensuring stability being completed, the skeleton is again removed to the Body Department, where the paneling follows; the floors are laid; the sides are built upon the proper curves; the seats and doors are introduced, and the body is ready for the painting room.

In this room, in the third story, eighteen or twenty coats of paint are given to the carriage; each being dried before the following coat is

36*

applied, and the whole surface repeatedly polished with pumice stone. These early coats are merely intended as the ground for future color, and are technically termed the "priming." White lead and litharge are used at first, and succeeded by coats of white lead and yellow ochre, upon which the selected colors of green, brown, black, &c., are properly applied.

The body is now removed to the Trimming Room, where considerable taste is employed in selecting the material, and adapting the color of the trimmings. Fine cloths, silks, carpetings, lace, oilskin, embossed leather, hair cushions, &c., are here employed in furnishing the coach. The coach now presents a very handsome appearance, but so far, instead of carrying others, it has been carried about itself. Leaving it upon the trestles, we return to the lower story.

The carriage manufacture is divided into two branches, those of body making and of carriage making; for though we have used the term "carriage" indiscriminately, in technical phraseology it applies only to axles and wheels, or the locomotive section of the vehicle. Wheel-making is usually a separate business; but Mr. Rogers prefers that all parts of his carriages should be of his own manufacture. Wheels are made in large quantities, and a stock is kept on hand. An exact proportion exists of the wheels to the body, and the average difference of diameter, between the fore and hind wheels, is two inches. In the lumber yard is an immense stock of hickory, of which the wheels and shafts are made; a material securing unusual lightness and strength. Ash is generally used for the body; oak, poplar, &c., are employed for various parts; and most of this wood is kept on hand two years before it is used.

The "carriage," like the body, passes through several successive stages, and after it is completed with axles, perches, shafts, &c., the body is hung upon the springs, and the coach is conveyed to the Finishing Room. Here it is polished, varnished, enameled, ornamented, finished: and only requires a pair of horses, a driver, a young lady, and a plank road, to display its comfort, durability, and speed.

In passing through this establishment, we were pleased with the arrangement of rooms, the facilities of transferring work from the ground floor to the fourth story; and, most of all, with the systematic management, evident in every department, as well as in the entire establishment. We saw much beautiful and elaborate work in process of manufacture, chiefly of light carriages; the building of heavy coaches commencing in July, for the fall trade. In the body room we saw a magnificent Brett, intended for St. Louis, in which by the use of curved iron work, bracing the wood, the ordinary perch is dispensed with, and

Rogers' Carriage Repository.—Exterior.

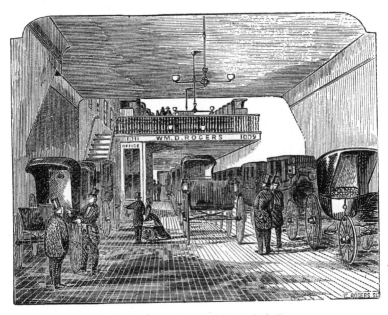

Rogers' Carriage Repository.—Interior.

the body is supported by its own firmness. Twelve shifting top wagons for the same city, of polished hickory, are not surpassed in our remembrance for excellence and beauty; and a Box Buggy for this city, ornamented with exquisite paintings, the order having been given to disregard expense, will be one of the handsomest light carriages ever built in Philadelphia or elsewhere.

Mr. Rogers is also building several light Buggies to order for gentlemen in Austria; and no doubt but many of these will be driven on the Prator at Vienna, ere this year closes.

The Repository, an exterior and interior view of which is given on the opposite page, is situated on Chestnut street above Tenth. This edifice is 46 feet front, and 178 feet deep, and three stories high, and is an ornament to the neighborhood. Here carriages of all kinds are kept for sale; and here all ordered work is deposited for delivery, after having received a careful test examination at the factory. Some of the handsomest carriages ever made in America are exhibited here; and persons more fond of examining beautiful results than inquiring into curious and complicated methods, are invited to pay a visit to the Repository.

VIII.
Remarkable Wagon-making Establishments.

In order to compensate in part for the compulsory brevity with which we were compelled to treat the important manufacture of Wagons, Carts, &c., we insert Mr. Young's description of two of these mammoth establishments.

WILSON, CHILDS & CO.'S WAGON MANUFACTORY.

The largest Wagon-making establishment in Philadelphia, and perhaps in the country, is that of D. G. WILSON, J. CHILDS & Co., whose office is at the corner of St. John and Buttonwood streets. Messrs. Wilson & Childs formed a copartnership in 1829, and commenced business on the premises now occupied by them, which fronts on three streets—Third, Buttonwood, and St. John—extending 230 feet on Buttonwood, and 113 on St. John. A four-story brick building, 86 by 45 feet, has recently been erected on the St. John street front. Their increased business requiring more room, they purchased, in 1850, a manufactory erected by Mr. Simons, also the adjoining property—comprising in all a square on Second street, and Lehigh Avenue; the whole containing 260,500 square feet, or over 6 acres. The square on the East side of Second street is 500 by 248 feet, and is used as a Lumber Yard, in which are piled plank and boards of various thicknesses, from one to five inches; also spokes and hubs. These are left to season from one to five years; one year being required for every inch in thickness. Spokes in the rough are also piled and seasoned. Timber for hubs is made chiefly from black locust trees of different sizes, sawed into suitable lengths, and before being stowed away to season, have the bark removed, and a hole bored in the centre, to facilitate the seasoning process. On the western side of Second street—the lot being 500 feet by 273, and containing over three acres—the various workshops are situated. The principal buildings are one brick Wheel and Body Shop, 100 by 45 feet, and three stories high; one brick Blacksmith Shop, 200 by 35 feet; the Saw-Mill, Engine House, and Machine Shop, 80 by 45 feet, and three stories high; Running-gear Shops, 100 by 45 feet; Paint Shop, Office, Stables, Sheds, &c.; the Boiler House is a separate building.

The variety and extent of the business may be learned by an inspection of the different successive operations required to make a wagon. Plank of three or four inches in thickness, for felloes and shafts, spokes, hubs of proper sizes—all sufficiently seasoned—are selected from the lumber yard, and removed into the saw-mill. Here there are two Upright and six Circular Saws; a machine for boring holes in the centre of the hub; another for boring holes for the spokes; four Drills (self-feeding) for drilling iron. In other shops there are a Planing Machine, a Mortising Machine, two machines for turning Spokes, machines for driving in the Spokes and for shaping the Felloes, and finishing them complete, and a machine for boring Hubs so as to put the boxes in properly, and to ensure accuracy and a solid bearing. The planks, after being marked, are sawed into shafts, and into felloes. Hubs are turned in a lathe, and both ends sawed at once by circular saws. They are then conveyed to other shops, where the wheels and running-gears are completed. The bodies are made by a number of men, each having his particular work assigned him; and, by such subdivision of labor, greater excellence and celerity is attained. In the Engine and Machine Shop the Iron Axles are turned, Screws cut, and a number of other operations performed. The Blacksmiths' Shop, a new brick building, is one of the best arranged I have ever seen. It is very high, and thoroughly ventilated, the gas being carried off, rendering it healthy and pleasant for the men; it has twenty-four fires, a machine for Punching, for making Bolts, Rivets, &c. Of course, all the Hoops or Bands, Tires, Straps, Bolts, Rivets, Staples—in short, all the iron-work for the wagon is made here. The upper story of the Body-shop, and also a separate building, are used as Paint Shops. Previous to this finishing operation, however, and when every defect can easily be discovered, the Wagons for the Government are minutely examined by an inspector, whose keen eyes are not easily blinded: and when passed by him their excellence and durability may be relied on.

To give an idea of the extent of business carried on in one of these great establishments, and of the large capital required, the stock on hand a few months since may be stated, viz.:—1,500,000 feet of hard-wood plank and boards, 30,000 hubs, 472,000 spokes, and 150 tons of iron. At the period of my visit, the second story of a building 273 by 23 feet, except twenty feet at one end, was filled with wheels; there must have been several thousands. The lower floor of the same building was filled with cart and wagon bodies. The number of hands employed averages 173.

The old establishment on St. John, Buttonwood, and Third streets, is still used by this firm, but it is eclipsed by the larger one above described. They also own the warehouse, 70 Carondelet street, New Orleans; and have an agency in Mobile.

D. G. WILSON, J. CHILDS & CO.'S WAGON FACTORY.

SIMONS, COLEMAN & CO.'S WAGON FACTORY,
1109 NORTH FRONT STREET.

This firm is also very extensively engaged in the manufacture of Wagons, Carts, Drays, Ox-Wheels, Timber-Wheels, Wheelbarrows, &c. Mr. Simons has been identified with the business for many years; and the other partners are sons of the late Mr. N. Coleman, a well-known manufacturer in this branch. The premises now occupied by Simons, Coleman & Co., consist of one brick building, 100 by 34 feet, four stories high; another 30 by 40 feet, four stories high; one frame building, 96 by 24 feet, two stories high; which are used for wood-working. Their Steam Engine is in a two-story building, 40 by 45 feet, in which also there are Iron Lathes, Wood Lathes, Rip Saws, Jig Saws, Drills, and a Screw-cutter. The last is on a new and improved plan, and will cut, it is said, more than any other now in use. Their Blacksmiths' Shops, which are divided into four apartments, occupy, in all, a space of about 600 by 45 feet, in which they have 25 forges, with Patent Bolt and Rivet machines, and Power Punch and Shears. Three yards are required for storing Lumber, Spokes, and Hubs; and they have now on hand about 1,200,000 feet of seasoned Oak Lumber, with Hubs and Spokes in proportionate quantities. Their increased business during the past year requiring additional room, they obtained a lot on North Second street, extending from Huntingdon to Cumberland street, 500 feet, and from Second to Washington, 273 feet. They have in course of erection a Wheelwright Shop, 105 by 66 feet, five stories high; a Blacksmiths' Shop, 106 by 51 feet, three stories high; an Iron Mill, 51 by 30 feet; and a Wood Mill, 51 by 27, both four stories high—all the Machinery to be driven by an Engine of sixty-horse power. These works, it is expected, will be in operation in September of this year (1858); and, in conjunction with the shops present, will afford employment to at least 250 workmen. When finished these buildings will, probably be the largest used for Wheelwright shops in the United States.

This firm have agencies at New Orleans, Mobile, Charleston, Savannah, Galveston, Indianola, and San Antonio, Texas.

In the two establishments just noticed, 550 Wagons for the Utah Expedition were completed in about five weeks, in addition to the large number which they make to keep good their stocks in their numerous depots throughout the South. The importance to Philadelphia of this branch of manufacture—affording, as it does, employment to various mechanics and artisans, and diffusing money among the proprietors of rolling-mills, saw-mills, &c., and through all these to hundreds of others—cannot be over-estimated; and the proprietors should, in common with other manufacturers, receive that consideration which their enterprise merits.   EDWARD YOUNG.

IX.

Lewis Thompson & Co.'s Steam Turning and Sawing Mills.

We have already stated that this establishment is one of the most remarkable of its kind in the country. It is remarkable both for the superior character of its Machinery, some of it not to be found in any similar manufactory in the United States, and for the variety of its departments, and the superior quality of its productions. With regard to the latter, our Reporter has furnished the following remarks.

BUILDERS' GOODS.—We were taken to one portion of the works occupied entirely by machinery for the manufacture of STAIR BALUSTERS, NEWEL POSTS, FANCY BRACKETS, &c., &c., and these are made in almost endless variety of size and style, and of every description of material; some of the *Balusters* being very richly carved, and so on down to the plainest and cheapest descriptions of these useful articles. The same remarks will apply to their variety of *Newel Posts* and *Fancy Brackets*, and *Scroll Work* for cornices, verandahs, &c. We suppose that architects, carpenters, builders, and contractors would find it a very difficult matter to find a similar establishment in the country, where their wants and requirements could be so readily and promptly supplied. Every description of this kind of work which cabinet-makers and carpenters have not the facilities for doing to advantage, is got out with dispatch.

FURNITURE DEPARTMENT.—Another, and very large portion of their Works are almost entirely devoted to the manufacturing of FANCY FURNITURE (in the white) for the trade. In this department are employed a very large number of workmen, and we were shown some fifty different patterns of *Sofa* and *Pier Tables*, many of them most richly and elaborately carved; fancy *Wash-Stands*, *Etageres*, *Hat Racks*, *Extension Tables*, &c., &c.—indeed, we shall not attempt even to name the different articles manufactured in this branch of their business. We will say, however, that we have been assured that these goods are all made with great care, and in the very best manner, in Rosewood, Mahogany, and Walnut; and, of course, are offered to the trade (to whom they are exclusively sold) at prices much below the cost of manufacturing by cabinet-makers, who have not the aid and appliances of the extensive and unique machinery which these gentlemen have in use.

MARBLE DEPARTMENT.—In still another portion of this establishment we were shown four large frames, each containing from twenty to thirty saws, and each frame covering a huge block of *Italian Marble*, which the saws were slicing up into slabs, as though they were so much wood. This department alone is well worth a visit from the curious, and from persons who admire the operations of fine machinery. We noticed here the pure white *Statuary Marble*, as well as the delicately-veined Italian, with several varieties of *Lisbon, Broccatella, Sienna, Irish Black, Egyptian,* and *Tennessee.* After being here sawed up into slabs of requisite thicknesses, these are taken to another portion of the premises, where skillful workmen are engaged in cutting and fashioning them, for *Table Tops* and similar uses, for which purposes alone this branch of their business is exclusively confined.

LUMBER DEPARTMENT.—In addition to their factories, these gentlemen deal also very largely in all descriptions of *Hard Woods*—usually denominated "*Albany Lumber.*" Their Lumber Yards occupy three squares of ground, and are filled with an immense quantity of choice and rare woods; consisting of Walnut, Cherry, Ash, Maple, and Cedar, of every variety of thickness; as well as *Rosewood* and *Mahogany*, in the logs, and cut to all sizes. We noticed here a log of Mahogany forty-nine inches wide! the largest, we believe, ever brought to this country. The most of this lumber is of their own direct importation and manufacture; and they are thus enabled to supply orders with much promptness and dispatch.

The proprietors are gentlemanly and courteous to strangers; and it is a pleasure to invite the especial attention of all likely to be interested to such an establish-ment.

X.

Abel Reed's Manufactory of Building Materials.

It has been remarked that in no branch of manufacture does the application of labor-saving machinery produce, by simple means, more important results than in the working of Wood. It was our misfortune to be compelled to defer any minute consideration of the wonderful machines that are in operation in the various Wood-working Establishments of Philadelphia; but, as a partial compensation, we would recommend those who take an interest in the subject to visit the establishment of ABEL REED, on Marshall Street, above Girard Avenue. Mr. Reed is one of the oldest established Wood-workers in the city. He commenced business some twenty-eight years ago, when those wonderful Planing and Moulding and Turning Machines of modern times were unknown. He was among the very first to introduce improved machinery into the business; and was one of those who were compelled to encounter the odium which was meted out to pioneers, and to endure the reproaches of journeymen whom it was supposed machinery would injure. The zeal which he manifested in behalf of improved machinery designated him as a suitable person to act as agent of the manufacturers of such machinery; and in this capacity he has supplied a greater part of the machines which are in use in the wood-working establishments of this city and the South. In his manufactory may at all times be seen a greater variety of remarkable machines for this purpose than can probably be seen anywhere else south of New England. Our reporter states: "Here we saw in operation *Gray & Wood's Planer*, which is adapted for all kinds of Shop Planing, and will plane 6,000 feet per day; Fay & Co.'s, and Ball & Ballard's *Tenoning Machines*, for all kinds of Tenoning, Coping, &c., and which will do the work of ten men; also *Sash* and *Moulding Machines* of the same makers, which will turn out 5,000 feet of ordinary moulding per day; and *Power Mortising Machines*, for doors, making two hundred cuts per minute; and *Foot Mortising Machines*,

37

for sash, &c., every one of which will do the work of three men ; Scroll Saws, for brackets and all kinds of scroll-work; and Circular Saws, for squaring and ripping.   But probably the most remarkable of all the machines was the *Circular Moulding,* or *Upright Shaping Machine,* adapted for any kind of circular work, or for any irregular cutting on the inside or outside of any crooked work requiring to be neatly smoothed. It may be made to work in any circle, from two inches and upward."

Mr. Reed, by means of his improved machinery, is enabled to supply Builders throughout the United States with Doors, Frames, Sash, Blinds. Shutters, Mouldings, &c., cheaper than they can make them in their own workshops, unless they possess similar facilities.

INDEX TO CONTENTS.

A COMPLETE ALPHABETICAL LIST

OF

ALL ARTICLES NOW MADE IN PHILADELPHIA,

WITH THE

ADDRESS OF ONE OR MORE MANUFACTURERS OF EACH.

☞ THIS LIST is intended to serve a double or triple purpose, as stated in the caption; but primarily we design to inform our readers *what articles are made in this city*, and incidentally to give the address of one or more manufacturers of each. This list, as respects articles, is far more minute than any ever before published; but the multitude embraced in the terms Philosophical Instruments, Hardware, Tinware, Housekeeping articles, Chemicals, and others, compelled us to make a selection of the more important. As respects manufacturers, it would afford us pleasure to give the name of every maker opposite to every article he produces; but, inasmuch as the products of a single establishment are often exceedingly numerous and diverse,—in one instance, as we stated, one hundred and fifteen distinct articles are made by one individual,—it is quite obvious that the repetition of names resulting from the adoption of such a plan would defeat the object of this book, by making it voluminous and unwieldy. We do not design to make a Directory, and therefore consider our duty both to readers and to manufacturers discharged, by mentioning the name of one maker of each article, trusting to the sagacity of purchasers to pursue the investigation for themselves. In preparing this List I have had the invaluable assistance of Dr. J. L. BISHOP, who has a more minute knowledge of the minor manufactures and mechanic arts of Philadelphia than probably any other man.

ARTICLES.	MANUFACTURERS.
Accordions.	JOSEPH SERVOSS, (sole agent for Fass' Patent Echo American Accordion,) 16 North Second street. C. M. ZIMMERMAN, (Patent American,) 238 North Second st.
Acids	(See Chemicals, page 207.)
Agricultural Implements.	(See page 141.)
Air Pumps.	GEO. C. HOWARD, (Submarine Armor,) 18th st., bel. Market.
Alcohol.	(See Alcohol, Camphene, and Burning Fluid, page 148.)
Ale and Porter.	(See Brewing, page 194.)
Almanacs.	KING & BAIRD, 9 Sansom street.

(455)

ARTICLES.	MANUFACTURERS.
Alum.	(See Chemicals, page 207.)
Ambrotype and Daguerreotype Tubes.	ADOLPH WIRTH, No. 704 Arch street, above Seventh.
Ammonia, Sulph. of.	HENRY BOWER, Gray's Ferry Road.
Anchors, Chains, and Cables.	WILLIAM H. REINHART, 522 Washington Av., below Brown.
Annunciators.	(See Bell Telegraphs.)
Apple Parers and Slicers.	E. L. PRATT, (Pratt's Patent,) 617 Sansom street.
Architectural Ornaments, in Iron.	HAGAR, SANSON & FARRAND, Willow st., bet'n 12th & 13th. W. P. HOOD, Broad and Ridge Avenue. E. W. SHIPPEN & CO., Market street, West Philadelphia.
———— in Plaster.	W. H. FRENCH, Eighteenth and Chestnut streets. THOMAS HEATH, southwest corner Eleventh and Arch.
———— in Wood.	THOMAS G. BERING, northwest corner of 10th and Ridge Av. CHARLES BUSHOR, 45 South Eighth street, and 200 Dock. R. S. SMITH, 234 South Second street. JOHN HARE OTTON, 306 S. Fifth street.
——— in Terra-Cotta.	LORENZE STAUDACHER, Chimney Tops, Caps, and Brackets, for churches and private dwellings, 1742 North Eleventh, above Columbia Avenue.
Artificial Flowers.	M. S. GERSTLE, 163 North Second street. HANLEY & BROTHER, 135 N. Third, below Race, up-stairs. A. M. HOSKINS, 15 South Fourth street. Mrs. M. J. VAN OSTEN, 29 South Fourth street. And *many others.*
Artificial Limbs.	B. F. PALMER, 1320 Chestnut street. JOHN F. ORD, 208 Dock, above Second street.
Artificial Teeth.	JONES, WHITE & McCURDY, 528 Arch st. ORUM & ARMSTRONG, 520 Arch street. CORYELL & ROBSON, (Dental Laboratory,) 7th & Sansom st.
Artists' Colors and Materials.	H. KAUSZ, (also Importer,) No. 802 Sansom street, second door above Eighth. THEODORE KELLEY, "Artists' Colorman," (Manufacturer and Importer of Materials, Fresco and Gilders' Tools,) No. 22 S. Eighth street. G. W. OSBORNE & CO., (Osborne's Superfine American Water Colors,) 104 North Sixth street.
Ash—Pot, Pearl, and Soda.	(See Chemicals, page 207.)
Ash Sifters.	O. L. BAKER, 315 North Front street.
Astronomical Instruments.	(See Instruments, Philosophical.)
Augers.	BENJAMIN PUGH, Oak street, West Philadelphia.
Awl Blades.	MICHAEL POTTER PARTRIDGE, (late from England,) Blair street, between Norris and Wood streets, Kensington.
Awnings, Tents, and Sacking Bottoms.	JOSEPH H. FOSTER, 443 North Third street, above Willow. GEO. W. FOX, No. 6 Hart's Buildings, 6th st., above Chestnut.
Awning and Canvas Printing.	C. E. FISK & SON, 13 South Sixth street, (basement). B. SACHS, 306 Market street

ARTICLES.	MANUFACTURERS.
Awning Frames.	HENRY BEAGLE, Willow street, above Fifth. J. E. SASS, 133 Elfreth's alley, below Second, first street above Arch.
Axes.	C. HAMMOND, 503 Commerce street.
Axe Handles.	CHIPMAN & WHITE, No. 11 South Front street.
Axles.	RICHARD FRENCH, Broad and Willow streets.
" Car	A. & P. ROBERTS, Broad street, near Vine.
Babbitt Metal.	JOSEPH BAKER & SON, 820 Rachel st., or 821 North Second. H. W. HOOK, Broad and Pleasant streets.
Baggage Checks.	WM. T. SCHEIBLE, 49 South Third street.
Bags.	JOSEPH H. FOSTER, No. 443 North Third street. GEORGE W. FOX, No. 6 Hart's Buildings, Sixth street, above Chestnut.
Bags, Paper.	JOHN H. LEWARS & CO., (Machine made,) 9th and Wallace.
Bakehouse Machinery.	R. J. HOLLINGSWORTH, No. 23 Coates alley.
Baking Powders.	BOHLER, TOMSON & WEIKEL, 248 North Third street, below Vine. M. GUGGENHEIM & CO., 1040 North Third st., corner George.
Balances.	(See Scales and Balances.)
Balusters.	LEWIS THOMPSON & CO., Eleventh and Ridge Road.
Bandages, Surgical.	(See Trusses and Bandages.)
Band Boxes.	(See Paper Boxes.)
Bark Mills.	J. YOCOM & SON, (Webster's,) 13 Drinker's alley.
Barometers.	(See Instruments, Philosophical.)
Barrels, Casks, and Kegs.	H. APPLE & SONS, 121 South Water street, and 230 North Water street. HENRY HAAS, 707 Brook street, between Third and St. John and Brown and Coates.
Baskets.	(See Willow Ware.)
Bath Tubs.	(See Cedarware and Tinware.)
Bay Water	JOHN GIBSON, SONS & CO., South Front, below Walnut.
Bed Coverlets.	PHILIP KOLMER, No. 1224 Germantown Road, above Franklin street. L. SIEFERT, No. 928 North Third street, between Poplar and Beaver.
Bedding, Beds, and Mattresses.	S. T. C. BELL, northeast corner Fourth and Callowhill. J. P. ERWIN, No. 525 Callowhill street, below Sixth. FISHER & BROTHER, northwest corner Fourth and Vine sts. E. FOLIOT, (Spring Mattresses,) No. 407 Walnut street. F. B. GILBERT, 936 and 938 Ridge Avenue. And *others*.
Beds, Spring.	JAMES RUGGLES, 333 Harmony Court, below Fourth. And *others*.
Bedsteads.	B. REEVES & SON, 441 St. John, above Willow. RIEBEL & LINCOLN, 409 and 411 Brown street.

37*

ARTICLES.	MANUFACTURERS.
Bedsteads, (Iron and Wire.)	S. MACFERRAN, No. 532 Arch street, corner of Sixth. SELLERS BROTHERS, 18 North Sixth street. M. WALKER & SONS, 535 Market street.
Beef, Packed.	JACOB T. ALBURGER & CO., 414 South Front street. N. H. GRAHAM & CO., 3 North Water street.
Bee-Hives.	P. J. MAHAN, (Langstroth's Patent,) 720 Chestnut street
Beer, Lager.	(See page 195.)
Bells.	JOSEPH BERNHARD & CO., No. 120 North Sixth street. GELBACH & METZGER, No. 1241 Howard st., above Franklin Avenue, Kensington.
Bell Hanging.	THOMAS H. AUROCKER, northwest corner 13th and Cherry sts. JULIUS BILLERBECK, No. 902 Ridge Avenue, above Vine st. H. HOCHSTRASSER, southwest corner 10th and George sts. J. B. SHANNON, No. 58 North Sixth street. SCHEERER & DIEHR, 264 Fourth street.
Bell Telegraphs.	H. HOCHSTRASSER, southwest corner 10th and George sts.
Bellows.	GEO. W. METZ & SONS, No. 813 Market street. And *others.*
Belts, (Waist and Money.)	P. HEALY & CO., No. 204 Walnut street, above Second.
Belting, Machine.	BARNETT & JENKINS, No. 112 North Sixth street. WM. ECKFELDT, No. 212 N. Thirteenth st., and No. 418 N. 3d. SELLERS & PENNOCK, 613 St. James street.
Bichromate of Potash.	BUCK, SIMONIN & CO., No. 127 Walnut street.
Billiard and Bagatelle Balls.	HARVEY & FORD, Goldsmith's Hall, Library street.

Billiard and Bagatelle Tables & Pools.	THOMAS DAVIS 242 Monroe street.
	SCHAFFER & ADLER, N. W. cor. Second and Callowhill.
	   (Cloths, Balls, Cues, Maces, and Bridges, constantly on hand,
	   and furnished at short notice.)
	M. SCHŒNHUT, S. W corner Franklin Avenue and Second st.
	   (See page 275.)

Billiard Table Mountings.	G. W. BRADFIELD, 525 North 12th st., near Ridge Avenue. PAUL SCHWARZKOPF, 519 Emlen, bet. Noble & Peg, ab. Front
Bindings.	JOHN DUDLEY, (Carpet, Mat, and Horse-sheet,) 207 Quarry st., above Arch, between 2d and 3d. JOS. SPITZ, (also Diamond Bed-Lace,) 928 N. 3d, ab. Poplar.
Bird Cages.	(See Wire Work.)
Bits and Braces.	BOOTH & MILLS, Little Washington st., below Ninth.
Bits, Bridle.	JOHN KLUFKEE, 1351 Marlborough, near West st., Kensing'n. PAUL SCHWARZ KOPF, 519 Emlen street, bet. Noble and Peg, above Front. E. HALL OGDEN, (Malleable Cast Iron,) 9th and Jefferson sts.
Blacking.	JOHN ANNEAR, No. 127 North Front street. WILLIAM CURRY, No. 208 Chestnut street. JAMES S. MASON & CO., No. 140 North Front street.
Blank Books. (See page 179.)	WM. B. EDWARDS, No. 152 S. 3d st., opposite the Exchange. JOHN GLADDING, No. 117 South Second street, & 529 S. 2d. HENRY HINKLE, (Improved English Flexible,) No. 1 Ranstead Place. DAVID M. HOGAN, (Stationer and Printer,) 418 Walnut street

ARTICLES.	MANUFACTURERS.
Blank Books.—*Continued.*	MOSS BROTHER & CO., No. 16 South Fourth street. C. P. PERRY, southwest corner Fourth and Race streets. PIERSON & DIAMENT, 136 and 138 N. Fourth, corner Cherry. THOMAS W. PRICE, 22 South Fifth street. JAMES B. SMITH & CO., 610 Chestnut street. JOHN B. SPRINGER, southeast corner Fifth and Walnut. And *many others.*
Blinds, Sashes, and Doors.	M. J. BRADY & CO., Beach, below Shackamaxon. COGILL & WILT, 421 North Front, below Willow. ABEL REED, 215 N. Second and Marshall st., above Girard Av. And *many others.*
Blinds and Shades.	A. BRITTON & CO., No. 44 North Second street. C. W. CLARK, No. 139 South Second street, above Walnut. L. M. HARNED, 139 North Sixth street. R. W. KENSIL, No. 939 Race street. B. J. WILLIAMS, No. 16 North Sixth street. And *many others.*
Block Letters.	WM. C. MURPHY, (Carved,) No. 47 South Third street.
Blocks and Pumps.	JAMES McCUSKER, No. 715 Swanson street. ALEXANDER RANKIN, No. 122 North Water street.
Blue, Prussian.	(See Chemicals, page 208.)
Boats.	ALBERTSON BROTHERS, Beach st., ab. Marlboro', Kensing'n. WOOD & BROTHERS, Penn street, above Poplar. And *others.*
Bobbins and Spools.	JOHN CUNDEY, 732 North Fourth street, below Brown. JOHN JACKSON & CO., 1089 Germantown Road, opp. Second— (various Turnings for Silk, Cotton, and Woolen Factories.)
Boilers.	(See Machinery.)
Bolts, &c.	E. & P. COLEMAN, Arch, above Twenty-first. HOOPES & TOWNSEND, 1330 Buttonwood, below Broad.
Bonnet Boards.	A. M. COLLINS & CO., 506 Minor street. HUCKEL & MILLER, 43 North Front street.
Bonnet Frames.	LINCOLN, WOOD & NICHOLS, 45 South Second street. JAMES TELFORD, 903 North Second, above Poplar.
Bonnet Wire.	M. BIRD, (Silk, Cotton, and Fancy,) northwest corner Franklin Avenue and Crown, Kensington. JOSEPH MOORE, 737 L. Crown st., rear of 522 North Fourth.
Bonnet Pressing Machines.	GEO. C. HOWARD, 17 South 18th street, below Market.
Bonnets and Straw Goods.	(See Straw Goods.)
Bonnets, Silk.	A. E. CARPENTER, 54 North Eighth street.
Bookbinding. (See page 177.)	J. ALTEMUS & CO., 62 North Fourth street. CRISSY & MARKLEY, Goldsmith's Hall, Library street. WM. B. EDWARDS, No. 152 South Third, opposite Exchange. EDWARD GASKILL, Post-office Building, Carter street. JOHN GLADDING, No. 117 South Second street, & 529 S. 2d. LANDENBERG & LEVER, Nos. 40 and 42 North Seventh st. J. B. LIPPINCOTT & CO., Fifth and Cherry. MILLER & BURLOCK, George, above Eleventh. JOSEPH MONIER, No. 10 North Sixth. PAWSON & NICHOLSON, No. 519 Minor street. THOMAS PRICE, No. 23 South Fifth. JOSEPH A. SPEEL, Cowperthwaite Building, Carpenter st And *many others.*

ARTICLES.	MANUFACTURERS.
Bookbinders' Muslin	N. M. ABBOTT & CO., 522 Minor street.
Bookbinders' Tools.	GASKILL, COPPER & FRY, No. 522 Minor street.
Book Mountings.	T. T. KINSEY, No. 207 Race street. EDWARD L. MINTZER, No. 23 North Sixth street. F. A. WAIT, No. 237 Spruce, below Third.
Book Publishers.	(See Books, &c., page 159.)
Boot Crimping Machines.	FETTER & CO., No. 31 South Sixth street.
Boot Crimping.	C. B. SYLVESTER, 338 Crown street, above Wood.
Boots and Shoes.	(See page 185.)
Boot Trees & Lasts.	PETER DEWEES, No. 113 Callowhill street, (old No. 31.) JACOB FOSTER, No. 305 Cherry street. J. HOWARD & CO., No. 112 Bread street. GEORGE MUNRO, (Anatomical,) No. 127 Callowhill.
Bottles.	(See Glass.)
Bottle Moulds.	EDWIN FAYLE, 205 Quarry street (Evans' Building.) GEO. H. MYERS, (established over 25 years,) No. 2 Lodge street, above Second. C. SOISTMAN, southwest corner York Avenue and Noble. STACY WILSON, 11 Drinker's alley. GEO. H. KOECHLEIN, (Wooden,) 526 North Front street.
Bottlers.	ANDREWS, JOHNSTON & CO., 1713 Cherry, West of 17th. P. CONWAY, No. 1021 Hunter street, below North 11th. EDWARD DUFFY & SON, No. 912 and 914 Filbert street. RICE & McKINNEY, 623 South Sixth street. And *others*.
Bottling Machines.	JOSEPH BERNHARD & CO., No. 120 North Sixth.
Boxes, Packing.	(See Packing Boxes.)
Boxes, Paper.	(See Paper Boxes.)
Boxwood for Engravers.	N. J. WEMMER, 205 and 207 (old No., 5) Pear street.
Bracelets.	J. T. MIDNIGHT, 145 North Fourth street.
Braces and Bits.	BOOTH & MILLS, Little Washington, below Ninth.
Brands and Stamps.	JOHN BUNTING, southwest corner Second and New streets.
Brass Cocks.	J. & S. AUSTIN, 1224 Clinton street, above Franklin Avenue, between Front and Second. J. & H. JONES, 243 Arch street. H. HOMER, 231 Race street. McCAMBRIDGE, FRY & CO., 317 Cherry street.
Brass Founding and Finishing.	J. & S. AUSTIN, 1224 Clinton street, above Franklin Avenue, between Front and Second. S. J. CRESSWELL, 816 Race street. M. A. DODGE, Fifteenth and Willow. JACOB FRICK, 614 North Third street. F. W. & G. A. KOHLER, 528 N. Second, above Buttonwood. GELBACH & METZGER, 1241 Howard st., above Franklin Av. THOMAS HARRINGTON, rear of 15 North Ninth street. And *many others*.
Brass Raised Letters, for Engines, Names, &c.	HENRY SINKLER, (also Card Plates, Engraved Names and Letters for Cemeteries, &c.) Pemberton st., south of Wallace.

ARTICLES.	MANUFACTURERS.
Brass Machine and Ship Castings.	GELBACH & METZGER, 1241 Howard st., above Franklin Av.
Brass Nails & Bands.	WILER & MOSS, 225 S. Fifth street, below Walnut.
Brass, Sheet.	SAMUEL CROFT, 22 Decatur street.
Brass Tubing, for Philosophical Instruments.	E. BORMAN, (Drawn) 523 Cherry street.
Bricks.	(See page 198.)
Brick Machines.	C. CARNELL, 1542 N. Sixth and Germantown Road. S. P. MILLER, No. 309 South Fifth street.
Bricklayers' & Plasterers' Tools.	JOSEPH RUE, (Ladders, Jacks, Hods, &c.) 845 Parrish street.
Britannia Ware. (See page 351.)	HENRY D. BOARDMAN, No. 245 Arch street. HENRY CALVERLEY, (Moulds, Coffin Trimmings, and Fancy Wares in Britannia Metal,) No. 205 Quarry street. ERNEST KAUFFMAN, No. 328 Noble street. G. ENGEL, No. 308 Chestnut street.
Bronzes.	ARCHER, WARNER, MISKEY & CO., 718 Chestnut street.
Bronze Powders.	WILLIAM EVETT, No. 28 North Fifth street.
Brooms and Whisk Brushes.	BERGER & BUTZ, No. 132 North Water st., below Race. CHIPMAN & WHITE, No. 11 South Front street. H. B. PENNOCK, Jr., 47 North Water street. ROWE & EUSTON, 157 and 159 North Third street.
Brush Blocks and Backs.	ALEXANDER BECRAFT, rear of 426 Walnut street. JOSEPH BUSCH, rear of 220 North Third. JOHN FUNSTON, (Bored,) 116 Edward st., near School.
Brushes. (See page 404.)	EDWIN CLINTON, No. 908 Chestnut, above 9th. CHARLES T. KERN, No. 58 North Second street. JACOB PEPPER, (Tooth,) Germantown Road, ab. Columbia. EDMUND W. P. TAUNTON, No. 14 Decatur street. And *many others.*
Buckets.	(See Wooden Ware.)
Buckskins.	McNEELY & CO., No. 64 North Fourth street. MORGAN & WELBANK, No. 402 N. Ninth, above Callowhill.
Buckskin Gloves, Drawers, Shirts, &c.	J. R. ASHFORD, (Gloves only,) 607 Callowhill street. P. HEALY & CO., No. 204 Walnut street. MORGAN & WELBANK, No. 402 N. 9th st., above Callowhill.
Bungs and Spigots.	GEO. H. KOECHLEIN, No. 526 N. Front street—(also, Mallets, Screws, Handles, and Moulds for Glass, &c., &c.) JOHN LOUIS & SON, 504 Vine st., above Fifth.
Burning Fluid and Camphene.	(See page 148.)
Burr Millstones.	SAMUEL M. MECUTCHEON, Haydock street, below Front J. E. MITCHELL, No. 310 York Avenue.
Buttons and Button Moulds, Bone.	J. C. NORTH, South Front street, below Reed. J. WITZEL, No. 1321 North Fifth, below Master.
Buttons, Covered.	GEIERSHOFER, LOEWI & CO., 907 Marshall & 5 Bank st.
Buttons, Horn.	LEWIS BUTZ, No. 978 Marshall st., below Franklin Avenue.
Buttons, Pearl.	EDWARD MARKLEW 15th street, bet. Hamilton and Willow

ARTICLES.	MANUFACTURERS.
Buttons, Silk and Fancy.	J. B. CHAMPROMY, 133 North Third street. JOHN C. GRAHAM, 607 Cherry street. H. W. HENSEL, 20 North Fourth street. W. H. HORSTMANN & SONS, 723 Chestnut street. CHARLES MEVES, 9 North Eighth street.
Cabinet Ware.	(See Furniture and Upholstery, page 273.)
Cabinet Makers' Findings.	NOBLITT, BROWN & NOBLITT, No. 222 South Second st. THOMAS THOMPSON, SON & CO., No. 238 South Second st.
Cables, Chain.	R. CLARK & SON, 763 South Front street.
Cables, Hempen.	(See Cordage.)
Candles.	FRANCIS CONWAY, 114 and 118 Relief street. WILLIAM CONWAY, No. 316 South Second st., below Spruce. G. DALLETT & CO., 122 and 1319 Market, and 10th & Callow'l. E. DUFFY & SON, No. 914 Filbert street. G. M. ELKINTON & SON, 116 Margaretta street. EPHRAIM WILSON, No. 1095 Germantown Road, opp. Second.
Candles, Adamantine and Stearine.	C. H. GRANT & CO., No. 126 South Delaware Avenue. DAVID THAIN & CO., Callowhill and Fairmount.
Candle, Wax.	F. SCHNEIDER, (also Importer and Dealer in Pictures and Religious articles,) 435 Franklin Avenue, below Fifth.
Candle Moulds.	HENRY CALVERLEY, 205 Quarry street, above Arch. JOHN CALVERLEY, 305 Race street.
Canes and Crutches.	AUGUSTUS BICKEL, No. 139 North Fourth street. GEORGE DOLL, No. 14 North Sixth street.
Canes and Ferrules.	FRANCIS KRAMER, No. 856 North Ninth st., below Poplar.
Cans and Canisters.	ARTHUR, BURNHAM & GILROY, (Self-sealing,) 119 S. 10th. HADDEN, CARLL & PORTER, 130 North Second street. REHN & EVERETT, 108 North Front street.
Caps, Cloth. (See p. 281.)	S. D. WALTON & CO., No. 127 North Third street.
Caps & Labels.	JOHN DAVEY, (Metallic,) rear 115 South Second street.
Capstans.	WILLIAM TEES, (also Windlasses,) Beach and Hanover.
Carboys.	H. B. & J. M. BENNERS & CO., 27 South Front street.
Cars.	(See page 311.)
Car Springs.	J. JEFFRIES & SONS, 811 Grape street.
Car Wheels.	A. WHITNEY & SONS, Sixteenth and Callowhill.
Cardboard.	A. M. COLLINS & CO., 506 Minor street. ROBINSON RITSON, corner Twenty-fourth and Green.
Cards, Playing.	SAMUEL HART & CO., 416 South Thirteenth.
Card Punching Machines.	CHAMBERS & RIEHL, 1033 N. Fourth, bet. Poplar and George. W. P. UHLINGER & CO., North Second, above Oxford.
Card Stamping.	JAMES MARTIN, 1128 North Second st., ab. Franklin Avenue.
Carpet & Traveling Bags.	GEO. B. BAINS, 302 Market street, and 6 North Fourth street. THOMAS W. MATTSON, 402 Market street, above Fourth.
Carpetings.	(See page 239.)
Carpets, List & Rag.	GEORGE ALBRECHT, 321 Callowhill, below Fourth. JAMES CRAWSHAW, No. 1516 North Second street. (Also, see page 239.)

ARTICLES.	MANUFACTURERS.
Carriages & Coaches. (See page 202.)	BECKHAUS ALLGAIER & PETRY, 1204 Frankford Road, ab. Frankliln Avenue. DUNLAP'S PHŒNIX WORKS, 490 York Avenue. LANE & CO., 1907 Market. WM. D. ROGERS, (see page 444,) 1009 Chestnut. GEO. W. WATSON, 1219 Chestnut. WENZLER, PFAFF & KROLL, Wallace, above Ridge Avénue. And *many others.*
Carriage Axles.	KIRCHNER & STICKEL, 309 Race street, above Third.
Carriage Bows, Poles and Shafts.	ALFRED RUHL, northwest corner Eighth and Wood streets.
Carriage Trimmings.	(See Dry Goods, page 245.)
Carts, Drays, &c.	(See Wagons and Carts, page 394.)
Carving, Furniture and Ornamental,&c.	THOMAS G. BERING, northwest corner Tenth and Ridge Av. CHARLES BUSHOR, 200 Dock street. EDWARD J. DELANY, corner Ridge Avenue and Eleventh. JOHN HARE OTTON, No. 306 South Fifth street. JOHN SCOTT, 452 North Twelfth, below Buttonwood. R. S. SMITH, 230 and 232 South Second st., below Walnut.
Carving, Pattern.	WILLIAM B. AITKEN, No. 203 Dock street. JACOB BEESLEY & CO., No. 424 Dillwyn st., ab. Callowhill. THOMAS G. BERING, northwest corner Tenth and Ridge Av. THOMAS W. MASON, No. 233 South Fifth street. SMITH & BROWN, 212 North Second street. JOHN HARE OTTON, 306 South Fifth st.
Carving, Ship.	FRANCIS FOX, 131 Dock street.
Cases, Card.	SAMUEL FISHER, (Pearl,) 1509 Linden st., west of Fifteenth, between Market and Chestnut. SAMUEL HART & CO., Thirteenth, above Pine.
Cases, Jewelers, Morocco and Velvet.	G. F. KOLB, 341 Harmony st., corner South Fourth PEACOCK & FICKERT, Fifth and Chestnut streets.
Cases, Mahogany, Morocco, &c	JOSEPH ELLIS, 101 South Eighth Street. G. F. KOLB, 341 Harmony st., corner South Fourth. JACOB LUTZ, 109 South Eighth street, (fourth story). PEACOCK & FICKERT, Fifth and Chestnut streets. JOSEPH LAUGHLIN, 300 Walnut street, near Exchange. NATHAN STARKEY, 116 South 8th, bet. Chestnut and Walnut.
Cask and Hogshead Shooks.	WM. M. COOPER & CO., 135 S. Water, and 71 Church. W. E. STEVENSON, 325 and 333 South Front.
Casters, French Furniture.	FRANCIS CLAVELOUX, 109 South Second, above Walnut.
Cedar Ware.	CHIPMAN & WHITE, 11 South Front. C. DREBY, 414 North Second street, above Callowhill. JOHN LOUIS & SON, 504 Vine street, above Fifth.
Chamois Leather.	McNEELY & CO., 64 North Fourth street.
Chandeliers.	(See page 352.)
Charcoal.	COOK & EMERICK, 410 Queen street, Kensington.
Chairs.	BENJAMIN H. BRAYMAN, Wholesale and Retail, No. 27 Front street, below Arch. (Chairs packed with care for shipping.) THOMAS L. PRICE, Franklin Avenue, below Eighth. W. D. REICHNER, (Cane Seat, of best materials and lowest cash prices,) 339 North Front street. NATHAN ROBBINS, (Boston, Rocking, Nurse, and Office; also, Night, Cabinet, Pivot, Office and Barber Chairs and Stools,) 243 North Front street.

ARTICLES.	MANUFACTURERS.
Chairs.—*Continued.*	WISLER & BROTHER, 223 and 225 North Sixth st., opposite Franklin Square. And *many others.*
Chair Findings.	JOHN McCULLOUGH, Willow street, between 12th and 13th.
Chemicals.	(See page 206.)
Chess Boards.	F. H. SMITH, 716 Arch street.
China Ware.	GLOUCESTER CHINA COMPANY, 17 North Sixth street.
Chocolate, Cocoa, &c.	(See Spices, and page 269.)
Chronometers.	WILLIAM E. HARPUR, 428 Chestnut street.
Churns.	G. SPAIN, southwest corner Ninth and Coates.
Cigar Boxes.	FREDERICK BRECHT, 406 Vine street. JOHN W. FARRELL, (and furnishing,) 336 Harmony Court. M. THALHEIMER, 417 Dillwyn street.
Cigar Cases.	H. A. HEUSSLER, 341 Harmony st., corner South Fourth.
Cigars.	(See page 388.)
Cloaks & Mantillas	(See Clothing, page 225.)
Clothing, Ready-made.	(See page 220.)
Cloths, Refinishing.	LOUIS SCHWARTSWAELDER, (also Sponging and Shrinking by a new process,) 8 Fetter Lane.
Coaches.	(See Carriages and Coaches.)
Coaches, Gigs, &c. (for Children.)	ASKAM & SON, No. 131 Dock street. CHARLES S. SWOPE, 753 South Third. T. W. & J. A. YOST, (successors to Bushnell & Tull,) 214 Dock street, above Second. WILLIAM QUINN, 416 Library st., back of Custom House.
Coffee, Essence of.	BOHLER, TOMSON & WEIKEL, 243 N. Third st., below Vine. M. GUGGENHEIM & CO., 1040 N. Third street, corner George. CHARLES KROBERGER, No. 458 Dillwyn street. McCULLOUGH & CO., (Concentrated Turkey,) Arch, bel. 3d. G. G. MILLER & CO., (Premium,) 1314 Crown st., Kensington.
Coffee Mills.	SELSOR, COOK & CO., Germantown.
Coffee Pots, Patent.	ARTHUR, BURNHAM & GILROY, ("The Old Dominion,") 117 and 119 South Tenth street.
Coffee, Roasting.	EAGLE SPICE MILLS, 244 North Front.
Coffins.	WM. H. MOORE, 1415 Arch st., and by Undertakers generally
Coining Machinery.	MORGAN, ORR & CO., Callowhill, below Broad.
Collars, Enameled.	WM. E. LOCKWOOD, (Hunt's Patent,) 236 Chestnut street.
Colors.	(See page 217.)
Combs, Brushmak'rs.	HENRY HUBER, Jr., North Fifth street, above Market.
Combs, Horn. (See page 407.)	Mrs. W. COOK, 1009 Randolph street, below Franklin Avenue. G. G. MILLER, 1314 Crown street, Kensington. REDHEFFER & SONS, 615 Coates street.
Combs, Shell & Buffalo Horn.	REDHEFFER & SONS, 615 Coates street. (See page 407.)

ARTICLES.	MANUFACTURERS.
Combs, Gold & Silver	GEORGE P. PILLING, No. 214 Gold street, above Second.
Concertinos.	C. M. ZIMMERMAN, (Premium awarded,) 238 North Second st.
Confectionery.	S. HENRION, No. 712 Market street, (Eagle Jujube Paste, Gum Drops, Chocolates, Syrups, &c.) WEISS & SCHELL, No. 831 Vine street, below Ninth. E. G. WHITMAN & CO., No. 102 South Second street. STEPHEN F. WHITMAN, No. 1210 Market street. (See Confectionery, page 226.)
Confection'rs Mo'lds.	JOHN GARDNER, 3418 Market st., West Philadelphia.
Cooking Ranges, &c.	(See page 290.)
Cooper Work.	H. APPLE & SONS, (also Gauging,) 121 South Water, and 230 North Water. C. DREBY, (Cedar,) 414 North Second, above Callowhill. HENRY HAAS, 707 Brook st., between Third and St. John, and Brown and Coates.
Cooper's Tools.	J. L. WILLIAMS, Fourth, above Columbia street.
Copper and Steel Plates.	J. KEIM, Eutaw street, near Race, above Seventh.
Copper Work.	JOSEPH OAT & SON, 232 and 234 Quarry street.
Copying Presses.	(See Presses.)
Cordage.	WEAVER, FITLER & CO., 23 North Water. And *many others.*
Cord, Cotton	HENRY CONKLE, No. 4 South Third street.
Corks.	GEO. M. FRIED & SON, No. 237 South Fifth street. GEORGE HAMMER, (By Patent Machine,) No. 823 N. Third, above Brown. JOSEPH MURPHY, No. 631 North Third, below Coates.
Cork Soles.	L. ZUEGNER, No. 278 North Third street.
Corsets. (See page 240.)	Mrs. C. BROWN, No. 329 Arch street. C. & E. HENSZEY, No. 521 Chestnut street. And *many others.*
Costumes.	COX & DESMOND, (for Balls, Theatres, &c.,) 917 Race street.
Cotton Machinery.	A. JENKS & SON, Bridesburg, (See page 299 and Appendix.)
Crackers & Biscuits. (See page 267.)	IVINS & ALLEN, No. 321 North Front street. P. MAISON, 134 North Front street. RICKETTS & CO., 153, 155, and 157 North Front street.
Cravats, Ties, &c.	C. A. BUTTS, No. 27 North Eighth street.
Cricket Bats.	W. J. WALKER, 819 Spruce street, above Eighth.
Crucibles, Clay.	J. & T. HAIG, 975 North Second street.
Curled Hair.	(See Glue, &c., page 218.)
Curriers' Knife-Blades.	J. M. EARNEST, No. 858 North Fourth street. JACOB ZEBLEY, 402 Cherry street.
Curry Combs.	W. BEACH, (Patent,) Willow street, between 12th and 13th.
Curtain Fixtures.	G. LYMAN MILLER, Fifteenth st., bet. Willow and Hamilton.
Curtains.	H. B. BLANCHARD & CO., 727 Chestnut street. WM. H. CARRYL & BRO., 719 Chestnut street.

38

ARTICLES.	MANUFACTURERS.
Cutlery. (See also " Edge Tools and Surgical Instruments.")	CLARENBACH & HERDER, (Shears, &c.) 634 Arch street. WM. GILCHRIST, (Razors,) No. 445 North Broad street. B. RICHARDSON, (Table,) 117 South Second street.
Daguerreotypes, &c.	GERMON, BROADBENT, and *many others.*
Daguerreotype Cases.	DABBS & BIRMINGHAM, (and Spectacle Cases,) 1 Ranstead Place, Fourth, above Chestnut. C. C. SCHLEUNES, Harmony Court, corner Fourth.
——— Chemicals.	(See page 209.)
——— Stock.	DABBS & BIRMINGHAM, 1 Ranstead Place, 4th, ab. Chestnut.
——— Tubes.	ADOLPH WIRTH, (also Ambrotype,) 704 Arch street.
Dental Instruments.	(See Surgical Instruments.
Dentists' Files.	J. M. EARNEST, No. 858 North Fourth, below Poplar. ROBERT MURPHY, 226 North Fourth, opposite Branch street.
——— Gold Foil.	CHARLES ABBEY & SONS, 228 Pear street, below Third.
——— Jewelry & Pearl Work, Mirrors, &c.	SAMUEL FISHER, 1509 Linden st., bet. Market and Chestnut. EDWARD J. JENKINS, 15 North Ninth street.
—— Gold Plate Work.	CORYELL & ROBSON, (and Dental Perfumery,) southeast corner Seventh and Sansom.
——— Moulds.	B. P. HOLLINGSWORTH, 123 Coates Alley.
Desks, Portable, &c.	W. T. FRY & CO., 15 North Sixth, above Market. NATHAN STARKEY, 116 South Eighth, above Sansom.
Diamonds, Cutting.	F. BOHRER, (Ruling, &c.) 20 Franklin Place.
Die Sinking.	A. & G. McCLEMENT, 321 Chestnut street.
Distilling. Grain.	(See " Distilling and Rectifying," page 231.)
Dolls and Toys.	LUDWIG GREINER, (Patent Doll Heads,) 414 North Fourth.
Drawing Instrum'ts.	THEODORE ALTENEDER, (Patent Joint,) also Importer of Optical Instruments, 229 New street.
Dressing Cases.	W. T. FRY & CO., (Wood and Leather,) 15 North Sixth.
Drug & Spice Mills.	C. VANHORN & CO., Fifteenth and Hamilton.
Drums, Military.	CHARLES M. ZIMMERMAN, (Patented,) 248 North Second st. (Also, Banjos, Tamborines, Wired Strings, &c.)
Drum Heads.	MORGAN & WELBANK, No. 402 North 9th, above Callowhill.
Dry Goods.	(See page 232.)
Dumb Bells.	STUART & PETERSON, Willow, above Thirteenth.
Dyeing.	(See Dry Goods, page 232.)
Dye Woods.	(See page 214.)
Earthenware. (Also, see page 201.)	J. & T. HAIG, No. 975 North Second street. HYZER & LEWELLEN, No. 952 North Ninth, above Poplar ISAAC SPIEGEL, Jr., Brown, bet. Cherry & Vienna, Kensing'n
Easels, Artists'.	THEODORE KELLEY, No. 22 South Eighth street.
Edge Gilding.	G. ECKENDORFF, 203 South Fifth.

ARTICLES.	MANUFACTURERS.
Edge Tools and Cutlery. (See page 336.)	BOOTH & MILLS, Little Washington, below Ninth WILLIAM CONAWAY, 402 Cherry street. C. HAMMOND, 503 Commerce street. JACOB ZEBLEY, 402 Cherry street above Fourth.
Electro-Plating and Gilding.	WM. BARBER, No. 110 South Eighth street. E. & G. EAKINS, (Howell's Building,) southwest corner Sixth and Chestnut. FRANCIS JAHN, No. 435 Race st., below Fifth street. T. T. KINSEY, No. 207 Race street, above Second. E. L. MINTZER, No. 23 North Sixth street. JOHN O. MEAD & SONS, Ninth and Chestnut.
Electrotyping.	E. & G. EAKINS, Howell's Building, Sixth and Chestnut. (Also, see Stereotyping.)
Embroideries.	E. GROSJEAN, Tenth, below Chestnut. And *many others.*
Emery Paper.	BAEDER, DELANEY & ADAMSON, No. 14 South Fourth.
Enameled Cloths.	THOMAS POTTER, 229 Arch, and 18th and Spring Garden.
Engines.	(See "Steam and Fire Engines," page 316.)
Engine Turning.	C. G. CROWELL, Dock, corner of Walnut.
Engraving. See page 180.)	BAXTER & HARLEY, (see page 180,) 35 South Sixth. BOERUM & NOBLE, (Wood,) 127 South Third. E. H. COGGINS, (also Printing,) 36 North Eighth. MARTIN LEANS, 402½ Chestnut street. SAMUEL MAROT, 434 Chestnut and Fifth. A. & G. McCLEMENT, (also Embossing,) 321 Chestnut street. E. ROGERS, (see page 180,) 132 South Third street. A. C. SUPLEE, 326 Chestnut street. And *many others.*
Engraving, Bk. Note.	AMERICAN BANK NOTE COMPANY, Philadelphia.
—— **Calico.**	MILLER, READER & CO., 6 Lagrange Place.
—— **Jewelry.**	E. F. BATON, 722 Chestnut street. WM. F. CAVENAUGH, 223 Dock street, (room No. 7, 3d story.)
—— **Seal.**	ROBERT LOVETT, 200 South Fifth.
Engravers' Tools.	GEORGE C. HOWARD, Eighteenth, below Market.
Envelopes.	SAMUEL H. BERRY, 33 South Sixth street. N. AMERICAN PAPER BAG AND ENVELOPE CO., northwest corner Ninth and Wallace. A. & G. McCLEMENT, 321 Chestnut street.
Extracts & Essences.	C. D. KNIGHT, 7 South Sixth street.
Eyelet Machines.	HYMEN L. LIPMAN, (Patent Improved Punch and Fastener combined,) 32 South Fourth street.
Fans.	* * * * * * * * * * *
Fan Blowers.	M. ALDEN, North Fifteenth and Willow.
Fashions, Publishers of.	A. T. WARD, (Tailors',) 333 and 335 Chestnut street.
Faucets.	(See Brass Cocks, &c.)
Felting, St'm Boiler.	H. W. MILLER, Sr., 1801 N. Sixth, and 5th & Germantown R.
Fertilizers.	(See page 145.)

ARTICLES.	MANUFACTURERS.
Files and Rasps. (See page 331.)	JAMES GILFEATHER, 1311 Germantown Road. J. M. EARNEST, (Dentists,)858 North Fourth street. R. MURPHY, (Jewelers and Watchmakers,) 226 North Sixth. JAMES B. SMITH, (See page 332,) 211 New st., ab. Second.
Fire Bricks, Tiles, & Clay Furnaces, &c.	NEWKUMET & MELICK, Vine, near Twenty-third street. ISAAC SPIEGEL, Jr., Brown, bet. Cherry & Vienna, Kensing'n. GEO. SWEENEY & CO., 1330 Ridge Avenue. (Also, see page 200.)
Fire Engines.	JOHN AGNEW, 922 Vine street.
Fire-Works.	SAMUEL JACKSON, Federal street, below 7th.
Fish Hooks.	SHADRACH HILL, 25 Bank street.
Fishing & Sporting Tackle.	CHARLES GODFREY, 104 North Second street. JOHN KRIDER & CO., northeast corner Second and Walnut.
Flags, Banners, &c.	WILLIAM G. MINTZER, 215 North Third.
Flasks, Pocket.	W. T. FRY & CO., 15 North Sixth, above Market.
Flour.	(See page 264.)
Flutes, &c.	KLEMM & BROTHER, 705 Market.
Fly Nets.	Mrs. R. MYERS, 1534 North Fourth, above Jefferson. GEO. TEILL, 38 North Third st., below Arch.
Force Pumps.	ARTHUR, BURNHAM & GILROY, (Burnham's Patent Double Acting,) 117 and 119 South Tenth street, corner George. A. C. BROWN, northeast corner Eighth and Buttonwood.
Forks, Steel.	(See page 330.)
Founding, Iron and Brass.	(See pages 287 & 189.)
Foundry Facings.	COOKE & EMERICK, 410 Queen street, Kensington. C. VANHORN & CO., 39 North Front, and 15th, cor. Hamilton.
Frames, Picture, &c.	(See Picture and Looking-Glass Frames.)
Fringes, Cord, Tassels, Gimp, &c.	J. B. CHAMPROMY, 133 North Third. JOHN C. GRAHAM, 607 Cherry street. H. W. HENSEL, 20 North Fourth. CHARLES MEVES, 9 North Ninth street. And *many others.*
Fruits, Preserved.	MILLS B. ESPY, 255 South Third.
Furnaces & Ranges.	(See page 291.)
Furnishing Goods, (Gentlemen's.) (See page 224.)	C. A. BUTTS, No. 27 North Eighth street. OLDENBERGH & TAGGART, 146 North Fourth st. JOHN C. REMINGTON, 217 North Fourth st., corner of Branch. WINCHESTER & CO., 706 Chestnut. Miss H. SOUDER, 44 North Sixth.
—— **Ladies.**	MADAME SEGORII, 628 Chestnut.
Furniture & Upholstery.	(See page 271.)
Furniture, Church.	(See page 274.)
Furniture, Iron.	SAMUEL MACFERRAN, Arch, below Sixth.
Furniture Polish.	SAMUEL RUE, 137 North Tenth.

ARTICLES.	MANUFACTURERS.
Furs, Plain & Fancy. (See page 282.)	P. GRAFF, No. 224 Arch street. EMANUEL C. PAGE, No. 325 Green st., near Fourth.
Gaiters, Over.	J. H. RICHELDERFER, Chestnut, above Ninth.
Galvanized Iron.	(See page 297.)
Gas Apparatus.	(See page 321.)
Gas Burners.	C. GEFRORER, 111 South Eighth street.
Gas Fitters Tools,&c.	EDWARD BORMAN, 523 Cherry street. MORRIS, TASKER & CO., Third, below Walnut.
Gas Fixtures.	(See page 352.)
Gas Flexible Tubing.	AMASA STONE, 207 Quarry street, above Second.
Gas Meters.	CODE, HOPPER & CO., (see page 323,) 1505 Jones st.
Gas Proving Apparatus.	C. GEFRORER, (for Gas-Fitters,) 111 South Eighth street.
Gas Stoves.	(See page 290.)
Gas Works, Portable	STRATTON & BROTHER, 719 Walnut.
Gauges, Steam, Vacuum and Water.	DAVID LITHGOW, (Grimes' Patent,) 305 Walnut st.
German Silver.	SAMUEL CROFT, 22 Decatur street.
—— **Silver Castings.**	THOMAS HARRINGTON, rear 15 N. Ninth st., above Market.
Gilt Mouldings.	EVERS & VOTTELER, 13 North Eighth st. FISHER & CO., 141 South Second st. F. GABRYLEWITZ, (Premium awarded by Franklin Institute,) No. 47 North Ninth st., below Arch. THIERRY & KRUSE, North Fourth, corner Branch.
Ginger, Essence of.	F. BROWN, Fifth and Chestnut.
Glass.	(See page 276.)
Glass Cutting.	WILLIAM BALL, 205 Quarry street.
Glass, Engraving on.	E. W. USSHER, Eighth, below Chestnut.
Glass Preserv'g Jars.	A. STONE & CO., 412 Race st., and 207 Quarry st.
Glass, Stained.	(See Stained Glass, page 278.)
Glass Syringes, Vials, and Tubing.	THOMAS BURNS, (Homœopathic Vials and Chemical & Philosophical Tube Ware,) 35 Walnut st.
Gloves, Mittens, &c. Buckskin.	P. HEALY & CO., 204 Walnut st. MORGAN & WELBANK, 402 North 9th st., above Callowhill.
Gloves, Kid.	J. R. ASHFORD, 607 Callowhill st.
Glue & Curled Hair.	(See Glue, &c., page 218.)
Glycerine.	HENNEL, STEVENS & CO., (page 209,) 4th st., ab. Market.
Gold Chains.	DREER & SEARS, (see page 344,) Goldsmiths' Hall. STACY B. OPDYKE, 610 Sansom st.
Gold Foil.	CHARLES ABBEY & SONS, (see page 347,) 228 Pear st.
Gold Pens.	PETER WALKER & BROTHER, 13 South Sixth.

38*

ARTICLES.	MANUFACTURERS.
Gold & Silver Leaf, &c.	WILLIAM EVETT, 28 North Fifth street. HASTINGS & CO., Fifth and Cresson's Alley. HENRY NELMS, 216 Pear street, bet. Third and Dock.
Gold Spectacles.	N. E. MORGAN, 610 Sansom street.
Grates and Fenders.	ARNOLD & WILSON, 1010 Chestnut st.
Greases, Rl'd Car, &c.	R. S. HUBBARD & SON, Office 107 Walnut st.
Grindstones.	JAMES E. MITCHELL, (and Plaster,) 310 York Avenue.
Grist Mills.	T. B. WOODWARD & CO., Germantown R. and New Market
Guitars.	J. BERWIND, 1513 George street.
Gun Cotton.	GARRIGUES & MAGEE, 108 North Fifth.
Guns, Pistols, and Rifles.	JOHN KRIDER & CO., northeast corner Second and Walnut. TRYON, SON & CO., 625 Market, 616 St. James street, and 220 North Second st. SHARP'S RIFLE FACTORY, (see page 337,) Fairmount. UNION RIFLE FACTORY, North Second, above Dauphin.
Gun Caps.	J. WURFFLEIN, (for Cannon,) 210 S. Third, below Walnut.
Gun Mountings.	THOS. HARRINGTON, rear of 15 North Ninth.
Gun & Pistol Stocks.	ELIZA BARRY, 1021 Melon street, near Eleventh.
Hair, Curled.	(See Glue, &c., page 218.)
Hair Dye.	GEORGE THURGALAND, 29 South Sixth street.
Hair Pins.	GEO. W. CARR & CO., (Whalebone,) 126 Willow street.
Hair Plaiting and Jewelry.	F. FROMHAGEN, 9 South Eighth street. Mrs. A. GREEN, 439 Arch street, near Fifth. F. SCHALCH, (also Hair Plaiting,) 336 South Fourth. CHARLES STUBENRAUCH, 140 North Fifth street.
Hair Stocks.	(See Stocks.)
Hames, Root.	HENRY BEAGLE, (Dray, Cart, Wagon, and Plow,) cor. Magnolia and Willow street, above Fifth.
Hames, Iron.	JOHN KLUFKEE, 135½ Marlboro', near West st., Kensington. PAUL SCHWARZKOPF, 519 Emlen st., ab. Front, bel. Noble.
Hammers, &c.	C. HAMMOND, 503 Commerce street.
Handles.	(See Tool Handles.)
Hardware.	(See Hardware and Tools, page 328.)
Harness.	(See Saddles, Harness, &c., page 375.)
Harness Mountings.	GELBACH & METZGER, (Brass,) 1241 Howard st., Kensington. WM. LITTLE & SONS, (Ornaments,) 452 Sansom's Alley, east of Third, between Noble and Willow.
Hats and Caps.	(See page 278.)
Hat & Bonnet Blocks.	JOHN AIKMAN, No. 141 Dock street, below Second. CHRISTIAN NONNENBERGER, 323 Race st., bel. Fourth. EDWARD DUFRENE, (Plaster,) South Seventh, bel. Chestnut.
Hat Cases.	BENJ. ANDREWS, (Leather,) 116 North Fourth. (Also, see Paper Boxes.)
Hats, Straw and Panama.	ADOLPH DESSART, (Washing, Dyeing, Pressing & Trimming of Felt, Panama, and Leghorn Hats,) 541 North Third. KLEINZ & FIELD, 36 North Second street.

ARTICLES.	MANUFACTURERS.
Hatters' Trimmings, &c.	DANIEL DOREY, 320 Chestnut, and 4 Hudson's Alley. J. C. KELCH, 506 Market.
Hat Tip Printing.	ISRAEL AMIES, (and Embossing,) 25 and 27 old, Minor.
Hinges, Brass Ship.	F. W. & G. A. KOHLER, 528 N. Second, above Buttonwood.
Hinges, Brass Butt.	G. W. BRADFIELD, (and Silver Plated,) 525 N. 12th, near Ridge Avenue.
Hinges, Iron.	W. H. McCALLA & CO., Montgomery, above Front.
Hoes.	GEO. GRIFFITHS, 1 Fetter Lane, ab. Arch, bet. 3d and Bread. PRINCE'S HOE FACTORY, Pennepack Creek.
Hoisting Machines.	ROBERT McCALVEY & CO., No. 602 Cherry street.
Hoops, Ladies.	SHARON SLEEPER, (for Skirts,) 1002 Market street.
Horse Collars.	M. & J. McCOLGAN, South Thirteenth, near Market. WM. R. SCOTT, (Patent Leather—Irish, Scotch, Draft, and all other,) 119 North Front.
Horse Nails.	FRANK DARDAS, Mechanic and Fourth street, Southwark. GEORGE JACKSON, Lilley Alley, above Green.
Horticultural Implements.	HENRY A. DREER, 327 Chestnut street.
Hose Couplings, &c.	THOMAS S. SMITH, (Branch Pipes,) Girard Av., W. of 12th.
Hose, Fire & Garden.	BARNETT & JENKINS, 112 North Sixth street. WM. ECKFELDT, 212 N. Thirteenth, and 418 N. Third st.
Hose, Hempen.	A. STONE, 207 Quarry street, above Second.
Hosiery.	(See pages 240–244.)
House Furnishing & Housekeeping Goods.	JOHN AMBLER, Jr., 711 Spring Garden street. ARTHUR, BURNHAM & GILROY, ("What Cheer" Yeast Cakes,) 117 and 119 South Tenth street. ISAAC S. WILLIAMS, 726 Market street. WM. J. WALKER, (Wooden Ware,) 819 Spruce street. J. LOUIS & SON, (Wooden Ware,) 504 Vine street.
Hubs.	(See Spokes and Hubs.)
Husks, (for Upholsterers.)	JONATHAN COLLINS, rear of 220 North Second.
Hydrant Cases.	BENJAMIN ESLER, 26 South Fifteenth.
Hydrants.	ARTHUR, BURNHAM & GILROY, (Cochrane's Patent Non-Freezing Non-Wasting,) 117 and 119 South Tenth, corner George.
Hydraulic Presses.	(See Presses.)
Hydraulic Rams.	A. C. BROWN, (Agent for Holliday's Windmill for Pumping Water,) northeast corner Eighth and Buttonwood.
Ice-Cream Freezers.	ARTHUR, BURNHAM & GILROY, (Masser's Patent Five-Minute Freezer,) 117 and 119 S. Tenth, corner George.
Ice Picks.	JAMES PATCHELL, 812 Race street.
Ice Tools.	LOWER & CO., 712 and 714 Washington Avenue.
Indigo Blue.	M. GUGGENHEIM & CO., 1040 North Third, corner George.
Indigo Paste.	J. ANDREYKOVICZ, (also Archil,) 108 Arch street.

ARTICLES.	MANUFACTURERS.
Inks. (See page 402.)	APOLLOS W. HARRISON, 26 South Seventh street. JOSEPH E. HOOVER, 416 Race street. J. S. MASON & CO., 138 North Front. SAMUEL SCHURCH, (Stationer,) 240 Race street. And *many others.*
Ink, Indelible.	W. C. BAKES, southwest corner Seventh and Buttonwood.
Inks, Printing. (See page 175.)	LAY & BROTHER, (Black and Colored,) 241 Dock street. L. MARTIN & CO., 215 Lodge street, above Second.
Ink-Stands.	ARTHUR, BURNHAM & GILROY, (Arthur's Patent Air-tight—also Paste Jars,) 117 and 119 South Tenth, corner George JACOB KIRCHEM, (Nock's Patent,) 8 South Seventh. HYMEN L. LIPMAN, (Nock's Patent Round Hinge,) 32 S. 4th.
Instruments, Draw'g	THEODORE ATTENEDER, (Patent Joint,) 229 New street.
—— **Engineers & Surveyors.**	EDMUND DRAPER, 226 Pear street, near the Post-office.
—— **Mathematical & Optical.**	SAMUEL L. FOX, (also Surveyors,) 537 North Second. McALLISTER & BROTHER, 728 Chestnut street. THEODORE MUELLER, 132 Noble st., corner New Market. ISAAC SCHNAITMAN, 225 North Fourth. ADOLPH WIRTH, 704 Arch, above Seventh. WILLIAM J. YOUNG & SON, 43 North Seventh street.
—— **Musical.**	(See page 408 ; also Pianos, Organs, &c.)
—— **Philosophical.**	L. C. FRANCIS, (and Chemical,) 100 S. Eighth, cor. Chestnut. THEODORE MUELLER, 132 Noble, corner New Market.
—— **Telegraphic.**	JAMES J. CLARK, 160 Dock street, corner South Second.
Iron.	And its MANUFACTURES, page 283.
Irridium.	J. BISHOP, 207 Pear street.
Isinglass.	BAEDER, DELANY & ADAMSON, 14 South Fourth.
Ivory or Bone Black.	E. M. SEELEY & SON, North Second, above Columbia street.
Ivory Turning.	HARVEY & FORD, (also Carving,) 422 Library street.
Jack Screws.	M. W. BALDWIN & CO., Broad and Hamilton.
Jacquard Machines.	CHAMBERS & RIEHL, 1033 N. Fourth, bet. Poplar & George. W. P. UHLINGER & CO., 919 and 921 North Second.
Japanning.	D. D. DICK, 625 St. James street. DAVID JONES, 413 Vine street, above Fourth.
Japan Ware.	ISAAC S. WILLIAMS, 726 Market st. And many others
Jet Ornaments.	H. OLIVER, southwest corner Eighth and Arch.
Jewelers' Tools.	H. H. SMITH, 201 Carter's Alley, corner Second street.
Jewelry. (See page 342); also, see Hair Jewelry, Masonic Marks, Silver Ware, &c.)	BAILEY & CO., 819 Chestnut. JACOB BENNET, (also, Diamond Setting,) 326 Chestnut, below Fourth. CARROW, THIBAULT & CO., (successors to Dubosq, Carrow & Co.,) 308 Chestnut street. DREER & SEARS, Goldsmiths' Hall, 418 Library street. S. B. OPDYKE, Sansom street Hall, rear Jones Hotel. PAYTON, HAWKINS & CO., 326 Chestnut street. GEO. W. SIMONS & BROTHER, 610 Sansom street.
Jewelry, Hair.	SCHMITT & STUBENRAUCH, (also, Hair Plaiting & Devices,) 928 Chestnut street.

ARTICLES.	MANUFACTURERS.
Jewelry and Silver Chasing.	C. J. SMITH, 1719 Lombard street.
Kindling Wood.	CHARLES RUMP, 37 Haydock street, below Front.
Knit Goods.	JOHN GADSBY & SONS, (Knit Jackets, Shirts, Drawers, Scarfs, Cravats, Mittens, and Guernsey Shirts,) 10 Fetter Lane.
Knitting Machines. (See page 303.)	JOHN LARARD, (every description of Needles for Knitting Machines,) 978 Marshall street, below Franklin. C. SHIRTCLIFF, 1226 Germantown Road. W. P. UHLINGER & CO., 919 and 921 North Second street.
Knives and Forks.	B. RICHARDSON, 117 South Second.
Knives, Mowing.	C. H. LAME, (and Reaping,) 407 Cherry street.
Lace, Coach.	(See Dry Goods, page 245.)
Laces, Embroideries, &c.	* * * * * * * *
Ladders.	JOSEPH RUE, 845 Parrish, below Ninth.
Ladies Dress Trimmings.	(See Trimmings.)
Lager Beer.	(See page 195.)
Lamp Black.	L. MARTIN & CO., 215 Lodge street, above Second.
Lamps, Chandeliers.	(See page 352.)
Lamps, Coach, &c.	E. W. USSHER, 109 South Eighth.
Lamp Shades.	V. QUARRE, 805 Race street, above Eighth.
Lamp Wick.	A. STONE, (for Solar, Fluid, and Astral, etc.,) 207 Quarry st
Lanterns.	J. H. ROHRMAN, 606 Cherry street. CHARLES WILHELM, (and Lamps,) 919 Race street.
Laps, Cotton.	(See Dry Goods, page 260.)
Lapidary's Work.	F. BOHRER, (Glass and Stone Cutting,) 20 Franklin Place.
Lasts.	PETER DEWEES, 113, (old, 31,) Callowhill street. JACOB FOSTER, 305 (old, 3,) Cherry street, above Third. J. HOWARD & CO., (Steam,) 112 Bread street. GEORGE MUNRO, (Anatomical,) 127 Callowhill, below Second.
Lathes.	(See Machinery, page 315.)
Laundry Work.	KOCHERSPERGER & CO., (Steam,) 645 North Broad.
Law Blanks.	JNO. B. SPRINGER, (and Parchment,) S. E. cor. Fifth & Walnut.
Leads for Pencils.	HYMEN L. LIPMAN, (see Pencils,) 4 Ranstead Place.
Lead Pipe, &c.	TATHAM & BROTHERS, Delaware Avenue, below South.
Leads, Sounding	JOSEPH BAKER & SON, 820 Rachel, and 821 North Second.
Lead, Sugar of.	SAMUEL GRANT, Jr., & CO., (see page 208,) 139 S. Water st.
Lead, Paints.	(See pages 208 and 215.)
Leather.	(Bookbinders', Glove, Morocco, &c., see page 357.)
—— English Kid.	G. R. CORRY, (Glove and Gaiter,) 455 N Third, ab. Willow

ARTICLES.	MANUFACTURERS.
Leather Belting and Hose.	BARNET & JENKINS, 112 North Sixth street. WM. ECKFELDT, 212 N. 13th, and 433 N. Third street.
Lenses and Prisms.	ADOLPH WIRTH, (Simple and Achromatic,) 704 Arch st.
Life Preservers.	JOHN THORNLEY, 311 Chestnut street
Lightning Rods.	THOMAS ARMITAGE, 1206 Vine street. (See page 341.) A. C. BROWN, (Copper and Iron,) N. E. cor. 8th & Buttonwood.
——— Points.	F. JAHN, (Plated and Solid Pointed Platina,) 435 Race street T. T. KINSEY, 207 Race street, above Second. F. W. & G. A. KOHLER, 528 North Second, ab. Buttonwood. EDWARD L. MINTZER, 23 North Sixth street.
Lime.	(Super-Phosphate of, see page 145.)
Lithography.	(See page 182.)
Lithographic Machines.	(See Machinery, page 321.)
Locks. (See page 332.)	THOS. H. AUROCKER, 134 North Thirteenth. JULIUS BILLERBECK, 902 Ridge Avenue, above Vine st. G. W. BRADFIELD, (Brass,) 525 North Twelfth street. JACOB KIRCHEM, (Nock's Patent,) 8 South Seventh. C. LIEBRICH, (see page 333,) 110 South Eighth. J. B. SHANNON, (see page 333,) 58 North Sixth street.
—— Bank.	LINUS YALE, Jr., & CO., (see page 333,) Front & New sts.
Locomotives.	(See page 305.)
Looking-Glass Frames.	(See Picture and Looking-Glass Frames.)
Looms.	(See Cotton and Woolen Machinery, page 299.)
—— Swivel & Hand.	CHAMBERS & RIEHL, 1033 N. Fourth, bet. Poplar & George. W. P. UHLINGER & CO., (page 303,) 919 & 921 N. Second st.
Lozenges.	GEORGE BATES, 1248 Hanover st., Kensington.
Macaroni, &c.	BOHLER, TOMSON & WEIKEL, 248 N. Third, below Vine.
Machine Cards.	JAMES SMITH & CO., Marshall st., cor. Willow, (see p. 302).
Machines and Machine Work. (See pp. 299 to 329.)	A. L. ARCHAMBAULT, (see p. 317,) Fifteenth and Hamilton. M. ALDEN, (see p. 324,) Fifteenth and Willow street. BEMENT & DOUGHERTY, (see p. 316,) 2029 Callowhill. F. CLAVELOUX, 109 South Second street. JAMES FLINN & CO., (see p. 325,) Sixth and Germantown R. J. J. HEPWORTH, (see p. 302,) cor. Edward and School streets. WM. H. HARRISON, 705 Lodge Alley. R. J. HOLLINGSWORTH, 23 Coates Alley. (See page 325.) HUNSWORTH, EAKINS & CO., Front, corner of Franklin HENRY HOWARD, (see p. 320,) Twenty-third and Hamilton. GEORGE C. HOWARD, (see page 321,) 15 S. Eighteenth. JOHN JACKSON & CO., 1089 and 1091 Germantown Road. KING & DORSEY, 233 S. Fifth, and Prune, below Fifth. E. KALLENBERG & CO., rear 220 N. Second st., and 441 N. 9th. JOHN L. KITE, (see page 317,) 13 Drinker's Alley, below 2d. MATTHEWS & MOORE, Sixteenth and Fairview. C. R. MELLOR, 448 North Twelfth, corner Pleasant st. MERRICK & SONS, (see page 327,) 430 Washington st. I. P. MORRIS & CO., (see page 326,) Office 125 Walnut street. REANEY, NEAFIE & CO., (see p. 317,) 135 Beach street. WM. SELLERS & CO., (see p. 313,) 16th and Pennsylvania Av. CHAS. SHIRTCLIFF, 1226 Germantown Road. CHAS. W. SMITH, 135 North Third street. STANHOPE & SUPLEE, Frankford.

ARTICLES.	MANUFACTURERS.
Machine and Machine Work.—*Cont'd.*	J. T. SUTTON & CO., 131 Franklin Avenue. J. & T. WOOD, (see page 301,) Wood street, near 21st. And *many others.*
Machine Tools.	(See page 314.)
Magnesia.	(See Chemicals, page 211.)
Magnetic Instru'mts.	WM. C. & J. NEFF, 5 South Seventh street.
Mahogany.	ALEX. BECRAFT, rear 426 Walnut street. JOHN EISENBREY & SON, Dock, corner Pear street. LEWIS THOMPSON & CO., Eleventh and Ridge Avenue.
Malt.	FREDERICK GAUL, New Market, corner Callowhill.
Mangles, Patent.	R. A. STRATTON, 1339 Cherry st., near Broad
Mantels, Marble.	(See Marble, page 363.)
—— **Enameled.**	ARNOLD & WILSON, 1010 Chestnut street.
Maps and Charts. (See page 183.)	RUFUS L. BARNES, 27 South Sixth. S. AUGUSTUS MITCHELL, 31 South Sixth. ROBERT P. SMITH, 517 Minor.
Marble.	(See page 360.)
Marbled Paper.	CHARLES WILLIAMS, (see page 178,) 605 Arch street.
Masts and Spars.	D. R. HUMPHRIES & SON, Vienna st. Wharf, Kensington.
Masonic & Odd Fellows Marks, Jewels, &c.	JACOB BENNET, 326 Chestnut street, below Fourth. MARTIN LEANS, 402½ Chestnut street.. SAMUEL MAROT, 434 Chestnut street, corner Fifth. GEO. P. PILLING, 214 Gold st., corner Dock. A. C. SUPLEE, 326 Chestnut street.
Matches.	JOHN S. HODGKINSON, (Friction and German Congreve,) 2 Fetter Lane, 3d street, between Arch and Race. JOHN SCHICK & CO., 919 St. John, above Poplar.
Match - Stands and Safes.	C. O. WILSON, (also Wax Matches and Cigar Lights,) 727 Race street.
Mathematical I'mts.	(See Instruments.)
Mattresses.	(See Beds and Mattresses.)
Mats, Door, &c.	JAMES CRAWSHAW, 1516 North Second street.
Meat Mauls.	WILLIAM BEACH, (Patent,) Willow st., bet. 12th and 13th.
Meat Safes.	W. J. WALKER, 819 Spruce st., above Eighth. J. LOUIS & SON, Vine street, above Fifth.
Medicines.	(See page 212.)
Medicinal Extracts.	EDWARD H. HANCE, (see page 211,) 627 Arch st. N. SPENCER THOMAS, (see page 211,) New Market, near Germantown Road.
Medicine Chests and Medical Saddle-Bags.	J. M. MIGEOD, (Medical Bags,) 27 South Eighth street. PEACOCK & FICKERT, northeast corner Fifth and Chestnut. NATHAN STARKEY, (also Pocket Cases,) 116 South Eighth, between Chestnut and Walnut.
Melodeons	MACNUTT & PRIOR, 108 North Sixth.
Metallic Letters.	HENRY SINKLER, (Raised Brass, &c. ; also, Card Plates and Engraved Names, for Engines and Cemeteries,) Pemberton street, South of Wallace.

ARTICLES.	MANUFACTURERS.
Military Feathers.	Mrs. MENCH, (Ladies Feathers' Dressed, Dyed, and Altered,) 510 North Second street, above Noble.
Military Goods.	W. H. HORSTMANN & SONS, 723 Chestnut. J. H. LAMBERT, 532 Callowhill. W. G. MINTZER, 215 North Third street.
Military Ornaments and Fancy Metal Work.	WM. H. GRAY, 4 Crockett's Court, Fifth, above Chestnut. WM. PINCHIN, 120 Jacoby street, between 12th and 13th. HENRY SINKLER, Pemberton st., S. of Wallace, bel. Ridge Av.
Milk Cans.	JOHN AMBLER, Jr., 711 Spring Garden street.
Millinery Goods.	(See page 413.)
Mineral Water.	J. & S. S. LIPPINCOTT, (in Fountains,) 916 Filbert street. And *many others.*
Mineral Water Apparatus.	JOSEPH BERNHARD & CO., 120 North Sixth.
Mittens and Gloves.	(See Knit Goods and Buckskin Gloves.)
Models, (for Patent Office.)	WM. B. AITKEN, (in Wood and Metal,) 203 Dock, corner 2d. EDWARD BORMAN, (and Small Machines,) 523 Cherry street. THEODORE MUELLER, 132 Noble, corner New Market.
Morocco.	(See Leather, &c., page 359.)
Mouldings, Wood.	BENJAMIN ESLER, 26 South Fifteenth. MATTHEW GRIER, (best quality always on hand or made to order,) Coates, West of Broad, south side.
Moulds.	(See Bottle Moulds, Candle do., Confectioners' do., Dentists' do.)
Mowing Machines.	SPANGLER & GRAHAM, 627 Market.
Mowing Knives.	C. H. LAME, (also Reaping,) 407 Cherry street.
Musical Instruments	(See page 408.)
Mustard.	(See page 269.)
Nails, Cut.	(See page 339.)
Nails and Spikes.	GELBACH & METZGER, (Composition,) 1241 Howard st., above Franklin Avenue. M. McFADDEN, Penn street, above Maiden, Kensington.
Needles, Mattress & Collar.	MICHAEL P. PARTRIDGE, (late from England,) Blair street, bet. Norris and Wood street, Kensington.
Nickel and Cobalt.	BUCK, SIMONIN & CO., (See page 208,) 121 Walnut street.
Nickel Silver Ware.	HARVEY FILLEY, (see page 351,) 1222 Market street.
Nitrate of Silver.	GARRIGUES & MAGEE, 108 North Fifth. (See page 209.)
Nuts, Bolts, &c.	HOOPES & TOWNSEND, (p. 334,) Buttonwood, bel. Broad.
Oars and Sculls.	ALBERTSON BROTHERS, Beach st., ab. Marlboro', Kensing'n WOOD & BROTHERS, Penn street, above Poplar.
Oils.	(Linseed, Lard, Resin, Sperm, &c. see page 369.)
Oil & Floor Cloths.	JAMES CARMICHAEL, 162 North Third street. THOMAS POTTER, (see p. 409,) 229 Arch street.
Oil Presses.	J. & T. WOOD, Wood st., near N. Twenty-first.
Omnibuses.	JOHNSON & ADARE, 329 Broad street.

ARTICLES.	MANUFACTURERS.
Optical Instruments.	(See Instruments.)
Ordnance.	SAVERY & CO., South Front and Reed.
Organ Pipes, Metal.	P. SCHENKEL, 922 Market street.
Organs.	J: C. B. STANDBRIDGE, 2107 Chestnut. (See Mus. Inst., p. 408.)
Oyster Knives.	S. P. MILLER, 309 South Fifth. JACOB SOUDER, 406 Noble.
Packing Boxes.	M. FIFE, 3 Elbow Lane, between Bank and South Third. SUPLEE & MYERS, 514 East North street, above Market.
Paging, Blank Book.	W. WILLARD, 439 Chestnut street.
Pails.	(See Cedar and Wooden Ware.)
Paints & Paint Mills.	(See page 217.)
Pans, Patent Bake.	WILLIAM BEACH, Willow street, between 12th and 13th.
Paper & Paper Mills.	(See page 176.)
Paper Bags, (Machine Made.)	N. AMERICAN PAPER BAG AND ENVELOPE MANUFAC-TURING CO., northwest corner Ninth and Wallace street.
Paper Boxes. (See page 402.)	JESSE BAKER, 21 Bank st., ab. Chestnut, between 2d and 3d. MERRICK BARNES, 102 Bread st., ab. Arch, bet. 2d and 3d. JOHN CROMPTON, 118 North Third street. EYRE & HARVEY, 531 East North street. RUDOLPH K. KNAPP, 230 North Third. GEO. W. PLUMLY, (See page 402,) 213 North Fourth. HENRY J. SEIBEL, 525 Commerce street.
Paper Folding Machines.	(See page 161.)
Paper Hangings.	(See page 371.)
Paper Machinery.	NELSON GAVIT, (see page 319,) 222 and 224 Broad street.
Paper, Metallic.	ROBINSON RITSON, corner Twenty-fourth and Green street.
Paper Ruling.	THOMAS W. PRICE, 22 South Fifth and Library. (See notice of Machine on page 179.)
Paper Staining.	W. H. PATTEN, 205 Arch.
Papier Mache Goods,	D. D. DICK, 625 St. James street.
Parasols,	(See Umbrellas, and page 391.)
Parchment, (and Vellum.)	McNEELY & CO., 64 North Sixth street.
Patent Leather.	GEORGE S. ADLER, 131 Margaretta street.
Pattern Makers.	W. B. AITKEN, 203 Dock street. J. BEESLEY & CO., (also Patterns for Needle-work,) 424 Dillwyn, above Callowhill. THOMAS G. BERING, northwest corner 10th and Ridge Av. THOMAS W. MASON, 233 South Fifth street. C. R. MELLOR, (and Models,) northwest cor. 12th & Pleasant. SMITH & BROWN, 215 North Second.
Pearl Studs & Ornaments.	EDWARD MARKLEW, (also Buttons,) 15th street, bet. Hamilton and Willow, and 1904 Hamilton.
Pearl & Shell Work.	SAMUEL FISHER, 1509 Linden st., W. of 15th, ab. Chestnut.

39

ARTICLES.	MANUFACTURERS.
Pearl Card Cases.	A. GREENHALGH, Manayunk.
Pearl Mountings.	SAMUEL FISHER, 1509 Linden st., W. of 15th, ab. Chestnut.
Pencil & Pen Cases.	GEO. W. SIMONS & BROTHER, (see p. 345,) 610 Sansom st.
Pencils.	HYMEN L. LIPMAN, (Mears' Propelling Pencil, Erasing Pencils; also, Lead and Erasing Pencil combined,) 32 S. 4th st.
Pens, Gold.	(See Gold Pens; also page 348.)
Penholders.	W. C. McREA, (Patent Adjustable,) 907 Chestnut.
Percussion Caps.	J. WURFFLEIN, 210 South Tenth.
Perfumery. (See page 410.)	XAVIER BAZIN, 917 Cherry. GLENN & CO., 720 Chestnut street. APOLLOS W. HARRISON, 26 South Seventh. A. HAWLEY & CO., 117 North Fourth. H. P. & W. C. TAYLOR, 641 and 643 N. Ninth, below Coates. R. & G. A. WRIGHT & CO., 35 South Fourth.
Philosophical I'nsts.	(See Instruments.)
Photographs.	(See Daguerreotypes.)
Physicians' Pocket Cases.	JOSEPH ELLIS, (Allopathic & Homœopathic Bottle,)101 S. 8th. W. T. FRY & CO., 15 North Sixth st., above Market. H. A. HEUSSLER, 341 Harmony Court, corner S. Fourth. JACOB LUTZ, (Allopathic and Homœopathic Bottle,) 109 S. 8th NATHAN STARKEY, 116 South Eighth st., above Sansom.
Piano Fortes. (See page 409.)	H. GOLDSMITH, 33 and 35 South Tenth street. HUNT & CO., 345 North Third street. CONRAD MEYER, 722 Arch street. AUGUST V. REICHENBACH, 1230 Chestnut. GEORGE VOGT, 628 Arch street. And *many others.*
Piano Hardware.	H. & E. GOUJON, 913 Marshall st., above Poplar.
Pickles.	(See Preserved Food.)
Picture & Looking-Glass Frames.	EVERS & VOTTELER, 13 North Eighth street. JAMES S. EARLE, 816 Chestnut street. FISHER & CO., 141 South Second street. F. GABRYLEWITZ, 47 North Ninth street, JACOB GRAEFF, Master st,, above Fifth GUNNING, ROGERS & MYERS, 814 Filbert. JOSEPH HILLIER, 65 North Second street. E. NEWLAND & CO., (Medal Awarded by N. Y. Exhib.,) 604 Arch street. PRICE & SANSOM, 227 Crown st., between Race and Vine. A. S. ROBINSON, 910 Chestnut street. DAVID WILLIAMS, (Packed and Insured from breakage free of charge,) 144 North Third, below Race. E. MASSE, (Black Oval and Passe Partout,) 201 N. Eighth.
Pile-Drivers.	A. L. ARCHAMBAULT, 15th, below Hamilton.
Pill Boxes.	GEORGE W. PLUMLY, 213 North Fourth.
Pills.	(See page 213.)
Pins.	HENRY MEHL, (Parisian Roundhead,) 353 N. Fourth st.
Pipes.	(See Gas Pipes, p. 322; Drain Pipes, p. 200.)
—— **Smoking.**	J. RICHARDS, Oxford st., ab. Frankford Road, Kensington.
Pistols.	(See Guns.)

ARTICLES.	MANUFACTURERS.
Planes.	S. H. BIBIGHAUS, 258 North Third street. B. SHENEMAN & BROTHER, 733 Market.
Planing Mills. (See page 395.)	NAYLOR & CO., 639 North Broad. GEO. B. SLOAT & CO., 1129 Beach street. And *many others.*
Plaster of Paris.	(See page 145.)
Plaster Ornaments.	E. DUFRENE, (also Figures,) Seventh, below Chestnut. THOMAS HEATH, (also Figures,) S. W. corner 11th and Arch. WM. H. FRENCH, (Architectural,) 18th and Chestnut.
Plated Ware.	(See Silver Plated Ware.)
Platina Work.	J. BISHOP, (Chemical Apparatus,) 207 (old No., 5,) Pear street.
Plows.	SAVERY & CO., South Front, below Reed.
Plumes.	Mrs. MENCH, 510 N. Second, bet. Noble and Buttonwood.
Pocket-Books.	PEACOCK & FICKERT, Fifth and Chestnut. JOSEPH LAUGHLIN, 300 Walnut street.
Porcelain Ware.	PORCELAIN FACTORY, Germantown Road.
Port Monaies, Cabas, &c.	H. A. HEUSSLER, (and Fancy Leather articles,) 341 Harmony street, corner Fourth. GEORGE FISCHER, Agent, 7 South Eighth. CHARLES RUMPP & CO., 118 North Fourth. And *many others.*
Porter.	(See page 192.)
Portfolios.	ARTHUR, BURNHAM & GILROY, (Arthur's Patent Elastic,) 117 and 119 South Tenth, corner George.
Potassium, Cyan. of.	BENJ. J. CREW & CO., (see p. 209,) N. W. cor. 5th & Callowhill.
Pottery.	(See page 201.)
Poudrette.	(See page 145.)
Preserved Food.	MILLS B. ESPY, (see page 269,) 255 South Third. J. L. WENDELL, 310 South Front.
Presses.	(Brick, Hydraulic, Oil: see Machinery, p. 302.)
Presses, Copying and Seal.	E. KALLENBERG, rear 220 N. Second, and 441 N. Ninth.
Presses, Printing.	(See page 320.)
Presses, Screw.	CHARLES DIEDRICHS, (and all other kinds,) 31 Vine street.
Printing, Book and Fancy.	(See page 173.)
Printing for Blind.	(See page 174.)
Printing, Plate. (See p. 180,)	HENRY QUIG, 115 South Seventh.
Printers' Furniture.	(See page 321.)
Print Works. w	(See Dry Goods.)
Prussiate of Potash.	CARTER & SCATTERGOOD, (see p 210,) 304 Arch street.
Publishing, Book, Newspaper and Magazine.	(See page 149)

ARTICLES.	MANUFACTURERS.
Pumps.	(See Machinery, Machine Work, Air Pumps, Force Pumps, &c.)
Putty.	(See Chemicals, page 217.)
Quills.	SAMUEL SCHURCH, (Stationer,) 240 Race street.
Quinine.	(See pages 207–208.)
Railing, Iron. (See page 295.)	BANCROFT, HAINES & CO., Christian, W. of Passayunk Rd. W. P. HOOD, (see page 295,) 684 Broad street. J. LANE, 224 Callowhill, below Third. JOSEPH E. SASS, (Plain and Fancy, Wrought and Cast Iron,) 133 Elfreth's Alley, below Second, first street above Arch. E. W. SHIPPEN & CO., Market street, West Philadelphia. And *many others.*
Railway Machinery.	(See page 304: also, Cars, Axles, Wheels, &c.)
Rakes, Hand, &c.	(See Agricultural Implements.)
Ranges, Cooking. (See page 291.)	JOHN ESTLIN, 231 North Fifth street. RAND & AYRES, 124 North Sixth street. And *many others.*
Rattan and Whalebone.	GEORGE W. CARR & CO., 126 Willow st., above Front.
Rat & Mouse Traps.	C. O. WILSON, 727 Race st., bel. Eighth.
Razors.	WILLIAM GILCHRIST, 445 North Broad.
Razor Strops.	C. Y. HAYNES, (Diamond Strop and Powder for Razors and Surgical Instruments,) No. 6 Lagrange Place.
Rectifying Whisky.	(See page 231.)
Reeds and Heddles.	M. MONGAN & SON, 114 S. Second, entrance 201 Carter street. JACOB SENNEFF, 230 Quarry, Office 28 North Front.
Refining, Gold and Silver.	DREER & SEARS, (see page 345,) Library street.
Reflectors, Daylight.	WILLIAM F. PULLINGER, 1 Carter's Alley
Refrigerators, Water Coolers & Filters.	GEORGE W. NICKELS, 606 Cherry street. JONES YERKES, 105 South Second street.
Regalia.	W. G. MINTZER, 215 North Third. JAMES KELLY, 146 North Sixth.
Revolvers.	(See Guns and Pistols.)
Ribbons.	W. H. HORSTMANN & SONS, 723 Chestnut.
Rifles.	(See Guns, &c.)
Rigging.	WILLIAM HUGG & SONS, 1053 Penn street.
Rivets.	PHILLIPS & ALLEN, (See page 334,) Penn'a Av., near 23d.
Roach & Fly Traps.	C. O. WILSON, 727 Race street, below Eighth.
Rolling Mills.	(See page 286.)
Roofing Composition.	H. M. WARREN & CO., Farquhar Buildings, Walnut st., bel. 3d. THOMAS, ALLEN & CO., N. Ninth and Girard Avenue
Roofing, Metallic.	R. S. HARRIS & CO., (Corrugated Iron,) S. E. cor. 11th & Pr.me. And *many others.*
Rope, Twines, &c.	(See page 373.)

ARTICLES.	MANUFACTURERS.
Rosin Oil.	(See page 370.)
Rotten Stone, Prep'd	GARRIGUES & MAGEE, 108 North Fifth street.
Rouge, Polishing, (for Jewelers.)	Mrs. M. S. WEST, Spruce street, above Third.
Rubber, India.	JOHN THORNLEY, 311 Chestnut street.
Ruches.	JOEL THOMAS, 26 S. Fifth street.
Rules, Gauging Instruments, &c.	J. E. CARPENTER, 862 North Fourth, below Poplar.
Saddle-Bags.	(See Medical Saddle-Bags.)
Saddlery & Harness. (See page 375.)	WM. S. HANSELL & SONS, 114 Market street. M. MAGEE & CO., 18 Decatur. LACEY & PHILLIPS, 30 and 32 South Seventh. W. R. SCOTT, (Collars,) 119 North Front street. And *many others.*
Saddlers' Tools.	HENRY HUBER, Jr., 4 North Fifth street.
Saddle-Trees.	C. PRUDDEN, Willow street, bet. Twelfth and Thirteenth.
Saddlery Hardware.	E. HALL OGDEN, (see page 340,) Ninth and Jefferson. KIRCHNER & STICKEL, (Steel Bitts, Hames, Rings, Eyes, Pad Plates, Hooks, Sterrets, &c.) rear of 309 Race. PAUL SCHWARTZKOPF, (Iron and Steel Chains, Buckles, &c.) 519 Emlen st., ab. Front, between Noble and Peg.
Safes and Fire Proof Chests. (See page 296.)	EVANS & WATSON, 26 South Fourth. FARREL, HERRING & CO., 130 Walnut. LINUS YALE, Jr. & CO., (Chilled Iron Burglar Proof,) Front & New sts. JONES YERKES, (Salamander,) 105 South Second street.
Sails.	McDONALD & LAUGHLIN, 16 N. Delaware Av. and 3 Dock. RICHARD F. SHANNON, 248 North Delaware Avenue.
Saleratus & Sal. Soda	BURGIN & SONS, 133 Arch street.
Salt, Dairy.	CONRAD KNIPE, 328 and 330 Noble street, below Fourth.
Salts, Bleaching, &c.	(See page 208.)
Sand, Writing.	JOHN W. CLOTHIER, 72 North Fourth street.
Sashes, Doors, Blinds &c.	COGILL & WILT, 423 North Front, above Callowhill. MATTHEW GRIER, Coates st., W. of Broad, south side. ABEL REED, Marshall and Girard Av., and 215 North Second. And *many others.*
Satchels, Fancy Leather.	J. T. MIDNIGHT, (also Leather Baskets, Bags, &c.) 145 North Fourth street.
Sausage Stuffers.	MICHAEL HEY, 637 North Third street.
Sawing, Scroll and Pattern.	MATTHEW GRIER, Coates st., W. of Broad, south side. JOSEPH RUE, 845 Parrish street. And *many others.*
Saw Mills.	GILLINGHAM & GARRISON (420, old No.), Queen street.
Saws.	(See page 330.)
Scaffolds, Portable.	A. C. FUNSTON, (Self-supporting,) Frankford Rd., opp. Master.
Scales and Balances.	(See page 335.)

39*

ARTICLES.	MANUFACTURERS.
Scales, Platform.	(See page 336.)
Scagliola.	E. DUFRENE, Seventh, below Chestnut. THOMAS HEATH, southwest corner Eleventh and Arch.
Scarificators and Spring Lancets.	F. LEYPOLDT, 508 E. North st., bet. 5th & 6th, ab. Arch,
Screws, Coffin.	HENRY CALVERLY, (also Studs, Plates, &c.) 207 Cherry st.
Screwdrivers, &c.	HORN & ELLIS, 307 Race street.
Screws, Wood.	H. & E. GOUJON, 913 Marshall st., above Poplar.
Screws, Wooden.	HARMAN BAUGH, (Bench, Hand, &c.) 125 Elfreth's Alley.
Screw Propellers.	REANEY, NEAFIE & CO., (See page 317,) 1365 Beach.
Scuttles, Coal.	GEORGE GRIFFITHS, 1 Fetter Lane, Third, above Arch.
Seeds, Garden, &c. (See page 142.)	HENRY A. DREER, 327 Chestnut. B. P. MINGLE & CO., 103 Market, and 4 North Front. C. B. ROGERS, 111 Market street.
Seines and Nets.	HENRY BEIDEMAN, Market st., bel. 2d, S. side, (basement.)
Settees.	BENJ. H. BRAYMAN, 57 North Front.
Sewing Machines.	PARHAM'S MANUFACTORY, George, below Tenth. GEORGE B. SLOAT & CO., 1229 Beach street. W. P. UHLINGER & CO., North Second, above Oxford.
Sewing—Machine.	W. H. TAYLOR, (also, Stitching and Quilting,) Chestnut street, above Fourth.
Sewing Silks.	B. HOOLEY & SON, (see page 249,) 16 Hudson's Alley, below Chestnut, between Third and Fourth.
Shawls.	(See page 238.)
Shears & Scissors.	CLARENBACH & HERDER, (Patent Cast Steel,) 634 Arch, and 235 Race.
Sheetings and Shirtings.	(See page 236.)
Sheet Iron.	ALLAN WOOD & CO., (Imitation Russia, &c.) 39 N. Front st.
Sheet Iron Ware.	GEORGE GRIFFITHS, 1 Fetter Lane, Third, above Arch.
Ship Bread.	(See Crackers, &c.)
Ship Building.	(See page 379.)
Ship Lanterns.	GEO. L. FLICK, (Bow, Signal, Binnacle and Fresnel,) 141 N. Front.
Ship Locks, Bolts, &c.	F. W. & G. A. KOHLER, (Brass,) 528 N. 2d, ab. Buttonwood.
Shirts, Collars, Bosoms, &c. (See p. 224.)	C. A. BUTTS, 27 North Eighth street. DAVIS & HOFF, 223 Church Alley. EDWIN A. KELLEY, (see page 224,) 16 Bank st. OLDENBERGH & TAGGART, 146 North Fourth. JOHN C. REMINGTON, 217 N. Fourth, corner Branch. H. SOUDER, 44 North Sixth st., above Market. WINCHESTER & CO., 706 Chestnut street.
Shoes, Gaiters, &c.	(See page 185.)
Shoe and Gaiter Uppers.	JACOB O. PATTEN, 52 N. Third st., below Arch. JACOB GILLER, northwest corner Eighth and Market.

ARTICLES.	MANUFACTURERS.
Shoemakers' Tools.	C. H. BLITTERSDORF, N. Fourth, above Callowhill. HORN & ELLIS, 307 Race street.
Shot.	THOMAS SPARKS, 121 Walnut street.
Shovels, Spades, &c. (See page 331.)	T. & B. ROWLAND & CO., Office, 501 Commerce. GEO. GRIFFITHS, 1 Fetter Lane, Third, above Arch. And *many others*.
Show Cards, Prem'm	STEELE & GLEASON, (Premium awarded by Penn'a Institute,) 109 North Seventh street.
Show and Counter Cases.	BEAL & FORMAN, (also Aquariums,)153 North Fourth. FRED'K HAFNER, Vine, below Second. JACOB LUTZ, 109 South Eighth.
Shuttles. (See page 303.)	E. JACKSON, 933 Charlotte st., above Poplar. HENRY SERGESON, 931 Charlotte st., above Poplar.
Shutter Bolts.	SELSOR, COOK & CO., Germantown. RIDGEWAY & RUFE, Germantown.
Sieves and Screens.	C. O. WILSON, (Coal Ash,) 727 Race street. (Also, see WIRE WORK, page 338.)
Signs, Letters, &c.	WM. C. MURPHY, 47 South Third street.
Silk Bonnets and Bonnet Frames.	AARON E. CARPENTER, (Sign of the Ostrich,) 54 N. 8th st. LINCOLN, WOOD & NICHOLS, 45 South Second. MORGAN'S CENTRAL BONNET FACTORY, 136 N. Ninth st. JAMES TELFORD, 903 North Second, above Poplar.
Silk Dyeing.	(See Dry Goods, page 249.)
Silk Moulds.	OLLIS & BROTHER, Beach and Shackamaxon.
Silver, Nitrate of.	(See page 209.)
Silver Ware. (See page 342.)	BAILEY & CO., (see page 348,) 819 Chestnut. W. FABER, (Forks and Spoons,) 214 North Fifth TAYLOR & LAURIE, 520 Arch street. W. WILSON & SON, corner Fifth and Cherry.
Silver-plated Ware. (See page 348.)	HARVEY FILLEY, (and Nickel Silver,) 1222 Market. EDWIN GUEST, 110 South Eighth street. G. ENGEL, 308 Chestnut st. J. S. JARDEN & BRO., 304 Chestnut street. ERNEST KAUFFMAN, 328 Noble street. JOHN O. MEAD & SONS, (see page 349,) 843 Chestnut, corner Ninth.
Silver Plating.	THOS. H. AUROCKER, northwest cor. Thirteenth and Cherry. G. W. BRADFIELD, 526 N. 12th st., corner Ridge Avenue. J. BILLERBECK, 902 Ridge Avenue. PHILIP CLINE, Fifth and Buttonwood. FRANCIS JAHN, 435 Race st., near Fifth. W. PAINE, Willow st., below Sixth. SCHEERER & DIEHR, 264 North Fourth street. SEDDINGER & BURWELL, 216 Arch st., above Second. J. B. SHANNON, 58 North Sixth street.
Skirts, Elastic.	M. BIRD, (also Spring Skirt Cord,) 227 Franklin Avenue.
Slates, Pat'nt School.	ZEBULON LOCKE, (see page 368,) 637 North Broad.
Slaw Cutters.	JOSEPH LEWIS, Germantown Road, below Diamond.
Sleighrunners.	ALFRED RUHL, (also Poles & Shafts,) N. W. corner Eighth and Wood.

ARTICLES.	MANUFACTURERS.
Smelting of Metals.	RICHARD WEST, 1247 Shackamaxon st., Kensington. (Always on hand and for sale the best quality of INGOT BRASS. Cash paid for Brass Turnings and Filings, Lead & Type Dross, &c.)
Snuff. (See p. 388.)	COOPER & WALTER, northwest corner 11th and Melon.
Soap and Candles. (See page 383.)	FRANCIS CONWAY, 114 and 118 Relief street. WILLIAM CONWAY, (see page 384,) 316 S. 2d, below Spruce. JOHN COOK, 1625 Market street. G. DALLETT & CO., 122 and 1319 Market, and northeast corner Tenth and Callowhill. E. DUFFY & SON, 912 and 914 Filbert street. G. M. ELKINTON & SON, (also Chemical Olive,) 116 Margaretta st., bet. Front and Second, above Callowhill. EPHRAIM WILSON, 1095 Germantown Rd,, opp. Second st. A. VAN HAAGEN & CO., (Magic Detersive,) Cadwalader st., above Columbia. And *many others.*
Soaps, Fancy & Toilet. (See page 410.)	GLENN & CO., 726 Chestnut. APOLLOS W. HARRISON, (see page 411,) 26 South Seventh. A. HAWLEY & CO., 117 North Fourth street. H. P. & W. C. TAYLOR, 641 and 643 N. 9th, below Coates. THOMAS WORSLEY, 518 Prune street, between 5th and 6th. R. & G. A. WRIGHT, (see page 411,) 35 South Fourth.
Soapstone. (See p. 367.)	EZRA PRATT, 127 North Sixth street.
Soda-Ash.	(See Chemicals, pp. 207–208.)
Soda Water Syrups.	EDWARD H. HANCE, 627 Arch street.
—— —— Apparatus.	JOSEPH BERNHARD & CO., 120 North Sixth.
Sofa Springs.	THOMAS THOMPSON, SON & CO., 238 South Second.
Solder.	JOSEPH BAKER & SON, (Spelter and Tinmen's,) 820 Rachel, or 821 North Second.
Sour-Crout Cutters.	JOHN LOUIS & SON, 504 Vine st., above Fifth.
Spectacles. w (See page 348.)	WM. BARBER, 110 South Eighth street, below Chestnut. McALLISTER & BROTHER, 728 Chestnut street. NATHAN E. MORGAN, 610 Sansom. SAMUEL FOX, 537 North Second, above Noble. ISAAC SCHNAITMANN, 225 North Fourth.
Spectacle Cases.	H. A. HEUSSLER, northeast cor. Fourth and Harmony Court.
Spices, &c.	BOHLER, TOMSON & WEIKEL, 248 N. Third, below Vine.
Spice Mills.	C. J. FELL & BROTHER, (see page 269,) 120 South Front.
Spirit & Plumb Levels.	J. E. CARPENTER, 862 North Fourth, below Poplar. WM. GOLDSMITH, northwest corner Green and New Market.
Spokes, Felloes and Hubs.	ELDRIDGE & FITLER, 1028 North Front. GEO. J. HENKELS, (Hubs,) North Sixth, above Thompson.
Spooling and Bobbin Machines.	JOHN JACKSON & CO., (also Warp Mills, Reels, and Temples,) 1089 and 1091 Germantown Road, opp. Second stree
Spoons & Forks, Silv. (See also Silver Ware.)	WILLIAM FABER, 214 North Fifth st., above Race. JABEZ E. WOOD, North Fifteenth, corner Hamilton.
Springs.	(See Car Springs and Carriage do., Sofa do.)
Stained Glass.	(See Glass, page 278.)
Stair Rods.	WILER & MOSS, (see page 191,) 225 South Fifth street.

ARTICLES.	MANUFACTURERS.
Starch.	SAML. T. STRATTON, Hancock & Phœnix streets.
Stationery. (Also see Blank Books, page 178.)	JOHN GLADDING, 117 and 529 South Second. DAVID M. HOGAN, 418 Walnut. HYMEN L. LIPMAN, 26 South Fourth street, (second story). PIERSON & DIAMENT, 136 and 138 N. Fourth, cor. Cherry. SAMUEL SCHURCH, 240 Race street JOHN B. SPRINGER, Fifth and Walnut. C. P. PERRY, (Bookbinder & Publisher,) S. W. cor. 4th & Race
Statues.	(Iron, see page 295 ; Marble, see page 365.)
Steam Engines.	(See Machinery, &c. ; also page 316.)
Steam Heating Apparatus.	JOHN L. KITE, (Kite's Patent Ventilating,) 13 Drinker's Alley, 147 North Second.
Steamers.	(See Ship Building ; also, see pp. 318–327.)
Stearin Candles.	(See Candles, page 384.)
Steel.	(See page 287.)
Steel Engraving.	(See page 181.)
Steelyards.	JOHN STEEL & CO., (Patent Balances,) Farmer, near 7th
Stencil Cutting.	C. E. FISK & SON, 13 South Sixth st., (basement.) B. SACHS, 306 Market st., above Third.
Stereotyping & Electrotyping. (See page 172.)	GEORGE CHARLES, 609 Sansom street. JOHN FAGAN, 623 St. James street. L. JOHNSON & CO., 606 Sansom street. MEARS & DUSENBERY, 322 Harmony street.
Stocks, Hair.	S. P. SMITH, (also Frames, Springs, Fasteners,) 1026 Chestnut.
Stocks, Ties, &c.	C. A. BUTTS, (also Cravats, Scarfs,) 27 North Eighth street.
Stockings.	(See Hosiery.)
Stockings, Elastic.	(See page 417.)
Stone-Cutters' Tools.	MILTON FORMAN, (also Awning Frames,) 668 Broad street, below Ridge Avenue, West side.
Stoneware.	(See page 201.)
Stoves. (See page 290.)	ABBOTT & LAWRENCE, 410 Brown street. NORTH, CHASE & NORTH, 209 North Second. (See also Iron Founding, Ranges, &c.)
Straw Goods	(See page 413.)
Street Lamps,	CHAS. WILHELM, (also, Hall, Hotel, &c.) 919 Race street.
Street-Sweeping Machines.	KING & HYNEMAN, 233 South Fifth, and Prime, below Fifth
Sugar, Refined.	(See page 386.)
Sugar Apparatus.	(See pages 192–328.)
Sulphuric Acid,	(See Chemicals, pp, 207–210.)
Surgical Instrum'ts.	(See page 415.)
Surveyors' Inst'mts.	(See Instruments.)
Suspenders.	J. J. HEBERSTECK, 1109 North Third street.

ARTICLES.	MANUFACTURERS.
Syringes, Glass.	THOMAS BURNS, 119 Walnut.
—— Metallic.	(See Britannia Ware.)
Swords & Side Arms.	WM. H. HORSTMANN & SONS, 720 Chestnut street.
Tables, Extension.	FRANCIS HOGUET, 307 Cypress. (See also FURNITURE.)
Tags, Tickets, &c.	WM. E. LOCKWOOD, (Sharp's Patent,) 236 Chestnut.
Tailors' Implements.	A. F. WARD, (Inch Measures, Crayons, Scales, Protractors, Squares, &c.) 333 and 335 Chestnut street. J. E. CARPENTER, (Squares, &c.) 862 N. 4th, below Poplar.
Tailors' & Sad Irons.	MORRIS L. KEEN, West Philadelphia.
Tanning & Currying.	(See Leather, page 357.)
Tape.	(See Bindings and Webbing ; also page 236.)
Taxidermy.	JAMES TAYLOR, 1916 Callowhill.
Tea Boxes and Canisters.	J. HALL ROHRMAN, (Japanned,) 606 Cherry street.
Tea Services.	JOHN O. MEAD & SONS, (Plated,) Ninth and Chestnut.
Teeth, Porcelain.	(See Artificial Teeth, and page 398.)
Telegraph Inst'mts.	JAMES J. CLARK, S. E. corner Second and Dock streets.
Telescopes.	ADOLPH WIRTH, 704 Arch, above Seventh.
Terra Cotta.	LORENZE STAUDACHER, (see page 201,) 1742 North Eleventh, above Columbia.
Thermometers and Barometers.	(See Philosophical Instruments.)
Thimbles, Gold and Silver.	GEORGE P. PILLING, 214 Gold street. GEO. W. SIMONS & BROTHER, (see p. 345,) Sansom st. Hall.
Threshing Machines.	DAVID LANDRETH & SON, 21 and 23 South Sixth.
Tickings.	(See Dry Goods, page 235.)
Tiles.	(See pages 200–365.)
Tin Boxes, Canisters, &c.	JOHN AMBLER, Jr., (Preserving Cans,) 711 Spring Garden st. JOHN S. HODGKINSON, (also Druggists' Tinware,) 2 Fetter Lane, Third, above Arch.
Tin, Crystals & Salts of.	(See Chemicals, pages 207–209.)
Tin Lamp Shades.	CHARLES WILHELM, (Patent Transparent Fancy Counting-House and Nursery Shades, and Patent Mica Shade Protectors,) 919 Race street.
Tin and Metallic Roofing.	JOHN AMBLER, Jr., 711 Spring Garden street. CUMMING & BRODIE, (and Composition,) 29 North Seventh. W. GILBERT, 7 South Seventh street. SAMUEL POWELL, 412 South Second, below Pine. CHARLES WILHELM, 919 Race street, above Ninth.
Tinware. (See page 417.)	JOHN AMBLER, Jr., (Improved Milk-Cans and Buckets, Ice-cream Freezers, &c. ; also, Gas-Consuming Sheet-Iron Stoves,) 711 Spring Garden street. W. GILBERT, (Plain and Planished ; also, Milk-Cans, Fancy Gas Lamps, &c.,) 7 South Seventh street. SAMUEL POWELL, (Milk, Cream, & Oil Cans, &c.,) 412 S. 2d.

ARTICLES.	MANUFACTURERS.
Tinware.—*Continued.* (See page 417.)	CHARLES WILHELM, (Models executed after designs,) 919 Race street. ISAAC S. WILLIAMS, (see page 417,) 726 Market.
Tobacco Manufact'rs	(See page 388.)
Tool Handles.	H. B. ANTRIM, 713 North Broad, corner Wallace. WM. G. BAMBREY, (Shoemakers' and Machinists',) 223 Crown street, above Race. JOSEPH RUE, (Hammer, Plane, Hand Saw, Axe, Pick, and Auger, &c.) 845 Parrish street.
Toys.	JOHN DOLL, (Checkers, Paper Babies, Snake Games, &c.) 144 North Second, above Arch. LUDWIG GREINER, 414 North Fourth, above Callowhill. HADDEN, CARLL & PORTER, (Tin,) 130 North Second. J. B. SHAW, (Drums, Boats, &c.) 1023 Chestnut.
Tops and Tubes, for Lamps.	J. T. VANKIRK, Agent, Frankford, (for Kerosene & Breckenridge Oils; also, Brass Fluid Tubes and Extinguishers.)
Transparencies.	M. ROBINSON & CO., (for Exhibitions; also, Sign Writing,) 19 North Second.
Traps.	(See Rat and Roach Traps.)
Trimmings, Carriage	(See Dry Goods, page 245.)
——, **Ladies Dress.** (See Dry Goods, page 244.)	J. B. CHAMPROMY, 133 North Third street. JOHN C. GRAHAM, 607 Cherry street. H. W. HENSEL, (see page 247,) 20 North Fourth. W. H. HORSTMANN & SONS, (see page 246,) Fifth and Cherry. CHARLES MEVES, 9 South Eighth street. J. G. MAXWELL & SON, Eleventh and Chestnut. And *many others.*
——, **Upholstery.**	G. T. BECHMANN, (Blind and Shade, &c.) northwest corner Third and Callowhill. CHARLES MEVES, 9 South Eighth street.
Trowels.	WM. ROSE & BROS., Market st., ab. Bridgewater, W. Phila.
Trucks, Store.	JONES YERKES, (and Packing Levers,) 105 South Second.
Trunks, Valises, &c. (See page 378.)	GEO. B. BAINS, 302 Market st., and 6 N. Fourth, ab. Market. DUNN & CO., Masonic Temple, 721 Chestnut street. THOMAS W. MATTSON, 402 Market street, above Fourth. J. M. MIGEOD, 27 S. 8th st., (2d story,) entrance on Lodge Al. WM. R. SCOTT, 119 North Front street. And *many others.*
Trunk and Valise Frames.	LAWRENCE M. POTTS, (Iron,) 9 St. James street.
Trusses & Bandages. (See page 417.)	HORN & ELLIS, 307 Race street. B. C. EVERETT, (Everett's Premium Patent Graduating Pressure Truss, &c., see page 417,) 14 North Ninth street. Dr. M. McCLENACHAN, (also Spinal Apparatus, Supporters, &c., by Mrs. McClenachan,) 50 North Seventh street.
Tubs.	(See Cedarware and Tinware.)
Tubes.	EDW'D BORMAN, (Drawn Metallic, for Philosophical Instruments, &c.,) 523 Cherry. MORRIS, TASKER & CO., (Wrought Iron,) 3d, below Walnut.
Turning, Bone, Ivory & Fancy Wood.	HARVEY & FORD, (see p. 393,) Goldsmiths' Hall, 422 Library THEODORE KANNEGIESER (also Metal,) 111 South Eighth. THOMAS SHAW, New st., cor. Front.
Turning, Wood.	WM. G. BAMBREY, 223 Crown st., above Race. HARMAN BAUGH, 125 Elfreth's Alley, bet. Arch and Race.

ARTICLES.	MANUFACTURERS.
Turning, Wood.—*Continued.*	JOHN CUNDEY, 732 North Fourth st., and 411 Brown. JOHN JACKSON & CO., 1089 and 1091 Germantown Road. GEO. H. KOECHLEIN, 526 North Front st., above Noble. JOSEPH RUE, 845 Parrish street below Ninth. JOHN LOUIS & SON, 504 Vine street.
Turn-Tables.	WM. SELLERS & CO., (see page 313,) 16th and Hamilton.
Twines and Lines.	(See Rope, page 373.)
Type Founding.	(See page 172.)
Type for Blind.	J. E. CARPENTER, (Pin Type,) 862 North Fourth.
Type Metal.	H. W. HOOK, (see page 171,) Broad and Pleasant street.
Ultramarine.	(See Chemicals, page 209.)
Umbrellas&Parasols. (See page 389)	WM. A. DROWN & CO., (see page 391,) 246 Market street. JOSEPH FUSSELL, 2 N. Fourth, northwest corner Market. SIMON HEITER, (see page 392,) S. W. corner 3d and Market. M. HINCKLEY, 905 Vine st., above Ninth. JOSEPH I. MATTHIAS, 555 North Second, above Noble. BENEDICT MILLER, 158 North Sixth st., near Race. W. H. RICHARDSON, (see page 392,) 418 Market street. SLEEPER & FENNER, (see page 391,) 336 Market street. SHARON SLEEPER, 1002 Market st., above Tenth. WM. S. TOLAND, 413 Coates, below Fifth. WRIGHT, BROTHERS & CO., (see page 391,) 324 Market. And *many others*.
Umbrella & Parasol Furniture.	SAMUEL FISHER, (Pearl,) 1509 Linden st., above Chestnut. HARVEY & FORD, Goldsmiths' Hall, Library st., (see p. 393.) THEODORE KANNEGIESER, 111 South Eighth. THOMAS SHAW, New st., cor. Front. J. T. VANKIRK, Agent, Frankford, (Metallic Mountings; also, Buggy Umbrella Handles and Mountings complete.)
Umbrella-makers' Tools.	WM. WEHRFRITZ, 242 North Fourth street.
Umbrella & Parasol Sticks.	BORIE & MACKIE, Frankford, (Carved in Imitation of French, Laurel, Hickory, &c.)
Upholstery. (See page 275.)	H. B. BLANCHARD & CO., 727 Chestnut street. FISHER & BROTHER, Vine, corner Fourth. ALFRED SMITH, 408 Spruce street. J. W. WINTER, 28 South Fifth street. And *many others*.
Varnishes.	(See page 219.)
Vats, Tanks, &c.	GEO. J. BURKHARDT & CO., (see p. 419) Broad & Buttonwood.
Velocipedes, Perambulators, &c.	ASKAM & SON, 131 Dock st., below Second. WM. QUINN, 416 Library street, (Spring and Lever Operating Four-wheel, for ladies, gentlemen, invalids, and juveniles.) CHARLES S. SWOPE, 753 South Third street. T. W. & J. A. YOST, 214 Dock st. and Franklin Av., cor. 3d.
Veneers and Fancy Woods.	ALEX'R BECRAFT, rear of 426 Walnut st. (See MAHOGANY.)
Veneers, Embossed.	ISRAEL AMIES, (Patentee,) 25 and 27 Minor, corner Sixth.
Venetian Blinds. (See p. 275.)	A. BRITTON & CO., 44 North Second street. C. W. CLARK, 139 South Second st., above Walnut street. R. W. KENSIL, 939 Race street. B. J. WILLIAMS, 16 North Sixth street.

ARTICLES.	MANUFACTURERS.
Venetian Blind Pulleys.	JOHN CUNDEY, 732 N. Fourth, and 411 Brown.
Verandahs Tents, &c.	JOSEPH H. FOSTER, 443 North Third, above Willow. GEORGE W. FOX, 6 Hart's Building, 6th, above Chestnut.
Ventilators.	ARNOLD & WILSON, 1010 Chestnut street.
Ventilating Chairs.	J. KAHNWEILER & BRO., (Patent,) Third, above Arch.
Vices.	M. MANSURE, Willing, near Arch, between 22d and 23d.
Vinegar and Cider. (See p. 270.)	EMIL MATTHIEU, 120 and 124 Lombard, below Second. JAS. G. PEALE, (Cider & White Wine,) N. E. cor. 3d & Noble.
Violins & Violoncellos. (See Musical Instruments, page 408.)	A. M. ALBERT, 303 Green street. JOSEPH NEFF, (Premium American,) 110 North Fourth street. JOSEPH WINNER, 148 North Eighth street.
Wagons, &c.	(See page 394.)
Wagon Boxes.	SAVERY & CO., Front street, below Reed.
Washing Machines.	DAVID LANDRETH & SON, 21 and 23 South Sixth street.
Watches.	(See Chronometers.)
Watch Cases.	(See page 345.)
Watch Case Springs.	CHARLES FARCIOT, (also Medallions,) 404 Library.
Waters, Saratoga.	J. & S. S. LIPPINCOTT, (in Bottles,) 916 Filbert street.
Watch Clocks, &c.	H. HOCHSTRASSER, 118 South Tenth, corner George.
Watch Springs.	CHARLES PRENOT, (also Clock,) 411 Merchant street.
—— **Guards.**	JOSEPH FRITZ, (French Leather, Fob Chains,) 1313 N. 4th.
Watchmakers' Tools.	CHARLES FARCIOT, Evans' Building, 404 Library street.
Water Colors.	G. W. OSBORNE & CO., (see page 209,) 104 North Sixth st.
—— **Coolers.**	JOHN AMBLER, Jr., (Improved Stone Jars,) 17 Spring Garden. W. GILBERT, No. 7 South Seventh street. GEO. W. NICKELS, (and XL–all Refrigerators,) 606 Cherry st.
Wax, Shoemakers'.	C. MOUSLEY, (also Channel Wax and Heel Balls, and Colors,) 1551 Germantown Road, below Oxford.
Wax Candles.	F. SNYDER, 435 Franklin Avenue, below Fifth.
—— **Flowers.**	Mrs. A. M. HOLLINGSWORTH, (also Fruit, and Material for Flowers, see page 414,) 48 North Ninth street.
Wax Taper Holders.	WILER & MOSS, (see page 191,) Fifth, below Walnut.
Weather Vanes.	A. C. BROWN, (Gilt,) northeast corner Eighth and Buttonwood.
Weaving.	(See Dry Goods, page 232.)
Webbing, (See Dry Goods, p. 256.)	JOHN DUDLEY, (Girth, Roller, Boot, Shoe, Trunk and Corded Rein, &c.,) 207 Quarry, above Arch, between 2d and 3d. JOS. SPITZ, (Rein, Girth, Boot and Trunk, &c.) 928 N. 3d st., above Poplar. AMASA STONE, (Chair and Sofa,) 207 Quarry, above Arch, bet. Second and Third. PHILADELPHIA WEBBING COMPANY, 9 Bank street.
Whalebone & Rattan	GEORGE W. CARR & CO., 124 and 126 Willow street. (See page 393.)

40

ARTICLES.	MANUFACTURERS.
Wheels, Ox & Timber.	SIMONS, COLEMAN & CO., 1109 North Front street. D. G. WILSON, J. CHILDS & CO., St. John and Buttonwood.
Wheelbarrows.	(See Wagons, &c., page 394.)
Whips and Canes.	CHARLES P. CALDWELL, (see page 378,) 4 North Fourth. PEARSON & SALLADA, 20 North Sixth.
Whip and Cane Mountings.	(See Umbrella Furniture.)
Whiting and Chalk.	CHARLES HASSE, Cadwalader, above Columbia street.
White Lead.	(See Paints, page 215.)
Wigs, Toupees, &c.	RICHARD DOLLARD, 513 Chestnut, opposite State House. GEORGE THURGALAND, 29 South Sixth, above Chestnut. And *many others.*
Willow-ware, Baskets, &c. (See page 400.)	H. COULTER, (also Importer of Fancy Baskets,) 17 N. Third, corner Church Alley. JOHN STINGER, (see page 400,) 511 Dickerson st., below Reed, Southwark. CHIPMAN & WHITE, 11 South Front street. ROWE & EUSTON, (and Cedar-ware,) 157 and 159 N. Third st.
Windlasses.	(See Capstans, &c.)
Window Shades,	M. ROBINSON & CO., (Buff Transparent, &c.,) 19 N. Second st. W. EARLE SMITH, 25 North Sixth street.
Wire-work. (See page 338.)	BAYLISS & DARBY, (see page 338,) 226 Arch street. JAMES P. FENNELL, 36 North Sixth st., corner Farmer. SELLERS BROTHERS, (see page 338,) 18 North Sixth street. WATSON, COX & CO., (see page 338,) 46 North Front. And *many others.*
Wire, Galvanized.	MARSHALL, GRIFFIN & CO., (see page 298,) 1142 N. Front.
——— Philosophical.	M. BIRD, (also Bonnet,) 227 Franklin Avenue.
Wood Manufactures.	(See page 395.)
Wooden-ware. (See p. 406.)	C. DREBY, 414 North Second, above Callowhill J. LOUIS & SON, 504 Vine street, above Fifth. W. J. WALKER, 819 Spruce st., above Eighth.
Woolen Goods.	(See Dry Goods, page 237.)
Woolen Machinery.	ALFRED JENKS & SON, Bridesburg, (see p. 299 & Appendix.)
Work Boxes.	(See Dressing Cases and Mahogany Cases.)
Yardsticks.	(See Wooden-ware.)
Yarns.	(See Dry Goods, page 232.)
Yeast Powders.	E. W. P. TAUNTON, (*Azumea*, the Premium Baking Powder, 14 Decatur street.
Yellow Metal.	RICHARD WEST, 1247 Shackamaxon street.
Zinc Manufactures.	(See TIN, ZINC, and SHEET-IRON WARE, page 417
Zinc Paints.	(See PAINTS, page 217.)

THE END.

Made in the USA
Lexington, KY
18 February 2015